Herbal Drug Development

Since ancient times, herbal remedies have been an essential component of human healthcare and are still used today to prevent and treat illnesses all over the world. Global interest in herbal and plant-based therapies has increased in recent decades due to growing concerns about the negative effects, high cost, and environmental impact of synthetic drugs. The need for safe, efficient, and reasonably priced herbal medications has increased due to the rising incidence of chronic illnesses, lifestyle problems, and antibiotic resistance. The scientific development, quality control, regulatory approval, and widespread commercialization of herbal medicines, however, continue to be difficult and complicated despite their enormous potential.

Herbal Drug Development: Insight into Techno-Entrepreneurial Opportunities aims to bridge this critical gap by integrating traditional medicinal knowledge with contemporary pharmaceutical science, emerging technologies, and entrepreneurial frameworks. The scope of the book spans the complete value chain of herbal drug development—from ethnomedicine and phytochemistry to industrial processing, regulatory affairs, and market translation. This book not only offers scientific insights but also strategic guidance for innovation-driven product development and enterprise creation in the herbal sector.

Key features:

- Overview of herbal, ethnomedicinal, and indigenous medical systems
- Comparative insights into Chinese, Indian, and Western medical practices
- Medicinal plant identification, phytochemicals, and pharmacological bioactivities
- Extraction techniques, optimization strategies, and formulation development
- Pharmacokinetics, pharmacodynamics, safety, toxicity, and clinical ethics
- Quality control and regulatory frameworks in India, China, the USA, and Europe
- Global market trends, commercialization challenges, and economic potential
- Innovation, technopreneurship, intellectual property, and business models
- Nanotechnology-based drug delivery systems and future perspectives

This book is particularly attractive to scientists, researchers, students, educators, entrepreneurs, and professionals in herbal medicine, pharmaceutical sciences, biotechnology, and allied disciplines. By blending scientific rigor with techno-entrepreneurial insights, it aims to inform, inspire, and empower readers to transform traditional herbal knowledge into sustainable, innovative, and globally competitive healthcare solutions.

Herbal Drug Development

Insight into Techno-Entrepreneurial Opportunities

Edited by
Mamoni Banerjee, Pawan Prabhakar,
and Sukanta Mondal

CRC Press
Taylor & Francis Group
Boca Raton London New York

CRC Press is an imprint of the
Taylor & Francis Group, an **informa** business

Designed cover image: Getty

First edition published 2027
by CRC Press
2385 NW Executive Center Drive, Suite 320, Boca Raton FL 33431

and by CRC Press
4 Park Square, Milton Park, Abingdon, Oxon, OX14 4RN

CRC Press is an imprint of Taylor & Francis Group, LLC

ISBN: 9781041034193 (hbk)
ISBN: 9781041034223 (pbk)
ISBN: 9781003623748 (ebk)

DOI: 10.1201/9781003623748

Typeset in Times
by KnowledgeWorks Global Ltd.

Dedicated to
Our beloved Parents and Families for making everything worthwhile and cheerfully tolerating many hours of absence involved in writing the book

Contents

About the Editors

Mamoni Banerjee, Dr. is currently working as Asst. Professor, Rajendra Mishra School of Engineering Entrepreneurship, Indian Institute of Technology, Kharagpur, India. She completed her PhD in Agricultural Science. Her major research interests are bioproduct development, pesticide formulation development, and problems related to the commercialization of biopesticides, rural technology development, agripreneurship, agrivalue chain optimization and management, etc. She has published over 35 publications in National/International journals and presented papers at 15 National/International Conferences. She has published 20 book chapters. She has communicated two patents. She is the Editor of National Journal and a reviewer of many reputed National/International journals. She is Life Member of several Academic Organizations, Societies, etc.

Pawan Prabhakar, Dr. is a nanotech and life science professional who has obtained his PhD degree from Rajendra Mishra School of Engineering Entrepreneurship, Indian Institute of Technology Kharagpur, Kharagpur, India, in "development of ethnomedicinal phytopharmaceuticals formulations" for diabetes mellitus. He did his master's research on "First order kinetics of nanoparticle drug delivery." Dr. Pawan earned his Integrated M. Tech degree in Nanotechnology in 2017 from the Centre for Nanotechnology, Central University of Jharkhand, Ranchi, India, with distinction. While continuing his M.Tech project, he joined a TATA Steel-funded project on the "Development of antibacterial graphene oxide materials" at the School of Biotechnology, KIIT University. After completing the project, he was selected for another project at nanomaterial synthesis laboratory, KIIT University, Bhubaneswar, on "Effect of High Energy Ball Milling on the Sintering Behavior and Thermal Properties of Lithium based tritium ceramic breeder" followed by "Development of Graphene based functional material for energy storage applications" funded by DAE-BRFST, GoI. After completing his M.Tech, he was selected as a Junior Research Fellow for a DST-funded project entitled "Mechanochemical milling and low temperature sintering of nanostructured MgA12O4" at the Department of Ceramic Engineering, NIT Rourkela, India, in 2018. As part of his area of specialization in nanobiotechnology, he has also built up a sound background in nanomaterial synthesis and its characterization, plant biotechnology, secondary metabolites extraction and characterization, herbal drug and food formulations, etc. He is also recipient of many awards and grants throughout his academic career including the Young Scientist Award, the Research Excellence Award, FEBS, and DST international travel grants. In addition, he is a member of different professional bodies and his activities and accomplishments include conferences, seminars, workshops, training and the publication of research articles, review papers, and book chapters in reputed international publishing houses.

Sukanta Mondal, Dr. is currently working as Principal Scientist, Physiology Division, ICAR-National Institute of Animal Nutrition and Physiology, Bangalore, India. His major research interests involve the impact of stress on maternal recognition of pregnancy, cloning, characterization and expression of genes regulating early embryonic loss, molecular characterization, and expression of hormone receptors. He has published over 200 research publications in various National/International journals and presented more than 30 papers at various National/International Conferences. He is pioneering and path-breaking contribution in hormone receptor as well as sheep and buffalo genomics has been recognized nationally and has received several prestigious awards like Prof. G. P. Talwar Midcareer Scientist Award (2014), Rashtriya Gaurav Award (2014), Prof. G. K. Pal Award (2013), Fellow of Indian Chemical Society (2013), Fellow of Society for Applied Biotechnology (2012), Fellow of Indian Association of BioMedical Scientists (2010), Dr. K. Anji Reddy Award (2011), Prof. P. B. Sen Memorial Oration Award (2008), Biotechnology Overseas Associateship Award (Long term, 2007–08), and ISSRF Young Scientist Award (2006). He has been bestowed with several fellowships including National Scholarship (Govt. of India), S. J. Jindal Trust Fellowship, Junior Research Fellowship (NDRI), Senior Research Fellowship (IVRI), and CRBR Postdoctoral Fellowship (Canada). He was Associate Editor of journals, viz., American Journal of Biochemistry and Molecular Biology, Asian Journal of Biotechnology, Asian Journal of Cell Biology, Current Research in Poultry Science and Biotechnology, etc.

Contributors

Trishna Bal
Biomaterial & Biomedical Research Lab
Department of Pharmaceutical Sciences & Technology
Birla Institute of Technology, Mesra
Ranchi, Jharkhand, India

Mamoni Banerjee
Bio-Research Laboratory
Rajendra Mishra School of Engineering Entrepreneurship
Indian Institute of Technology, Kharagpur
Kharagpur, West Bengal, India

Shanaz Banu
College of Pharmaceutical Sciences
Dayananda Sagar University
Bengaluru, Karnataka, India

P. L. Bharati
Ayurveda
Central Ayurveda Research Institute
Guwahati, Assam, India

Niharika Sahoo Bhattacharya
Rajiv Gandhi School of Intellectual Property law
Indian Institute of Technology, Kharagpur
Kharagpur, West Bengal, India

Sukanta Shekhar Bhattacharya
Department of Biosciences
JIS University
Kolkata, West Bengal, India

Devanjal Bora
Botany Section
Central Ayurveda Research Institute
Guwahati, Assam, India

Manajit Bora
Pharmocology Section
Central Ayurveda Research Institute
Guwahati, Assam, India

Naveen Chandrakar
Bio-Research Laboratory
Rajendra Mishra School of Engineering Entrepreneurship
Indian Institute of Technology
Kharagpur, West Bengal, India

Dimple Chauhan
Biomaterial & Biomedical Research Lab
Department of Pharmaceutical Sciences & Technology
Birla Institute of Technology, Mesra
Ranchi, Jharkhand, India

Mousumi Das
School of Pharmacy
Techno India University
West Bengal, Kolkata, India
and
P.G. Institute of Medical Sciences
Dhurabila, West Bengal, India

Sagar Surendra Deshmukh
School of Agro and Rural Technology
Indian Institute of Technology, Guwahati
Guwahati, Assam, India

Santanu Dhara
Biomaterials & Tissue Engineering Laboratory
School of Medical Science & Technology (SMST)
Indian Institute of Technology, Kharagpur
Kharagpur, West Bengal, India

Laxmi Ashok Kumar Dubey
Rajiv Gandhi School of Intellectual Property Law
Indian Institute of Technology, Kharagpur
Kharagpur, West Bengal, India

Sonal Dubey
College of Pharmaceutical Sciences
Dayananda Sagar University
Bengaluru, Karnataka, India

Pankhi Goswami
School of Agro and Rural Technology
Indian Institute of Technology, Guwahati
Guwahati, Assam, India

Swapnil G. Jaiswal
Department of Agricultural Engineering
Maharashtra Institute of Technology
Chhatrapati Sambhajinagar, Maharashtra, India

J. Josephine Leno Jenita
College of Pharmaceutical Sciences
Dayananda Sagar University
Devarakaggalahalli, Bengaluru, Karnataka, India

Deepayan Kar
School of Pharmacy
Techno India University
Kolkata, West Bengal, India
and
P.G. Institute of Medical Sciences
Dhurabila, West Bengal, India

Ankit Kumar
ICAR-National Research Centre on Litchi
Muzaffarpur, Bihar, India

Anshu Kumari
Biomaterial & Biomedical Research Lab
Department of Pharmaceutical Sciences & Technology
Birla Institute of Technology, Mesra
Ranchi, Jharkhand, India

Padmavati Manchikanti
Rajiv Gandhi School of Intellectual Property law
Indian Institute of Technology, Kharagpur
Kharagpur, West Bengal, India

Satyanarayan Naik
Supercritical Fluid Extraction Laboratory
Centre for Rural Development and Technology
Indian Institute of Technology, Delhi
New Delhi, India

Ashok Nimesh
Department of Anthropology and Tribal Studies
Central University of Jharkhand
Jharkhand, Ranchi, India

Jhansi Lakshmi Parimi
School of Pharmacy
Techno India University
Kolkata, West Bengal, India

Pawan Prabhakar
Bio-Research Laboratory
Rajendra Mishra School of Engineering Entrepreneurship
Indian Institute of Technology, Kharagpur
Kharagpur, West Bengal, India

Sonam Prabhakar
Department of Anthropology and Tribal Studies
Central University of Jharkhand
Jharkhand, Ranchi, India

Pavuluri Srinivasa Rao
Department of Agricultural and Food Engineering
Indian Institute of Technology, Kharagpur
Kharagpur, West Bengal, India

Seema S. Rathore
College of Pharmaceutical Sciences
Dayananda Sagar University
Bengaluru, Karnataka, India

Rahul Kumar Rout
ICAR-RCER Research Centre for Makhana
Darbhanga, Bihar, India

Shalini Shreya
Biomaterial & Biomedical Research Lab
Department of Pharmaceutical Sciences & Technology
Birla Institute of Technology, Mesra
Ranchi, Jharkhand, India

Puneet Kumar Singh
Chemistry Section
Central Ayurveda Research Institute
Guwahati, Assam, India

Manonayan Singha
School of Pharmacy
Techno India University
Kolkata, West Bengal, India
and
P.G. Institute of Medical Sciences, Dhurabila
Dhamkuria, Chandrakona Town, Paschim Medinipur, West Bengal, India

Prem Prakash Srivastav
Department of Agricultural and Food Engineering
Indian Institute of Technology, Kharagpur
Kharagpur, West Bengal, India

Soubhagya Tripathy
Department of Agricultural and Food Engineering
Indian Institute of Technology
Kharagpur, West Bengal, India

Preface

The domain of Herbal Medicine, also known as Botanical Medicine or Phytomedicine, encompasses all medicinal products derived from plants, plant parts, or plant extracts that undergo minimal or no chemical alteration. Since antiquity, natural herbs have played a crucial role in the treatment and prevention of a vast array of ailments. Over recent decades, extensive research into traditional systems of medicine has led to the development of diverse herbal formulations that are environmentally sustainable, biologically safe, and cost-effective.

In the twenty-first century, herbal medicines are gaining remarkable momentum in mainstream healthcare as individuals increasingly seek safe and natural therapeutic alternatives. The global demand for herbal medicines, health products, pharmaceuticals, nutraceuticals, food supplements, and cosmetics continues to rise, largely due to their perceived non-toxic nature, minimal side effects, physiological compatibility, and affordability. Medicinal herbs have long formed the foundation of ancient healing systems – with evidence of their documentation and use in Indian, Chinese, Egyptian, Greek, and Roman medical traditions dating back over 5000 years. Despite this resurgence of interest, the pharmaceutical industry has historically paid limited attention to herbal drug development, largely due to the inherent complexities of drug discovery, quality assurance, safety and efficacy evaluation, regulatory compliance, and market promotion. Additionally, market and regulatory research on herbal medicines remains underdeveloped, hindered by funding limitations and complex administrative frameworks. However, in recent years, entrepreneurship in the herbal drug sector has emerged as a rapidly growing field, offering substantial opportunities for innovation and business development. Addressing this evolving need, ***Herbal Drug Development: Insight into Techno-entrepreneurial Opportunities*** presents a distinguished compilation of contemporary research exploring the multifaceted realm of herbal medicine development. This volume serves as an invaluable resource for academics, entrepreneurs, and policymakers, offering critical perspectives on the intersection of science, technology, and business in the herbal industry. It is particularly beneficial for scientists, researchers, and graduate students specializing in traditional and herbal medicine, pharmacognosy, and pharmaceutical sciences.

This comprehensive work elucidates the synergy between entrepreneurship and herbal drug innovation, detailing market trends, regulatory frameworks, intellectual property rights, and strategic business approaches essential for success in this dynamic field. Beyond its scientific depth, the book also highlights the drug development process – from the identification and extraction of key phytocompounds to purification, characterization, and formulation using novel technological approaches. This collection serves as both a scholarly reference and a practical guide for aspiring innovators and professionals. It stands apart as a holistic resource that seamlessly integrates scientific principles, pharmacological research, and regulatory standards, illuminating the rigorous criteria governing herbal drug production and paving the way for future advancements in this promising domain.

This book comprises twelve chapters, each addressing key aspects of ethnomedicine, herbal medicine, and modern approaches to herbal drug development. Chapter 1 explores ethnomedicine practices, their contemporary relevance, challenges, and future outlook, emphasizing their importance in healthcare and socioeconomic development. Chapter 2 presents a comparative overview of herbal medicine within Chinese, Indian, and Western medical systems, highlighting their unique philosophies and therapeutic approaches. Chapter 3 compiles the taxonomical, therapeutic, and commercial profiles of 25 high-value medicinal plants. It also examines the role of biologically active secondary metabolites in treating various ailments, providing a detailed understanding of the pharmacological applications of each plant. Chapter 4 discusses the therapeutic benefits of phytocompounds, emphasizing their immense potential in natural product drug discovery through the integration of modern technologies. Chapter 5 focuses on bridging the gap between the identification of bioactive compounds and their effective application in promoting human health. It also highlights their relevance in the development of functional foods and disease prevention, contributing significantly to overall well-being. Chapter 6 delves into advanced extraction concepts, with a particular emphasis on modern extraction techniques that ensure efficient and selective separation of target compounds. The chapter further explores contemporary chromatographic methods for phytochemical identification and isolation, alongside formulation strategies for herbal drug development. Additionally, it provides a comprehensive analysis of drug release behaviour using *in vitro*, *in vivo*, and *in silico* approaches. Chapter 7 illustrates current laws and regulations governing herbal medicines in India and discusses the challenges and opportunities within this regulatory landscape. Chapter 8 examines the global herbal medicine market, detailing its size, segmentation, key players, growth drivers, challenges, and limitations. Chapter 9 evaluates patenting activity and prosecution trends within India's herbal sector, emphasizing the importance of fostering innovation and improving intellectual property management to accelerate industry growth. Chapter 10 highlights innovation and entrepreneurship in herbal drug development, showcasing strategies for translating research into commercially viable products. Chapter 11 explores advancements in herbal drug delivery systems powered by nanotechnology, which have enhanced the biological activity, solubility, bioavailability, and potency of therapeutically active molecules. Chapter 12 covers the critical insights presented throughout the book and outlines a strategic vision for the future of herbal drug development. It underscores the integration of traditional medicinal knowledge with modern scientific approaches, advocating for robust phytochemical standardization, clinical validation, and regulatory harmonization.

This book serves as a comprehensive guide for readers and entrepreneurs seeking opportunities in the herbal drug market. It is also an excellent teaching resource for academic institutions offering courses in pharmacology, herbal medicine, and entrepreneurship. The volume will be of particular interest to scientists, researchers, students, educators, and professionals in traditional medicine, herbal medicine, and pharmacy. The 35 contributors to this book are internationally recognized experts representing various esteemed institutions across India. We express our deep gratitude to all contributors for their valuable efforts and insights that made this work possible. We also extend heartfelt thanks to our families for their constant encouragement and patience during the completion of this project. Special appreciation is due to **Renu Upadhyay, Jyotsna Jangra,** and **the publishing team** for their exceptional care, dedication, and perseverance in bringing this book to publication.

Mamoni Banerjee,
Pawan Prabhakar,
Sukanta Mondal

Acknowledgements

It is a pleasure to formally acknowledge our profound gratitude to all those whose contributions have materially supported the preparation of this book. We extend our sincere thanks to everyone who assisted directly or indirectly through their encouragement, thoughtful guidance, and countless valuable suggestions. We wish to offer special appreciation to the editorial and production teams at CRC Press, Taylor and Francis Group, USA, for their exemplary work. Their professionalism, responsiveness, and unwavering support have been vital throughout the development of this book. Our heartfelt thanks also go to Renu Upadhyay, Jyotsna Jangra, and the entire publishing team for their exceptional dedication, patience, and meticulous care in bringing this book to publication. Above all, we express our deepest appreciation to our families for their enduring support, understanding, and patience. Their encouragement has been a constant source of strength throughout this endeavour.

Mamoni Banerjee,
Pawan Prabhakar,
Sukanta Mondal

1 Ethnomedicine in Tribal and Indigenous Cultures

Practices, Popularity, and Future Prospects for Entrepreneurial Development

Sonam Prabhakar and Ashok Nimesh

1.1 INTRODUCTION

Plants have been an essential component of our traditional therapeutic system since ancient times. Since the inception of human civilisation, plants have supplied all vital necessities as well as life-saving treatments. The persistent relationship between humans and plants dates back to the beginning of humanity and civilisation on Earth. Ethnomedicine comprises traditional ideas and practices associated with healing, frequently regarded as unscientific, yet possessing significant insights within societies. These methods encompass empirically grounded natural treatments, frequently sourced from flora, and rituals possessing a supernatural dimension (Alum 2024; Sultana, Muhammad, and Zakaria 2018). Ethnomedicine is an area of ancient medical practices of various cultures that attracted considerable attention from the scientific community for its essential role in documenting and preserving ancestral knowledge related to medicinal plant usage. It is the examination of traditional medical practices and knowledge across diverse cultures, frequently passed orally through generations (Teall 2014; Duche-Pérez et al. 2024). It comprises a multidisciplinary domain that integrates anthropology, pharmacology, ethnobotany, and public health, particularly concerning sustainable development, biodiversity protection, and healthcare accessible in tribal and ethnic communities. It also discusses the ways in which various ethnic groups observe health, sickness, and healing, including cultural interpretations and beliefs related to these notions. A wide variety of medicinal plant parts, including herbs, shrubs, and trees, is used in the treatment of diseases and conditions such as neurological, anthelmintic, diaphoretic, inflammatory, and diuretic problems. Therefore, ethnomedicine is one of the most important fields, as it safeguards ancestral information regarding therapeutic plants and has emerged as a crucial field of study for its potential in drug discovery and insights into cultural health

DOI: 10.1201/9781003623748-1

views. It is the scientific study of the traditional medical practices and knowledge of ethnic communities, which have been taught orally over centuries and evolved over millennia (Chen and Bu 2019). The World Health Organisation (WHO) characterises a medicinal plant as one that has compounds in one or more of its components that can be utilised for therapeutic purposes and serve as precursors for chemo-pharmaceutical semi-synthesis (Kasilo and Nikiema 2014). Thus, plants are the living pharmacies with a variety of medicinal qualities that promote mental, emotional, and spiritual health rather than just being biological entities. Utilisation of these plants is closely associated with tribal indigenous or ethnic communities having long history of practising these medicines for treatment of various illnesses (Tripathi 2019; Paul and Das 2025). Tribal and ethnic communities globally, especially in biodiversity-rich areas like India, Southeast Asia, Africa, and Latin America, have developed an extensive and profound knowledge system of ethnomedicinal plants, cultivated over centuries through intimate engagement with their natural surroundings (Prabhakar and Banerjee 2020; Tripathi 2019; Anju and Kumar 2024).

India's tribal population, constituting about 8.6% of the country's total population, is unevenly distributed across states and union territories, with significant variations in their share of the local demographic. The highest proportions are found in the north-eastern region and certain union territories, with Mizoram having 94.4% tribal population, Lakshadweep 95.3%, Nagaland 86.5%, Meghalaya 86.2%, and Arunachal Pradesh 68.8%. Other north-eastern states like Manipur (35.1%), Tripura (31.8%), and Sikkim (33.8%) also have substantial tribal presence. In central India, where some of the largest tribal groups reside, Chhattisgarh has 30.6%, Jharkhand 26.2%, and Odisha 22.8%, while Madhya Pradesh stands at 21.1% and Gujarat at 14.8%. Rajasthan records 13.5%, Maharashtra 9.4%, and Andhra Pradesh 7.0%. Assam, though not predominantly tribal, has a considerable 12.5% share, while Karnataka has around 7% and West Bengal 5.8%. States with smaller proportions include Bihar at 1.3%, Uttar Pradesh 0.6%, Tamil Nadu 1.1%, Kerala 1.5%, and Goa 10.2% (Census of India, 2011; Ministry of Tribal Affairs, 2021; Press Information Bureau, 2025). These variations reflect diverse geographical, historical, and cultural contexts, with tribal communities concentrated in forested, hilly, and border areas, and their presence linked closely to traditional livelihoods, unique cultural identities, and distinct social structures. The indigenous populations in India, including *the Santhals, Gonds,* Bhils, Nagas, Mundas, Pahari, and Birhors, sustain a comprehensive and symbiotic relationship with forests, depending on flora for sustenance, shelter, and traditional medicinal techniques. Here, ethnomedicinal plants are integral to their cultural, spiritual, and healthcare systems, addressing problems such as fever, gastrointestinal disorders, reproductive concerns, and chronic illnesses using plant-based therapies (Reddy, Subedi, and Guite 2023; Surendran et al. 2023).

The expertise regarding these medicinal plants, encompassing their identification, collection, preparation, dosage, and application, is frequently conveyed orally from one generation to the next by tribal healers, elders, and medicine practitioners, rendering it an essential component of intangible cultural heritage (Pradeep 2016). Some of the well-recognised ethnomedicinal plant species including Neem (*Azadirachta indica*), Tulsi (*Ocimum sanctum*), Ashwagandha (*Withania somnifera*), Amla (*Phyllanthus emblica*), Giloy (*Tinospora cordifolia*), and Harad (*Terminalia chebula*) are integral and important to indigenous healing practices and constitute the foundation of Ayurveda and various traditional medicinal systems (Mistry et al. 2014; Upasani et al. 2017). Here, Table 1.1 shows different ethnomedicinal plants well known for their therapeutic actions in tribal communities in India.

TABLE 1.1
Ethnomedicinal Plants Used by Tribal Groups in India

S. No.	Local/Vernacular Name	Scientific Name	Part of the Plant Used	Name of the Disease/Ailment	References
1.	Medshingi	*Dolichandrone falcata (Wall. ex DC.)*	Leaf	Leprosy	Dushing and Patil (2010)
2.	Latumoni	*Abrus precatorius L.*	Root	Diuretic	Buragohain (2011)
3.	Jorot goch	*Bixa orellana L.*	Stem	Dysentery and kidney trouble	
4.	Lamkhuang	*Artocarpus integrifolia Hook f.*	Leaves, fruit	Anti-microbial, fungal, inflammatory, wound, skin diseases	Shankar and Rawat (2013) Laldinsanga et al. 2019
5.	Bara Gokhru	*Pedalium murex Linn.*	Fruits	Gonorrhoea and dysuria	Kapoor et al. (2013)
6.	Saadpatti	*Flemingia fruticulosa Wall.*	Root	Epilepsy	Sharma et al. (2013)
7.	Madar	*Calotropis gigeantean*	Leaf, seed, bark	Wound healing, treatment of skin diseases, snake venom, leprosy, dysentery, treatment for helminthiasis, and ulcers, anticancer and anti-inflammatory, anti-diabetic, antioxidant, cough, haemolytic	Chandrakar et al. (2014)
8.	Chandrabhaga	*Rauwolfia serpentina*	Root, leaf	Snake bike, antihypertensive	
9.	Khus	*Vetriveria zizaniodes*	Root	Stomach	
10.	Simloo	*Berberis lyciumRoyle*	Root	Intermittent fever and weakness	
11.	Malabar nut	*Justica adhatoda*	Leaves, Stem, Flower	Stomach ache and fever	Singh et al. (2017)
12.	Pindi kura	*Avera lanta jun*	Seeds	Wound healing	Sridhar et al. (2018)

(Continued)

TABLE 1.1 ***(Continued)***
Ethnomedicinal Plants Used by Tribal Groups in India

S. No.	Local/Vernacular Name	Scientific Name	Part of the Plant Used	Name of the Disease/Ailment	References
13.	Khanghu	*Acacia pennata L.*	Leaves and bark	Bronchitis, stomach ache, gum bleeding, snake bite, headache, fever, cholera	Laldinsanga et al. (2019)
14.	Falsa/Phalsa	*Grewia asiatica L.*	Fruits	Sun stroke, acid reflux	Kuar et al. (2024)
15.	Sylo	*Tectona grandis Linn.*	Leaf	Kidney stones	Gudulkar et al. (2016)
16.	Kalajeerige	*Baccharoides anthelmintica (L.) Moench.*	Seed	Paralysis	Yogeesha and Krishnakumar (2023)
17.	Adalodakam	*Justicia adhatoda L.*	Whole plan	Asthma	Sasidharan et al. (2024)
18.	Seetha	*Annona squamosa L.*	Leaf	Hysteria	Raj et al. (2024)

The application of medicinal plants can be categorised into two primary groups: first, their traditional use by local practitioners for treating diseases, and second, their incorporation by pharmaceutical companies as active ingredients in their products (Beyene et al. 2016; Prabhakar et al. 2023b). The WHO estimates that 80% of the global population relies only on regional medicinal plants for basic healthcare due to poverty and limited access to modern drugs in numerous rural and tribal areas (Pathak, Gupta, and Pandey 2024). India is acknowledged as one of the four 'megadiversity' countries in Asia mostly due to two significant biodiversity 'hotspots': the Himalayas, encompassing the north-eastern hills along the northern boundary, and the Western Ghats in peninsular India. The Himalayas encompass 18% of India's physical area, including 12 states, and are broadly classified into Northern, North-Eastern, and Central Himalayas (Pullaiah 2018; Negi et al. 2002). In remote tribal regions, where modern healthcare facilities are limited or unattainable, these ethnomedicines provide efficient and economical healthcare alternatives (Biswas et al. 2023). These ethnomedicinal plants are an integral part of their lifestyle and very popular in their culture, and they are very efficient in the treatment of fever, bronchitis, malaria, vomiting, pain and inflammation, skin diseases, diabetes, arthritis, constipation, and snakebite. These pharmacological attributes are due to the presence of a number of bioactive constituents called secondary metabolites or phytocompounds belong to different classes like alkaloids, phenolics, and terpenoids (Ahmed 2018, Prabhakar and Banerjee 2020; Shukla et al. 2022; Dean 2024). Alkaloids constitute a heterogeneous category of nitrogenous organic molecules, predominantly located in plants, but also present in certain fungi, bacteria, and animals. Alkaloids exert a variety of significant physiological impacts on humans and other animals, and some of the notable alkaloids encompass morphine, strychnine, quinine, ephedrine, and nicotine (Kukula-Koch and Widelski 2017; Prabhakar et al. 2024). Phenolics are one of the major classes of secondary metabolites present in most plant tissues, encompassing fruits and vegetables. The predominant classification categorises phenolics into two principal groups: flavonoids (e.g., anthocyanins, flavanols, flavanones, flavonols, and isoflavones) and non-flavonoids (e.g., phenolic acids, coumarins, stilbenes, lignans, lignins, and tannins. These derivatives confer a broad spectrum of biological and possible therapeutic effects, including antioxidant, anti-inflammatory, and anticancer properties, among others (De La Rosa et al., 2019; Khoddami, Wilkes, and Roberts 2013; Prabhakar et al., 2025a; de Oliveira et al. 2025). Terpenoids are the most extensive class of natural compounds, exhibiting a wide variety of structures and functions. Numerous terpenoids exhibit therapeutic capabilities, including antibacterial, anti-inflammatory, immunomodulatory, and chemotherapeutic effects, rendering them highly significant in the medical domain (Ludwiczuk, Skalicka-Woźniak, and Georgiev 2017; Prabhakar et al. 2024a).

The ethnic group possesses its own medicinal plant preparations and treatment methodologies, and they also safeguard and cultivate diverse therapeutic plant species while transmitting ethno-pharmacological knowledge between generations. The provision of medicinal plants has been linked to rituals, religious convictions, and traditional practices, suggesting an animist or pantheistic perspective. From an ethnobotanical standpoint, most medicinal plants are regarded as omnipotent, ubiquitous, and intangible entities capable of remedying diseases that would typically require physical intervention (Quinlan 2011). It is important to note that traditional and ethnomedicine were disregarded by physicians and biomedical practitioners owing to several circumstances. The advancement of efficacious medicinal plants for managing chronic illnesses, especially those not immediately associated with significant public health emergencies, frequently encounters obstacles pertaining to health behaviour and accessibility. A significant concern is the people's demand for, coupled with restricted access to, recognised medical settings, and phototherapeutic goods. This scenario is exacerbated by other causes, including the absence of scientific validation for herbal practitioners, minimal institutional endorsement of recognised herbal therapies, and a general inadequacy in incorporating medicinal plants into conventional healthcare practices (Bussmann 2013; Feeney 2016). The pharmacological effects of many of these plants have been increasingly confirmed by scientific study, which has found active chemicals with anti-inflammatory, anti-cancer, anti-microbial, antioxidant, and anti-diabetic qualities. Examples of chemicals that have been identified from ethnomedicinal plants that are now being investigated for medication development include curcumin, tannins, flavonoids, alkaloids, and saponins (Prabhakar et al. 2024; Shahrajabian and Sun 2023).

Despite its importance, ethnomedical knowledge of plants is quickly dwindling because of urbanisation, cultural deterioration, deforestation, and a lack of documentation. Furthermore, the commercial exploitation of plant resources and bio-piracy frequently takes place without the appropriate acknowledgement or benefit-sharing of indigenous and tribal populations. Proper acknowledgement and benefit-sharing with indigenous and tribal groups are essential for guaranteeing ethical, egalitarian, and sustainable utilisation of their traditional knowledge and biological resources. These groups have maintained extensive collections of ethnomedicinal, ecological, and cultural knowledge through generations of experiential learning and oral traditions. (Prabhakar et al., 2023a; Saini, Lather, and Gahlawat 2022; Gitima et al. 2025). Ethnomedicine presents a powerful and promising avenue for entrepreneurial development among tribal communities, offering them a sustainable path to economic empowerment rooted in their own traditional knowledge systems. Tribal populations, who have long been custodians of biodiversity and rich repositories of plant-based healing wisdom, possess unique insights into the medicinal value of herbs, roots, barks, and forest resources. These insights can be transformed into viable micro- and small-scale enterprises focused on the cultivation, processing, packaging, and marketing of herbal medicines, wellness products, nutraceuticals, and organic personal care items (Kurniadi 2024; Prabhakar et al. 2023). Therefore, it is essential to preserve ethnomedicinal knowledge through documentation, community involvement, entrepreneurial setups, sustainable harvesting, and the protection of intellectual property rights. Thus, innovative, reasonably priced, and long-lasting healthcare solutions can result from combining this old knowledge with contemporary medical and scientific systems through courteous, cooperative methods. In addition to being a rich source of therapeutic potential, these plants are also part of a cultural legacy that prioritises holistic health and balance with the natural world (Mgbeoji 2012).

1.2 ETHNOMEDICINE: HISTORICAL AND CULTURAL SIGNIFICANCE

The history of ethnomedicine is closely associated with the development of human civilisation, originating thousands of years ago when early cultures engaged closely with their natural surroundings to identify methods for healing and survival. Prior to the emergence of contemporary science and allopathic medicine, ancient societies worldwide from indigenous tribes in Africa, Asia, and the Americas to early civilisations such as Mesopotamia, Egypt, India, and China were documenting the impacts of flora, minerals, and animal derivatives on health and illness (Singh et al. 2017). These observations established the basis of ethnomedicine, a sophisticated system of knowledge that integrates botany, spirituality, empiricism, and tribal belief systems. The Ayurvedic system in India originated from Vedic traditions over 3,000 years ago and constitutes one of the first documented types of ethnomedicine, relying significantly on botanical compositions and lifestyle concepts (Ravishankar and Shukla 2007). In China, traditional Chinese medicine (TCM) evolved over thousands of years, integrating herbal remedies, acupuncture, and notions of energy balance such as yin and yang. In the tribal and oral cultures of Africa and Latin America, healing knowledge was transmitted over generations by shamans, tribal healers, and medicine practitioners, who were frequently regarded with high social esteem. These practitioners employed indigenous plants, frequently found in forests and fields, and possessed profound expertise in its combinatorial application to address diverse diseases, including infections, injuries, mental disorders, and spiritual disarray (Wang, Wong, and Liao 2018). The ethnomedicinal traditions characterise therapeutic plants as divine offerings, and consequently, the apprehension of supernatural consequences from improper usage has been employed to maintain, protect, and prevent the misuse of medicinal herbs. With the expansion of colonialism worldwide, numerous ethnomedicinal systems were marginalised or suppressed in favour of Western biological paradigms. The colonial explorers and botanists documented various medicinal plants from indigenous civilisations, which subsequently contributed to the discovery of significant pharmaceutical medications (Curbishley 2015; Singh et al. 2020). Quinine, utilised in the treatment of malaria, was extracted from the bark of the Cinchona tree, which was historically employed by indigenous South Americans. Aspirin similarly originates from the utilisation of willow bark by ancient Greeks and Native Americans. In the twentieth and twenty-first centuries, ethnomedicine has undergone a revival as individuals pursue alternative, holistic, and culturally grounded healthcare methods (Gachelin et al. 2017). The preservation of ethnomedicinal traditions is jeopardised by urbanisation, habitat destruction, the decline of indigenous languages, and the deterioration of oral knowledge systems. Consequently, it is imperative to preserve, document, and integrate ethnomedicine with scientific research and public health policy to safeguard this valuable heritage and harness its extensive potential for future generations. The use of flavour enhancers in everyday life is strongly supported by the history of the oldest civilisations, which include a variety of cultures in Africa, China, Egypt, and beyond the Indus valley (Queiroz, Nogueira, and Ribeiro 2024). Notably, the Rigveda and other texts, which were produced between 4500 BC and 1600 BC, include most of the primary records describing several medicinal plants in India. With the profiles of 67 plants recorded, this era is considered the oldest reservoir of human knowledge. This collection of writings, which is respectful of the Rigveda, is a major source of inspiration for the study of life and shapes Hindu culture. Native Americans were already aware of the utilisation of plants. These plants were already well documented by 2500 BC, which was a crucial step in the spread of knowledge (Süntar 2020). Plants became the most common therapeutic agents used by humans in the middle of the nineteenth century, and this function is still important in contemporary medicine. Hansen's illness, also known as Eubacteria, was the causal organism of a terrible infection that swept over the plains' inhabitants during this time. Ethnobotany started to develop into a scientific field in the late nineteenth century, providing a new method for analysing pharmaceuticals (Kimta et al. 2024).

1.3 EVIDENCES FROM ETHNOMEDICINAL PRACTICES

Ethnomedicinal practices, deeply rooted in cultural traditions and indigenous knowledge systems, provide an expansive body of evidence that demonstrates the therapeutic value of phytocompounds in the prevention and treatment of diseases, and this knowledge often accumulated over centuries of trial, observation, and intergenerational transmission has laid the foundation for much of modern pharmacognosy and drug discovery. Across Asia, Africa, the Americas, and Europe, plant-based remedies have been intricately woven into daily healthcare practices, with communities relying on herbal formulations for conditions ranging from common infections and inflammatory disorders to complex chronic illnesses such as diabetes, cardiovascular disease, and cancer, and the increasing scientific validation of these phytocompounds reveals the molecular mechanisms by which they act, thus bridging the ancient and the contemporary in therapeutic science (Suman et al. 2019). In the Indian subcontinent, Ayurveda represents one of the oldest codified systems of medicine, and perhaps the most iconic plant in its pharmacopoeia is Curcuma longa (turmeric), traditionally applied to wounds, used in decoctions for respiratory infections, and prescribed for inflammatory conditions such as arthritis. Its primary bioactive compound, curcumin, has been extensively studied for its capacity to inhibit nuclear factor kappa B (NF-κB), cyclooxygenase-2, and other pro-inflammatory mediators, while also functioning as a potent antioxidant that neutralises reactive oxygen species, and these findings provide mechanistic substantiation for its historical reputation as a universal healer (Aggarwal and Harikumar 2009). Alongside turmeric, *Azadirachta indica* (neem) has long been revered in Indian ethnomedicine as a 'village pharmacy.' where every part of the tree is used against ailments such as skin infections, fever, intestinal worms, and dental disorders. Contemporary research reveals that neem contains diverse limonoids, triterpenoids, and azadirachtin, which exert antibacterial, antiviral, antifungal, and antiparasitic effects, validating centuries of traditional applications (Biswas et al. 2002). Similarly, *Withania somnifera* (Ashwagandha), classified as a Rasayana or rejuvenating tonic in Ayurveda, has been traditionally prescribed to improve vitality, fertility, and resilience against stress. Modern pharmacological investigations have demonstrated that withanolides, its active constituents, regulate the hypothalamic-pituitary-adrenal axis, reduce cortisol levels, promote synaptic regeneration, and enhance immunomodulatory responses, providing evidence that aligns with its long-established roles in strengthening both physical and psychological endurance (Singh et al. 2011). Another well-known Indian medicinal plant, *Tinospora cordifolia* (Guduchi), often described as 'Amrita' or the nectar of immortality, has been utilised for febrile illnesses, metabolic disorders, and immune deficiencies, and scientific studies confirm that diterpenoid lactones and alkaloids present in Guduchi stimulate macrophage activity, modulate cytokine production, and exhibit antidiabetic properties, thereby corroborating its ethnomedicinal relevance (Singh and Chaudhuri 2017).

Chinese traditional medicine offers equally compelling illustrations of ethnomedicinal wisdom. *Panax ginseng*, considered a superior tonic for longevity, energy, and mental clarity, has been documented in Chinese medical texts for over two millennia. Its ginsenosides exhibit adaptogenic properties, balancing the central nervous system, enhancing immune surveillance, improving glucose homeostasis, and exerting cardioprotective effects, thereby providing scientific validation for ginseng's revered role in restoring systemic harmony (Attele et al. 1999). Another landmark example is *Artemisia annua* (qinghao), used historically for intermittent fevers and malaria in TCM. The isolation of artemisinin, a sesquiterpene lactone, from *A. annua* in the 1970s revolutionised malaria treatment worldwide, earning a Nobel Prize for its discoverer, Tu Youyou, and demonstrating in the most direct way how ethnopharmacological knowledge translates into modern life-saving drugs (Klayman 1985). In addition, Chinese medicine frequently prescribes *Scutellaria baicalensis* (Huang Qin), used traditionally for respiratory infections, hepatitis, and inflammatory disorders. Its bioactive flavonoid, baicalin, has been shown to possess antiviral, anti-inflammatory, and hepatoprotective effects, confirming the mechanistic rationale behind its ethnomedicinal use (Li-Weber 2009).

In African ethnomedicine, *Vernonia amygdalina* (bitter leaf) remains a vital multipurpose plant employed against malaria, helminthiasis, gastrointestinal problems, and diabetes. Scientific analysis has revealed the presence of sesquiterpene lactones, flavonoids, and alkaloids, which exhibit antimalarial activity through interference with parasite mitochondrial function and also demonstrate hypoglycaemic and anticancer effects, thereby providing concrete pharmacological evidence for its widespread use among indigenous populations (Ijeh and Ejike 2011). Another African species, *Harpagophytum procumbens* (devil's claw), used by Khoisan and Bantu peoples for musculoskeletal disorders, arthritis, and digestive issues, contains iridoid glycosides such as harpagoside, which have been shown to suppress inflammatory mediators and reduce pain, aligning closely with its ethnomedicinal reputation as a potent anti-rheumatic agent (Brendler et al. 2021). Garcinia kola (bitter kola), commonly chewed in West Africa for stamina, respiratory ailments, and infections, is another key ethnomedicinal species; its biflavonoid kolaviron exhibits antioxidant, hepatoprotective, and antimicrobial effects, providing scientific validation for its cultural importance (Farombi, Akanni, and Emerole 2002). Furthermore, *Moringa oleifera*, widely cultivated and consumed across Africa and Asia, has been traditionally used for nutritional supplementation, lactation support, and treatment of hypertension and diabetes. Rich in polyphenols, glucosinolates, and isothiocyanates, *M. oleifera* exhibits strong antioxidant, anti-inflammatory, and antihypertensive properties, offering modern confirmation of its long-standing role as both food and medicine (Anwar et al. 2007).

In South America, particularly among Amazonian communities, *Uncaria tomentosa* (cat's claw) has been employed for immune support, arthritis, gastritis, and infections. Scientific studies have confirmed that oxindole alkaloids from *U. tomentosa* modulate immune function by enhancing phagocytosis and reducing tumour necrosis factor-alpha (TNF-α), while also exerting antioxidant effects, thus explaining its broad ethnomedicinal reputation (Sandoval et al. 2000). Another South American plant, *Erythroxylum coca*, historically used by Andean populations to combat altitude sickness, hunger, and fatigue, yielded the alkaloid cocaine, which, although associated with abuse, also provided the prototype for local anaesthetics, demonstrating both the benefits and complexities of ethnopharmacological discovery (Plowman 1984). *Paullinia cupana* (guaraná), traditionally consumed by indigenous Brazilians as a stimulant and tonic, contains high levels of caffeine, tannins, and saponins, which improve endurance, alertness, and oxidative balance, thereby confirming traditional beliefs about its energising and protective properties (Bempong and Houghton 1992). Additionally, *Croton lechleri* (dragon's blood), used in Amazonian ethnomedicine for wound healing, diarrhoea, and ulcers, has been found to contain the alkaloid taspine, which promotes tissue regeneration and exhibits antiviral activity, further supporting its cultural significance (Jones 2003).

European folk medicine also provides abundant ethnomedicinal evidence. Allium sativum (garlic) has been traditionally employed across Mediterranean societies for cardiovascular protection, infections, and as a general health enhancer.

Its sulphur-containing compounds, notably allicin, have been shown to reduce cholesterol levels, inhibit platelet aggregation, and exert antimicrobial activity, thereby confirming its broad ethnopharmacological legacy (Rahman and Lowe 2006). Another significant European herb is *Hypericum perforatum* (St. John's Wort), used in folk medicine for centuries to alleviate melancholy and nervous conditions. Its bioactive constituents, hypericin and hyperforin, modulate neurotransmitter activity, including serotonin and dopamine reuptake, thereby providing a mechanistic rationale for its antidepressant properties, which are now recognised in modern psychiatric practice (Müller 2003). *Silybum marianum* (milk thistle), used in European traditions for liver disorders, contains silymarin, a flavonolignan complex that exerts hepatoprotective effects through antioxidant activity and regulation of hepatocyte regeneration, aligning well with traditional knowledge (Flora et al. 1998). In addition, *Valeriana officinalis* (valerian), employed for centuries as a sedative and sleep aid, contains valerenic acid and related sesquiterpenes that modulate gamma-aminobutyric acid (GABA) receptors, thereby providing a biochemical explanation for its calming effects (Houghton 1999). Beyond regional examples, certain plants illustrate the universality of ethnomedicinal practices. *Zingiber officinale* (ginger), utilised in Indian, Chinese, African, and Middle Eastern traditions for nausea, digestive issues, and respiratory conditions, contains gingerols and shogaols that display antiemetic, anti-inflammatory, and antioxidant properties, providing a pharmacological foundation for its global use (Ali et al. 2008). *Camellia sinensis* (green tea), widely consumed in East Asia for health and longevity, contains catechins such as epigallocatechin gallate (EGCG), which demonstrate anticancer, neuroprotective, and cardioprotective activities, supporting its ethnomedicinal role as a health-promoting beverage (Cabrera et al. 2006). *Ocimum sanctum* (holy basil or tulsi), revered in Indian ethnomedicine as a sacred plant with protective functions, has been shown to possess eugenol and ursolic acid, which modulate immune responses, reduce oxidative stress, and exhibit antimicrobial activity, thus confirming its spiritual and medicinal significance (Prakash and Gupta 2005).

The accumulated evidence from ethnomedicinal practices therefore highlights the profound role of phytocompounds in disease mechanisms. These compounds act through diverse biological pathways: as antioxidants reducing oxidative stress, as anti-inflammatories suppressing cytokine cascades, as antimicrobials disrupting pathogen viability, as metabolic regulators modulating insulin and cholesterol levels, and as neuroactive agents influencing neurotransmitter systems. The translation of ethnomedicinal knowledge into modern medicine, as exemplified by artemisinin from *Artemisia annua*, aspirin's origin in willow bark (*Salix* spp.), or the continued development of nutraceuticals from ginseng, garlic, and green tea, demonstrates that indigenous knowledge is not only historically valuable but also a critical resource for contemporary drug discovery (Desborough and Keeling 2017). Moreover, the continuity between traditional observations and modern validations demonstrates that ethnomedicine is not mere superstition or anecdote, but a sophisticated empirical science rooted in centuries of experimentation and cultural practice. The future of global healthcare lies in preserving and integrating this ethnopharmacological wisdom with modern biomedical research, ensuring that plant biodiversity and indigenous knowledge systems remain vital sources for therapeutic innovation, particularly in the context of rising chronic diseases, antimicrobial resistance, and the urgent need for affordable and accessible medicines worldwide (Kaur et al. 2020).

1.4 MODALITIES AND PROSPECTS IN ETHNOMEDICINE

Medicinal plants have been utilised throughout human history for diverse objectives, including illness prevention, treatment, and enhancement of human health. Twenty per cent of global plant species possess the capability to properly remedy numerous human problems. A plethora of medicinal plants exists globally, although emerging nations, particularly in Africa, Asia, and South America, are especially abundant in both the variety of beneficial species and the traditional knowledge of their utilisation. The ecological significance of these raw resources is notable, as the included species are frequently endemic. Over fifty per cent of all plant species inhabit tropical climates, many of which remain unclassified. This is particularly accurate in rainforests. Nevertheless, these species are predominantly impacted by habitat degradation and climate change. The correlation between the availability of medicinal plants and ecological degradation is highly intricate (Kim, Lim, and Kwon 2016; Sofowora, Ogunbodede, and Onayade 2013).

Various diversity-related criteria indicate that plant groupings indigenous to a place hold significant importance for the region's biodiversity and frequently possess relevance for the local human population. Consequently, the obliteration of rainforests can facilitate the accessibility of species that were previously concealed beneath the canopy. A disease or abnormal condition in humans is typically attributed to the presence of a specific pathogen, often a bacterium or virus. Consequently, the quest for effective treatment has historically focused on identifying a toxic agent, a substance capable of eliminating or incapacitating the pathogen. Drugs have been, and continue to be, derived from plants, animals, minerals, and synthetic processes. Generally, these methods are not recognised as safe and effective preventive strategies against diseases by the biomedical model. Contemporary pharmaceuticals are frequently unaffordable for developing nations (Prabhakar et al., 2023a; Reyes-García et al. 2022; Behera et al. 2025). Consequently, the significance of medicinal plants within the healthcare systems of these nations is substantial. Epidemiological surveys indicate a significant reliance on traditional healers among people in Africa, Asia, and Latin America, with 80% of respondents in perceptual studies seeking their services. Nevertheless, access to this medical practice is frequently challenging. There are limited practitioners, and they are typically remote from the mass of the populace (Ahmed et al. 2023). Moreover, inadequate infrastructure and training may result in unsanitary behaviours that worsen the diseases. Deeply ingrained in the social, cultural, spiritual, and ecological fabric of indigenous and ethnic communities worldwide, ethnomedicine is a comprehensive and multidimensional system of traditional healing that includes a wide variety of elements and techniques. Health is viewed as a dynamic equilibrium among the body, mind, spirit, and environment in ethnomedicine, in contrast to modern biomedicine, which frequently isolates symptoms or diseases as the use of chemicals of natural origin and medicinal plants, which serve as the pharmacological foundation of traditional healing, is a fundamental aspect of ethnomedicine (Prabhakar et al., 2024; Rizvi et al. 2022). Pharmacological research, medication discoveries, and the confirmation of traditional practices have resulted from the scientific community's recognition of ethnomedicine's therapeutic potential in recent decades. However, there are still a lot of obstacles to overcome, such as a lack of documentation, the possibility of biopiracy, the deterioration of traditional knowledge, and insufficient legal recognition. Ethnomedicine frequently occurs in unofficial or alternative healthcare systems and is not institutionally supported, particularly in national health programs and policymaking. The gap between traditional and modern medicine is being closed through programs such as ethnopharmacology research, community-based ethnobotanical surveys, and integration with primary healthcare practices (Pirintsos et al. 2022). To sum up, ethnomedicine's elements and practices are numerous, intricate, and deeply ingrained in culture. It provides not just medical treatments but also a philosophy that encourages communal harmony, sustainable living, and overall well-being. In addition to being crucial for cultural heritage, acknowledging and protecting this rich knowledge system has enormous promise for biodiversity conservation, innovative global healthcare, and robust health systems that can withstand contemporary difficulties (Mahapatra et al. 2019).

The conservation of biodiversity and global healthcare is intricately linked through ethnomedicine, which relies significantly on the sustainable utilisation of plant and animal species within various ecosystems worldwide. Ethnomedicine, grounded in traditional ecological knowledge, depends on the extensive variety of medicinal plants, herbs, fungi, animal products, and minerals found in forests, grasslands, wetlands, and mountainous areas, many of which are inhabited by indigenous and tribal populations. These societies have, over generations, developed intricate healing systems grounded in observation, experimentation, and spiritual comprehension of nature, rendering them stewards of both cultural heritage and ecological resources (Chen et al. 2016). The conservation of biodiversity is essential for sustaining ecological equilibrium and for safeguarding traditional medicinal knowledge systems that provide alternative and frequently effective methods for health and illness management. According to the WHO, more than 80% of the population in underdeveloped nations continues to depend on traditional medicine as their major healthcare source, underscoring the global significance of ethnomedicine (Caballero-Serrano et al. 2019).

A substantial percentage of contemporary pharmaceutical drugs such as aspirin (derived from willow bark), quinine (extracted from cinchona bark), and artemisinin (sourced from sweet wormwood) stem from ethnobotanical knowledge, highlighting the importance of biodiversity as a repository of biochemical innovation. The devastation of natural ecosystems caused by deforestation, climate change, urbanisation, and industrialisation jeopardises not only plant and animal species but also the capacity of traditional healers to access and utilise their medicinal knowledge. This ecological degradation leads to the loss of biological and cultural variety, which are intrinsically connected within the realm of ethnomedicine. Consequently, conservation initiatives must encompass not only species protection but also the preservation of traditional knowledge, community rights, and the environments that sustain these healing systems (Faurant 2011; Petrovska 2012;

Shah et al. 2015). Ethnomedicine advocates for the use of indigenous, natural, and sustainable resources for therapeutic purposes, closely aligning with the principles of ecological sustainability and environmental ethics. Moreover, when global healthcare systems encounter escalating challenges such as antibiotic resistance, chronic illnesses, and health disparities, ethnomedicine offers a supplementary approach through affordable, culturally relevant, and preventative health practices (Kumar et al. 2021). Integrating ethnomedical practices into national health systems, honouring intellectual property rights, and investing in community-driven conservation and healthcare models can promote biodiversity preservation and equitable health outcomes globally. Ethnomedicine functions as a significant conduit, connecting the knowledge of historic healing practices with the contemporary necessity for sustainable, inclusive, and accessible global healthcare systems, while also emphasising the ethical obligation to preserve the natural environment that supports them (Awoke and Cosendey 2025).

1.5 CHALLENGES IN PRESERVING AND PROMOTING ETHNOMEDICINE

Preserving and promoting ethnomedicine face a multitude of challenges that span cultural, scientific, legal, ecological, and socio-economic dimensions, making it a complex task that requires urgent and coordinated global attention. One of the most pressing challenges is the rapid erosion of traditional knowledge systems, which is primarily due to modernisation, urbanisation, and the declining interest among younger generations in indigenous practices. As older custodians of this knowledge pass away without documented transmission, centuries-old wisdom about medicinal plants, healing rituals, and diagnostic methods risks being permanently lost. This is further exacerbated by the lack of formal documentation and academic validation of ethnomedical practices, which are often transmitted orally, making them vulnerable to distortion, neglect, and misappropriation (Alum 2024; Kumar et al. 2021). Moreover, biopiracy and intellectual property rights violations remain a serious concern, as multinational corporations and researchers often extract traditional knowledge and medicinal resources without proper consent, recognition, or benefit-sharing with the indigenous communities who are the rightful knowledge holders. There is also a lack of clear legal frameworks and international protocols that protect the intellectual and cultural property rights of tribal and ethnic groups, resulting in exploitation and commercial misuse. Furthermore, ethnomedicine often suffers from scientific scepticism and lack of standardisation, as many practitioners do not follow evidence-based protocols recognised by modern science, which makes it difficult to integrate these practices into mainstream healthcare systems (Farhat 2021; Imaran et al. 2021).

Furthermore, commercial pressure impacts the sustainability and ethical use of therapeutic herbs. The rising popularity of specific herbs in the worldwide market often results in overharvesting and unsustainable extraction practices, which jeopardise species and degrade ecosystems. This endangers biodiversity and deprives local communities of vital resources for healthcare and cultural traditions

(Wanzala and Minyoso 2024; Prabhakar et al. 2025). The worldwide increase in demand for Hoodia, a traditional appetite suppressant utilised by the San people of Southern Africa, resulted in its overexploitation and commercialisation by multinational pharmaceutical corporations, with no advantages accruing to the originating group. To resolve these issues, it is essential to implement equitable benefit-sharing frameworks, as promoted by the Convention on Biological Diversity (CBD) and the Nagoya Protocol, which guarantee that indigenous communities obtain fair compensation and acknowledgement when their traditional knowledge is commercialised (Landor et al. 2015). Furthermore, ethical commercialisation necessitates obtaining free, prior, and informed consent (FPIC) from knowledge holders and ensuring their involvement in every phase of product creation, branding, and revenue allocation. In conclusion, although the commercialisation of traditional medicine presents opportunities for economic advancement and increased accessibility to natural healthcare, it should not occur at the expense of cultural degradation, spiritual contempt, or environmental damage. Safeguarding the integrity of traditional knowledge systems necessitates not only legal frameworks but also ethical accountability, intercultural reverence, and the proactive empowerment of indigenous voices in determining the utilisation, representation, and preservation of their knowledge within the global marketplace (Mgbeoji 2007; Quinlan 2011). Preserving their knowledge is essential for global efforts in biodiversity conservation and climate resilience. Hence, preserving tribal heritage and ethnomedicine is essential for safeguarding cultural diversity, improving health equity, advancing scientific research, and protecting natural ecosystems. This requires community-led documentation, educational integration, policy support, legal protections, and genuine respect for the rights and wisdom of tribal people. It is not just about preserving the past but securing a wiser, more inclusive future (Beyene et al. 2016).

1.6 PROSPECTS FOR ENTREPRENEURIAL DEVELOPMENT

Ethnomedicine offers significant and unexplored business prospects for tribal and ethnic communities, especially in areas with abundant biodiversity and well-preserved traditional knowledge systems. With the rising global interest in natural, plant-based, and holistic healthcare solutions, there is a growing acknowledgement of the economic potential of traditional healing methods and medicinal flora. This transition presents a favourable opportunity for tribal and ethnic communities to convert their ancestral knowledge into sustainable economic enterprises, thus enhancing their socio-economic standing while safeguarding their cultural heritage (Ota 2020; Sonam et al. 2021). The cultivation and commercialisation of therapeutic plants is one of the most immediate potentials. Indigenous tribes that historically gather plants from forests can now participate in scientific cultivation, enhance product value, and obtain organic certification to comply with national and international market norms. Herbal medications, nutraceuticals, essential oils, and wellness products based on traditional knowledge such as Ashwagandha, Giloy, Tulsi, Amla, Arjun bark, and Turmeric have experienced significant increase in demand in both local and international markets. With suitable training in agro-processing, packaging, branding, and marketing, tribal businesses can transition from mere raw material suppliers to value chain participants and brand proprietors (Chaachouay and Zidane 2024; Nasim, Sandeep, and Mohanty 2022).

The increasing demand for natural cosmetics and Ayurveda-inspired personal care products offers tribal tribes an opportunity to enter the beauty and wellness sector, particularly when paired with eco-friendly packaging and ethical sourcing stories that attract conscientious customers. A significant opportunity exists in the advancement of community-oriented wellness tourism. Ethnomedicinal expertise, alongside tribal culture, forest ecosystems, and traditional therapeutic modalities (including massages, herbal baths, and spiritual therapies), can entice travellers in search of genuine, nature-oriented healing experiences (Reddy, Subedi, and Guite 2023). Homestays, nature trails, and wellness resorts operated by tribal entrepreneurs provide immersive experiences while generating jobs in hospitality, guiding, and handmade crafts. Government initiatives in eco-tourism and tribal tourism development can facilitate connections between conservation and entrepreneurship. Furthermore, tribal kids can emerge as ethno-entrepreneurs by participating in documentation, digital storytelling, and product development (Bhwomick and Prabhakar, 2023; Mishra et al. 2025). Digital technology and smartphone applications enable communities to catalogue their local medicinal plant diversity, record healing traditions, and disseminate this knowledge via interactive platforms. This content can facilitate research, education, and awareness, while also constructing intellectual property portfolios that protect traditional knowledge and provide benefit-sharing systems (Chirima 2025). Moreover, tribal women who are frequently the principal custodians of traditional therapeutic knowledge can be empowered as micro-entrepreneurs through the establishment of Self-Help Groups (SHGs) and cooperatives dedicated to the production of herbal teas, dietary supplements, and environmentally sustainable healthcare kits. Government, non-profit, and academic policy assistance can expedite these opportunities (Prabhakar and Mamoni 2021; Prabhakar et al., 2023; Sarawagi and Singh 2024). Initiatives such as TRIFED (Tribal Cooperative Marketing Development Federation of India), Van Dhan Yojana, AYUSH programs, and Start-Up India for Tribal Entrepreneurs are actively establishing supportive ecosystems through training, financial accessibility, market connections, and incubation assistance (Gupta et al. 2022; Naik and Panda 2023; Prabhakar and Nimesh 2022; Sonam et al., 2020). Following are the areas where tribal people can work in setting up new enterprises and ventures in ethnomedicinal products.

1.6.1 Herbal Product Manufacturing

The potential for herbal product manufacture by tribal communities is extensive and consistently growing, propelled by the increasing global desire for natural, chemical-free, and sustainable healthcare alternatives. Tribal areas exhibit abundant biodiversity and a profound repository of traditional medical knowledge, affording them a distinctive advantage in the procurement and processing of high-value medicinal plants, roots, and herbs. As consumer interest for herbal cures, nutraceuticals, and organic cosmetics increases, tribal communities might exploit several product segments, including herbal medications, immune enhancers, digestive aids, essential oils, skincare items, and herbal teas. The growing popularity of Ayurveda, naturopathy, and holistic wellness in India and elsewhere presents profitable potential for domestic sales and exports.

1.6.2 Cultivation and Sustainable Harvesting of Ethnomedicinal Plants

Indigenous communities, having coexisted with forests and biodiversity for generations, possess profound knowledge regarding the identification, seasonal cycles, growth patterns, and therapeutic properties of medicinal flora, including Giloy, Safed Musli, Ashwagandha, Shatavari, Tulsi, Kalmegh, and Aloe Vera, alongside aromatic species such as lemongrass, citronella, and eucalyptus. Transitioning from exclusive wild harvesting to systematic cultivation enables the assurance of a reliable, high-quality supply of raw materials for the flourishing herbal, nutraceutical, cosmetic, and wellness sectors, while alleviating pressure on endangered species and forest ecosystems.

1.6.3 Traditional Healing Services

Ethnomedicinal traditional healing services constitute a distinctive and culturally significant business potential for tribal and ethnic communities, providing a synthesis of ancestral knowledge, natural cures, and holistic wellness practices that are increasingly in demand in the contemporary health-conscious landscape. Grounded in generations of observation, experimentation, and oral tradition, these services include phytotherapeutic treatments, ritualistic healing, massage techniques, detoxification strategies, and preventative healthcare practices that cater to both bodily and spiritual wellness.

1.6.4 Ethnomedicine in Food and Beverage Sector

The food and beverage sector in ethnomedicine provides tribal and ethnic communities with a dynamic commercial opportunity by converting traditional medicinal plants, herbs, and natural components into edible goods that merge health advantages with cultural authenticity. Leveraging generations of indigenous wisdom regarding the nutritional and medicinal attributes of native flora, tribal entrepreneurs can develop an array of value-added products, including herbal teas infused with Tulsi, lemongrass, or Giloy; fortified beverages derived from Amla, Jamun, or Aloe Vera; and functional foods such as immunity-enhancing snacks, herbal jaggery, or spiced health powders. These products address the rising global demand for natural, chemical-free, and wellness-focused consumables while aligning seamlessly with the rapidly expanding nutraceutical and health drink sectors.

1.6.5 Entrepreneurship in Ethnomedicine Tourism

Entrepreneurship in ethnomedicine tourism provides tribal and ethnic groups with a distinctive, high-value business model that integrates traditional medical practices with the flourishing global wellness and cultural tourist sectors. Utilising their extensive knowledge of medicinal flora, herbal treatments, and holistic practices, tribal entrepreneurs may create immersive tourist experiences that draw people in search of genuine, nature-oriented wellness options. This may encompass guided medicinal plant trails, herbal identification excursions, practical workshops for crafting traditional remedies, and involvement in indigenous therapeutic practices such as herbal steam therapy, detoxification diets, or spiritual medicine processes. Entrepreneurs may incorporate eco-lodges, homestays, and healing retreats into their offers, so crafting a comprehensive cultural and wellness package that attracts domestic visitors, international travellers, researchers, and environmentally aware wellness enthusiasts.

Together with this collaborative research with institutes specialising in ethnopharmacology, biotechnology, and traditional medicine can augment the credibility, safety, and efficacy of products, thereby assisting tribal entrepreneurs in securing broader customer trust. Intellectual Property Rights (IPR) mechanisms, like Geographical Indications (GI) and the Traditional Knowledge Digital Library (TKDL), serve to safeguard indigenous resources and knowledge against exploitation. Notwithstanding these encouraging prospects, obstacles persist, including inadequate formal education, infrastructural deficiencies, restricted financial access, feeble market linkages, and exploitation by intermediaries (Goel and Srikanth 2023; Radha, Kosuri, and Kumar 2023). Therefore, the future approach must prioritise community engagement, participation, and capacity development. Tribal entrepreneurship in ethnomedicine encompasses not only economic advancement but also cultural revitalisation, ecological balance, and the respect for knowledge systems that have benefited humanity for millennia. By investing in this domain with respect, resources, and acknowledgement, we can promote inclusive development, guarantee sustainable healthcare solutions, and enable tribal and ethnic communities to assume leadership roles in the green economy.

1.7 ETHNO-TOURISM AND HEALING RETREATS

Ethno-tourism and Healing Retreats represent two of the most profound convergences between culture, nature, and spirituality in contemporary travel, offering experiences that transcend recreation and instead focus on transformation, self-realisation, and intercultural understanding. In the twenty-first century, as globalisation compresses distances yet widens the gap between cultural authenticity and commercial uniformity, the rise of ethno-tourism embodies a reawakening of curiosity towards indigenous lifeways, traditional knowledge, and ecological ethics that have long been overshadowed by industrial modernity. Ethno-tourism and healing retreats increasingly converge in contemporary travel, blending interest in local/Indigenous culture with structured programs designed to improve psychological, spiritual, and physiological well-being; in practice, this fusion spans yoga and Ayurveda retreats in the Himalayas, mindfulness residencies in Buddhist centres, and community-based programs that share ethnobotanical knowledge, all marketed through the experience economy yet grounded at their best in host agency and reciprocal benefit-sharing (Liu, Li, and Kim 2023; Reyes-García et al. 2022). Evidence suggests retreats can produce measurable short-term and near-term gains across stress, mood, sleep, and anthropometrics, indicating that the retreat 'container' diet, nature exposure, community, and guided practices—functions like a therapeutic micro-ecology rather than a single intervention (Cohen et al. 2017). Parallel tourism studies show wellness experiences spill over into higher life satisfaction via satisfaction-mediated pathways, helping to explain why travellers report transformation that persists after the trip (Liu, Li, and Kim 2023). From a place-based perspective, 'therapeutic landscapes' research adds that restorative natural settings and meaning-rich cultural environments amplify perceived quality of life and spiritual well-being, illuminating how landscape, memory, and flow interact during rural/spiritual stays (Si, Guo, and Zhang 2024). Yet, the same forces that make ethno-healing travel attractive—novelty, authenticity, and embodied knowledge—also create ethical tension: when Indigenous healing is detached from its lifeworlds, commercialised without consent, or scaled via global markets, asymmetries of power and benefit can widen (Marcus 2022; Reyes-García et al. 2022). In ethnomedicine-linked circuits, demand can reframe sacred practice as consumable spectacle, compress timelines (e.g., weekend 'quick-fix' intensives), or privilege visitor expectations over community protocols; scholarship on neo-shamanic ayahuasca networks, for instance, documents real benefits for some participants alongside safety, consent, and cultural-appropriation risks (Manuel-Navarrete 2024; Marcus 2022). A more durable model centres host control, clear health screening and integration support, fair pricing, and knowledge governance that recognises collective rights (Reyes-García et al. 2022). In South Asian settings, retreat content often layers codified systems (yoga, meditation, Ayurveda) with local ethnobotanical practices; ethnobiology and ethnopharmacology studies in the Indian Himalayas and Kashmir show active, evolving plant knowledge embedded in socio-ecological relations an orientation that aligns conceptually with wellness travel's turn towards preventive, lifestyle-based care, but requires careful curation to avoid extractive 'sampling' of TEK (traditional ecological knowledge) for touristic novelty (Batool et al. 2023; Manzoor et al. 2023). Methodologically, the field benefits from triangulating health-science outcomes (e.g., pre/post-biomarkers and psychometrics), tourism-behaviour models (e.g., experience → satisfaction → well-being), and critical policy analysis (e.g., Indigenous rights, FPIC, and benefit-sharing), because retreat impacts are co-produced by program design, guide competence, safety protocols, and socio-environmental context (Cohen et al. 2017; Liu, Li, and Kim 2023; Reyes-García et al. 2022). Looking forward, researchers highlight three priorities: (1) stronger prospective designs with follow-ups to test durability and boundary conditions of retreat effects across populations; (2) integration of therapeutic landscape constructs into quantitative models of well-being and spirituality to capture how nature, culture, and ritual co-shape outcomes; and (3) governance frameworks that operationalise Indigenous agency and knowledge stewardship within tourism value chains (Cohen et al. 2017; Reyes-García et al. 2022; Si, Guo, and Zhang 2024). When these strands are woven together—rigorous health evaluation, culturally grounded facilitation, and rights-based collaboration—ethno-tourism and healing retreats can move beyond wellness consumerism toward ethically robust, place-affirming practice that supports visitor flourishing while strengthening the communities and landscapes that make the healing possible (Liu, Li, and Kim 2023; Manuel-Navarrete 2024; Reyes-García et al. 2022; Si, Guo, and Zhang 2024).

1.8 START-UPS AND E-COMMERCE MODELS IN ETHNOMEDICINE

Start-ups and E-commerce models in ethnomedicine represent one of the most fascinating intersections of ancient knowledge systems and modern digital capitalism, blending the epistemological depth of indigenous healing traditions with the technological scalability of contemporary entrepreneurship. As the twenty-first century witnesses a global rediscovery of holistic health and sustainable consumption, ethnomedicine—defined broadly as the collective medical wisdom embedded in the socio-cultural fabric of indigenous, tribal, and traditional communities—has evolved from being a subject of anthropological curiosity into a vibrant domain of commercial innovation. Ethnomedicine is moving rapidly from field notes and local dispensaries into digitally intermediated value chains in which discovery, validation, branding, compliance, logistics, and after-care are orchestrated by start-ups across e-commerce, telehealth, and data platforms; the leading patterns visible since 2021 show three mutually reinforcing tracks: (i) e-pharmacy and marketplace models that list verified herbal formulations, practitioner-made preparations, and standardised phytopharmaceuticals; (ii) tech-enabled service models (tele-Ayurveda/tele-traditional medicine) bundling consultation, personalisation, subscription refills, and remote monitoring; and (iii) infrastructure plays (quality, traceability, and data commons) that use AI and blockchain to assert provenance, reduce adulteration, and document benefit-sharing obligations across the medicinal-plant supply chain (Apte et al. 2024; D'Cruz and Karuvanthodi 2022; Nguyen et al. 2024; Si, Guo, and Zhang 2024; Wen et al. 2025).

In practice, e-pharmacy is the commercial beachhead because it already solves payments, last-mile delivery, and Rx/OTC compliance at scale; Indian and African compliance audits, however, underscore that online sellers variably meet labelling, prescription, and consumer-information norms, particularly with herbal and traditional products that fall into regulatory grey zones, so start-ups increasingly differentiate on regulatory by design (e.g., automated prescription validation, pharmacist chat, batch-level CoAs, and tamper-evident packaging) (Apte et al. 2024; D'Cruz and Karuvanthodi 2022). Parallel to retailing, tele-Ayurveda and 'digital Ayurveda' services expand access to expert care, especially outside urban centres; platforms integrate scheduling, video consults, structured SOAP-style documentation adapted to traditional nosologies, and follow-up care, while emergent e-learning ecosystems (virtual patients, case libraries) upskill practitioners and standardise triage pathways that can plug into e-commerce fulfilment for personalised formulations (Nedungadi et al. 2025; Puthiyedath, Nair, and Jayan 2025). These service layers create defensible moats (data, protocols, practitioner networks), but they also raise data-protection and ethical-AI questions around culturally embedded diagnostic knowledge; early reviews argue for transparent model cards and participatory co-design with traditional clinicians (Rathor, Gupta, and

Singh 2024). On the sourcing side, e-commerce is constrained by raw-material volatility, species misidentification, and substitution; systematic ethnopharmacology and ethnobotany studies from South Asia illustrate both demand growth and conservation threats, making traceability a core differentiator (Batool et al. 2023; Manzoor et al. 2023). Here, blockchain and IoT propositions target three failure points: (1) geo-origin and species provenance, (2) quality drift along aggregation/processing, and (3) fair compensation for collectors/communities; pilot architectures log farm/forest coordinates, community permits, drying/processing events, and lab results on distributed ledgers, with hash-anchored certificates accessible through QR codes on consumer packs (Bandara, Jayasena, and Karunanayaka 2024; Han, Huang, and Chen 2025; Yang et al. 2025). Strategically, these stacks let start-ups run tiered brands (mass standardised vs. small-batch terroir) and unlock B2B APIs for insurers, hospitals, and cross-border marketplaces that demand high verifiability; operationally, they reduce recalls and counterfeit risk, which improves unit economics for D2C subscription models (Wen et al. 2025). Demand-side evidence suggests wellness and 'therapeutic landscapes' tourism feed funnel activity to digital ethnomedicine brands: travellers discover ingredients/practices *in situ*, then continue care digitally via consults and replenishment plans, creating an O2O loop with higher lifetime value and well-being effects mediated by satisfaction and meaning (Si, Guo, and Zhang 2024). Yet, the translation from use tradition → marketed product is not automatic; marketing science shows that credible ethnomedical evidence (documented indications, dosage traditions, safety narratives) materially improves adoption, and recent work from West Africa demonstrates how curated ethnomedical dossiers can raise conversion without over-claiming (Sangho et al. 2024). Still, two grand challenges persist. First, rights and equity: the CBD/Nagoya ethos demands recognition of Indigenous and local communities' agency, prior informed consent, and benefit sharing—principles often missing from fast-moving D2C funnels; policy scholarship warns that commercialisation surges without community governance risk reproducing extractive patterns even when products are 'natural' (Reyes-García et al. 2022). Second, safety and claims: compliance studies flag gaps in labelling, contraindications, pregnancy warnings, and drug–herb interaction advice online; platforms that embed pharmacist review, structured adverse-event capture, and post-market surveillance can reduce risk and strengthen trust with regulators and payers (Apte et al. 2024). A pragmatic blueprint for start-ups blends: (a) evidence-bearing content (ethnographic context + pharmacology + contemporary RCTs where available); (b) closed-loop personalisation (symptom scores, sleep/stress metrics, and lab markers feeding recommendation engines supervised by clinicians); (c) chain-of-custody tech (blockchain, IPFS, and GS1 identifiers connected to third-party labs) for raw materials and finished goods; and (d) rights-respecting sourcing (contracts with community bodies, audited revenue shares, and open ledgers for benefit-sharing flows) (Bandara, Jayasena, and Karunanayaka 2024; Nguyen et al. 2024;

Reyes-García et al. 2022; Yang et al. 2025). Market-structure research from pharmaceutical e-commerce suggests multi-channel (manufacturer → retailer → platform) games converge on equilibria where authenticity/traceability premiums can be sustained, especially when counterfeit penalties and recall costs are internalised by platforms with reputation at stake; this supports platform-as-insurer models (escrows, quality bonds, and warrantees) that are beginning to appear in herbal categories (Wen et al. 2025).

The next generation of ethnomedicine start-ups looks less like 'online herb stores' and more like vertical operating systems that couple community-anchored knowledge with compliant telehealth, high-integrity supply chains, and longitudinal outcomes data; if entrepreneurs meet the dual test of cultural legitimacy and regulatory rigor, digital ethnomedicine can improve access and livelihoods while protecting biodiversity—otherwise, growth will stall under distrust, regulatory pushback, and supply-side degradation (Apte et al. 2024; D'Cruz and Karuvanthodi 2022; Reyes-García et al. 2022; Si, Guo, and Zhang 2024). Looking ahead, future trends in ethnomedicine and wellness tourism point towards hybrid innovations that merge technology with ancestral wisdom. AI-guided wellness diagnostics are increasingly being paired with tribal herbal pharmacopoeia to personalise treatment protocols and expand evidence-based ethnomedicine (Rathor, Gupta, and Singh 2024). Simultaneously, blockchain-based fair-trade certification systems are emerging to ensure transparency, authenticity, and equitable benefit-sharing within cultural and wellness experiences (Bandara, Jayasena, and Karunanayaka 2024; Han, Huang, and Chen 2025). The rise of carbon-neutral retreats supported by local cooperatives underscores how sustainability and community ownership can reinforce both ecological integrity and socio-economic resilience (Si, Guo, and Zhang 2024). Universities and innovation hubs are expected to establish 'living labs' within healing destinations, integrating plant medicine research with architecture, ecology, and psychology—bridging empirical science and holistic health (Nedungadi et al. 2025). Policy frameworks could further accelerate this transformation by providing visa incentives for long-stay wellness residencies, developing heritage-linked tourism corridors, and offering tax benefits for community-owned resorts, aligning with the United Nations' vision for inclusive, sustainable tourism (Reyes-García et al. 2022). Crucially, international intellectual property frameworks must safeguard Indigenous knowledge systems against biopiracy and exploitative franchising, ensuring that local communities retain agency and economic participation in global ethnomedicine markets (Nguyen et al. 2024; Reyes-García et al. 2022). In spiritual terms, both ethno-tourism and healing retreats reaffirm that the Earth itself is the ultimate healer and teacher, echoing eco-psychological findings that nature immersion enhances emotional balance and social empathy (Si, Guo, and Zhang 2024). Whether one drinks herbal tea under a tribal thatched roof or practices pranayama facing the Himalayas, the underlying lesson is humility before the biosphere. Tourism—often blamed for ecological degradation—here becomes a vector of restoration, as travellers plant trees, fund schools, learn permaculture, and carry the message of interconnectedness back to urban life. The convergence of anthropology and wellness thus signals an evolution in human consciousness—from conquering nature to conversing with it. Ultimately, ethno-tourism and healing retreats invite humanity to rediscover its roots while envisioning a sustainable future. They serve both as a mirror to the past reflecting how ancestral cultures lived in rhythmic harmony with seasons and spirits and a window to the future, where technology and tradition coexist to nurture both planet and psyche. The journey is circular: one travels outward to remote cultures or mountains only to return inward to one's own forgotten wholeness. In this reciprocity lies their enduring allure and civilisational importance.

1.9 CONCLUSION

Ethnomedicine in tribal and indigenous cultures embodies a significant amalgamation of ecological knowledge, cultural identity, and holistic health practices that have developed over generations of intimate engagement with nature. These therapeutic systems imbued with spiritual, botanical, and communal wisdom have not only preserved the health of indigenous populations but are also garnering global acclaim for their natural, sustainable, and culturally grounded methodologies to well-being. The global transition towards organic, herbal, and preventative healthcare is increasing the prominence of ethnomedicine, providing both recognition and prospects for these traditional systems. Ethnomedicine transcends its role as a medical modality and now occupies a pivotal position in the entrepreneurial advancement of tribal and ethnic communities. Therefore, with appropriate support, encompassing ethical commercialisation, legal safeguards, scientific confirmation, and community-driven initiatives, traditional healers and knowledge custodians can convert their history into sustainable economic opportunities. These initiatives not only safeguard cultural heritage but also economically empower communities, particularly in marginalised areas. This future must be constructed with a foundation of cultural integrity, equitable benefit-sharing, and sustainability. Encouraging tribal-led enterprises in ethnomedicine is a chance for the world to adopt a development paradigm that respects tradition, enhances biodiversity, and facilitates inclusive progress for some of the most resilient and resourceful populations globally.

CONFLICT OF RESEARCH INTEREST

The authors declare no conflict of research interest.

ACKNOWLEDGEMENT

The authors are thankful to the Department of Anthropology and Tribal Studies, Central University of Jharkhand, Ranchi, India, for providing necessary support for the research work.

REFERENCES

Aggarwal, B. B., & Harikumar, K. B. 2009. "Potential Therapeutic Effects of Curcumin, the Anti-inflammatory Agent, against Neurodegenerative, Cardiovascular, Pulmonary, Metabolic, Autoimmune and Neoplastic Diseases." *The International Journal of Biochemistry & Cell Biology*, 41 (1):40–59.

Ahmed, G.E.M., et al. 2023. "Prevalence and Reasons to Seek Traditional Healing Methods Among Residents of Two Localities in North Kordofan State, Sudan 2022: A Cross-Sectional Study." *Health Science Reports* 6 (8): e1487.

Ahmed, H.M. 2018. "Ethnomedicinal, Phytochemical and Pharmacological Investigations of *Perilla frutescens* (L.) Britt." *Molecules* 24 (1): 102.

Ali, B. H., Blunden, G., Tanira, M. O., & A. Nemmar. 2008. "Some Phytochemical, Pharmacological and Toxicological Properties of Ginger (Zingiber Officinale Roscoe): A Review of Recent Research." *Food and Chemical Toxicology*, 46(2):409–20.

Alum, E.U. 2024. "The Role of Indigenous Knowledge in Advancing the Therapeutic Use of Medicinal Plants: Challenges and Opportunities." *Plant Signaling & Behavior* 19 (1): 2439255.

Anwar, F., Latif, S., Ashraf, M., & A. H. Gilani. 2007. "Moringa Oleifera: A Food Plant with Multiple Medicinal Uses. Phytotherapy Research: An International Journal Devoted to Pharmacological and Toxicological Evaluation of Natural Product Derivatives." 21(1):17–25.

Anju, T., & A. Kumar. 2024. "Traditional Ecological Knowledge and Medicinal Plant Diversity Usage Among the Mullu Kuruman Tribes of Wayanad District of Kerala, India and Its Implications for Biodiversity Conservation in the Face of Climate Change." *Trees, Forests and People*, 16: 100595.

Attele, A. S., Wu, J. A., & C. S. Yuan. 1999. "Ginseng Pharmacology: Multiple Constituents and Multiple Actions." *Biochemical Pharmacology*, 58(11):1685–93.

Apte, A., R. Swain, A. Rai, N. Ajmera, S. Kulkarni, and A. Mehendale. 2024. "Facilitators, Barriers, and Potential Impacts of Implementation of E-Pharmacy Guidelines in India and Kenya." *JMIR Public Health and Surveillance* 10 (1): e51080. https://doi.org/10.2196/51080

Awoke, A., and B.N. Cosendey. 2025. "Investigate the Ethnomedical Practices of Different Indigenous Communities: A Literature Review." *International Journal of Ethnoscience and Technology in Education* 2 (1): 64–89.

Bandara, K., M. Jayasena, and D. Karunanayaka. 2024. "Exploring the Blockchain's Green Revolution in Medicinal Plant Supply Chains: Blossoming Trust—A Review." *Journal of the Association of Public Health Inspectors of Sri Lanka* 53 (2): 35–54.

Batool, Z., M. Zahoor, I.U. Rahman, and R.W. Bussmann. 2023. "Medicinal Plants Traditionally Used in the Health Care System of Northern Pakistan and Kashmir: Current Therapeutic Practices and Implications." *Journal of Ethnopharmacology* 314: 116763. https://doi.org/10.1016/j.jep.2023.116763.

Behera, D., Menon, D., Wilson, V. K., & N. Ayyappan. 2025. "Tree Diversity, Community Structure and Aboveground Biomass of a Lowland Dipterocarp Forest of Western Ghats, India." *Trees, Forests and People*, 101048.

Bempong, D. K., & P. J. Houghton. 1992. "Dissolution and Absorption of Caffeine from Guarana." *Journal of Pharmacy and Pharmacology*, 44(9):769–71.

Beyene, B., B. Beyene, and H. Deribe. 2016. "Review on Application and Management of Medicinal Plants for the Livelihood of the Local Community." *Journal of Resources Development and Management* 22 (1): 33–39.

Bhowmick, B., & S. Prabhakar. 2023. "Phytochemicals for Drug and Food Development: A Techno-Entrepreneurial Opportunity for Herbal Industry." *International Journal of All Research Education and Scientific Methods*, 11(10):528–39.

Biswas, K., Chattopadhyay, I., Banerjee, R. K., & U. Bandyopadhyay. 2002. "Biological Activities and Medicinal Properties of Neem (*Azadirachta indica*)." *Current Science*, 82(11): 1336–45.
Biswas, G., B. Sarkar, S.R. Sarkar, N. Sharma, S. Sultana, and N.R. Madhu. 2023. "Exploring the Ethnobotanical Resources of Indian Sundarbans: An Insight into Medicinal Plants." In *The Basic Handbook of Indian Ethnobotany and Traditional Medicine*. India: International Academic Publishing House.
Brendler, T. 2021. "From Bush Medicine to Modern Phytopharmaceutical: A Bibliographic Review of Devil's Claw (*Harpagophytum* spp.)." *Pharmaceuticals*, 14(8): 726.
Buragohain, J. 2011. "Ethnomedicinal Plants Used by the Ethnic Communities of Tinsukia District of Assam, India." *Recent Research in Science and Technology* 3 (9): 31–42.
Bussmann, R.W. 2013. "The Globalization of Traditional Medicine in Northern Peru: From Shamanism to Molecules." *Evidence-Based Complementary and Alternative Medicine* 2013 (1): 291903.
Caballero-Serrano, V., B. McLaren, J.C. Carrasco, J.G. Alday, L. Fiallos, J. Amigo, and M. Onaindia. 2019. "Traditional Ecological Knowledge and Medicinal Plant Diversity in Ecuadorian Amazon Home Gardens." *Global Ecology and Conservation* 17: e00524.
Cabrera, C., Artacho, R., & R. Giménez. (2006). "Beneficial Effects of Green Tea—A Review." *Journal of the American College of Nutrition*, 25(2):79–99.
Census of India. 2011. *Primary Census Abstract: Scheduled Tribes*. New Delhi: Office of the Registrar General & Census Commissioner, Ministry of Home Affairs, Government of India.
Chaachouay, N., and L. Zidane. 2024. "Plant-Derived Natural Products: A Source for Drug Discovery and Development." *Drugs and Drug Candidates* 3 (1): 184–207.
Chandrakar, K., T. Satapathy, M. Nayak, G. Dhalendra, and B. Kumar. 2014. "Ethnomedicinal Plants in Chhattisgarh Origin." *Journal of Applied Pharmaceutical Research* 2 (3): 28–41.
Chen, H., and Y. Bu. 2019. "Anthropocosmic Vision, Time, and Nature: Reconnecting Humanity and Nature." *Educational Philosophy and Theory* 51 (11): 1130–40.
Chen, S.L., H. Yu, H.M. Luo, Q. Wu, C.F. Li, and A. Steinmetz. 2016. "Conservation and Sustainable Use of Medicinal Plants: Problems, Progress, and Prospects." *Chinese Medicine* 11 (1): 37.
Chirima, R. 2025. "Harnessing Digital Technologies to Preserve and Promote Indigenous Knowledge Systems in Africa." *International Journal of Innovative Science and Research Technology* 10 (5): 4411–16.
Cohen, M.M., F. Elliott, L. Oates, A. Schembri, and N. Mantri. 2017. "Do Wellness Tourists Get Well? An Observational Study of Multiple Dimensions of Health and Well-Being after a Week-Long Retreat." *Journal of Alternative and Complementary Medicine* 23 (2): 140–8. https://doi.org/10.1089/acm.2016.0333
Curbishley, L.S. 2015. "Destabilizing the Colonization of Indigenous Knowledge: In the Case of Biopiracy." Master's thesis, [institution not specified].
D'Cruz, A.C., and M. Karuvanthodi. 2022. "The Rise of E-Pharmacy in India: Benefits, Risks, and Regulatory Concerns." *Indian Journal of Public Health Research & Development* 54 (4): 31–6.
Dean, M. 2024. "Exploring Ethnobotanical Knowledge: Qualitative Insights into the Therapeutic Potential of Medicinal Plants." *Golden Ratio of Data in Summary* 4 (2): 154–66.
De La Rosa, L.A., J.O. Moreno-Escamilla, J. Rodrigo-García, and E. Alvarez-Parrilla. 2019. "Phenolic Compounds." In *Postharvest Physiology and Biochemistry of Fruits and Vegetables*, 253–71. Cambridge: Woodhead Publishing.
de Oliveira, I., Santos-Buelga, C., Aquino, Y., Barros, L., & S. A. Heleno. 2025. "New Frontiers in the Exploration of Phenolic Compounds and Other Bioactives as Natural Preservatives." *Food Bioscience*, 68:106571.

Desborough, M. J., & D. M. Keeling. 2017. "The Aspirin Story–From Willow to Wonder Drug." *British Journal of Haematology*, 177(5):674–83.

Duche-Pérez, A., O.A. Gutiérrez Aguilar, J.L. Valero Quispe, A. Serruto Castillo, and L.J. Mamani Daza. 2024. "Systematic Review of Ethnomedicinal Knowledge: Documentation, Evaluation, and Conservation of Medicinal Plants and Their Therapeutic Applications." *F1000Research* 13: 1324.

Dushing, Y.A., and D.A. Patil. 2010. "Studies on Ethnomedicine in Buldhana District of Maharashtra (India)." *Journal of Phytology* 2 (12): 35–41.

Farhat, C. 2021. *Biopiracy: Business vs Traditional Knowledge and Biodiversity*. SSRN 3933062.

Faurant, C. 2011. "From Bark to Weed: The History of Artemisinin." *Parasite:* Journal de la Société Française de Parasitologie 18 (3): 215.

Farombi, E.O., Akanni, O.O. and G.O. Emerole. 2002. "Antioxidative and Scavenging Activities of Kolaviron In Vitro." *Pharmaceutical Biology*, 40:107–16.

Flora, K., Hahn, M., Rosen, H., & K. Benner. 1998. "Milk Thistle (Silybum Marianum) for the Therapy of Liver Disease". *The American Journal of Gastroenterology*, 93(2):139–43.

Feeney, K.M. 2016. "Peyote and the Native American Church: An Ethnobotanical Study at the Intersection of Religion, Medicine, Market Exchange, and Law." PhD diss., Washington State University.

Gachelin, G., P. Garner, E. Ferroni, U. Tröhler, and I. Chalmers. 2017. "Evaluating Cinchona Bark and Quinine for Treating and Preventing Malaria." *Journal of the Royal Society of Medicine* 110 (2): 73–82.

Gitima, G., Gebre, A., Berhanu, Y., & T. Wato. 2025. "Exploring Indigenous Wisdom: Ethnobotanical Documentation and Conservation of Medicinal Plants in Goba District, Southwest Ethiopia." *Scientific African*, 27:e02571.

Goel, S., and N. Srikanth. 2023. "Local Health Traditions and Preservation of National Heritage: Emphasis on Intellectual Property Rights Protection, Benefit Sharing and Mainstreaming." In *Ethnomedicine and Tribal Healing Practices in India: Challenges and Possibilities of Recognition and Integration*, 307–23. Singapore: Springer Nature Singapore.

Gudulkar, S., Rajbhar, K., Dawda, H., & U. Mukundan. 2016. "Plant system as a Tool for Validating Ethnobotanical Claims for Kidney Stone Treatment." *World Journal of Pharmaceutical Research*, 5(6), 9.

Gupta, S., Satpathy, B., & S. K. Baral. 2022. "A study on start-up India: a governmental initiative to boost entrepreneurship, growth and employment generation of tribal youth of TSP Region of Southern Rajasthan." In Proceedings of the 2nd Indian International Conference on Industrial Engineering and Operations Management (pp. 476–86).

Gupta, V., Jamwal, G., Rai, G. K., Gupta, S. K., Shukla, R. M., Dadrwal, B. K., ... & C. N. Aguilar. 2024. "Biosynthesis of Biomolecules from Saffron as an Industrial Crop and Their Regulation, with Emphasis on Chemistry, Extraction Methods, Identification Techniques, and Potential Applications in Human Health and Food: A Critical Comprehensive Review." *Biocatalysis and Agricultural Biotechnology*, 59, 103260.

Han, M., Z. Huang, and Y. Chen. 2025. "Blockchain Technology in Herbal Medicines: Applications, Opportunities, and Challenges." *NanoImpact* 34: 100487. https://doi.org/10.1016/j.impact.2025.100487

Houghton, P. J. 1999. "The scientific Basis for the Reputed Activity of Valerian." *Journal of Pharmacy and Pharmacology*, 51(5):505–12.

Ijeh, I. I., & Ejike, C. E. C. C. 2011. "Current Perspectives on the Medicinal Potentials of Vernonia Amygdalina Del." *Journal of Medicinal Plants Research*, 5(7):1051–61.

Imran, Y., N. Wijekoon, L. Gonawala, Y.C. Chiang, and K.R.D. De Silva. 2021. "Biopiracy: Abolish Corporate Hijacking of Indigenous Medicinal Entities." *The Scientific World Journal* 2021 (1): 8898842.

Jones, K. 2003. "Review of sangre de drago (Croton lechleri)-a South American Tree Sap in the Treatment of Diarrhea, Inflammation, Insect Bites, Viral Infections, and Wounds: Traditional Uses to Clinical Research". *The Journal of Alternative and Complementary Medicine: Paradigm, Practice, and Policy Advancing Integrative Health*, 9(6):877–96.

Kapoor, B.B.S., and S. Lakhera. 2013. "Ethnomedicinal Plants of Jodhpur District, Rajasthan Used in Herbal and Folk Remedies." *Indian Journal of Pharmaceutical and Biological Research* 1 (4): 71–5.

Kasilo, O.M., and J.B. Nikiema. 2014. "World Health Organization Perspective for Traditional Medicine." In *Novel Plant Bioresources: Applications in Food, Medicine and Cosmetics*, 23–42.

Kaur, J., R. Kaur, and A.K. Nagpal. 2020. "Traditional Use of Ethnomedicinal Plants Among People of Kapurthala District, Punjab, India." *Pharmacognosy Magazine* 16 (68): 123–31.

Kaur, J., Kaur, R., & A. Nagpal. 2020. "Traditional Use of Ethnomedicinal Plants Among People of Kapurthala District, Punjab, India." *Pharmacognosy Magazine*, 16(68):69–80.

Kaur, S., Shams, R., Dash, K. K., Pandey, V. K., Shaikh, A. M., Harsányi, E., & B. Kovács. 2024. "Phytochemical and Pharmacological Characteristics of Phalsa (Grewia asiatica L.): A Comprehensive Review." *Heliyon*, 10(2).

Keshari, D.S., and M. Bishnupriya. 2022. "Ethono-Medico-Botanical Survey of Netravali Villages Nearby Wildlife Sanctuary of Goa." *International Ayurvedic Medical Journal* 10 (1): 57–61.

Khoddami, A., M.A. Wilkes, and T.H. Roberts. 2013. "Techniques for Analysis of Plant Phenolic Compounds." *Molecules* 18 (2): 2328–75.

Kim, H.P., H. Lim, and Y.S. Kwon. 2016. "Therapeutic Potential of Medicinal Plants and Their Constituents on Lung Inflammatory Disorders." *Biomolecules & Therapeutics* 25 (2): 91.

Kimta, N., A.F. Majdalawieh, G.K. Nasrallah, S. Puri, E. Nepovimova, K. Jomova, and K. Kuča. 2024. "Leprosy: Comprehensive Insights into Pathology, Immunology, and Cutting-Edge Treatment Strategies, Integrating Nanoparticles and Ethnomedicinal Plants." *Frontiers in Pharmacology* 15: 1361641.

Klayman, D. L. 1985. Qinghaosu (Artemisinin): An Antimalarial Drug from China. *Science*, 228(4703):1049–55.

Kukula-Koch, W.A., and J. Widelski. 2017. "Alkaloids." In *Pharmacognosy*, 163–98. Cambridge: Academic Press.

Kumar, M., S. Rawat, B. Nagar, A. Kumar, N.A. Pala, J.A. Bhat, et al. 2021. "Implementation of the Use of Ethnomedicinal Plants for Curing Diseases in the Indian Himalayas and Its Role in Sustainability of Livelihoods and Socioeconomic Development." *International Journal of Environmental Research and Public Health* 18 (4): 1509.

Kurniadi, R. 2024. "Integrating Ethnomedicine Practices into Family Economic Education: Exploring Health and Economic Empowerment Synergies." *International Journal of Ethnomedicine and Nutraceutical* 1 (1): 36–44.

Laldinsanga, H.S., T. Jahan, A. Kumar, and H.K.S. Goswami. 2019. "Traditional Anti-Malarial Drugs from Serchhip and Lunglei Districts of Mizoram." *Current Trends in Pharmaceutical Research* 6 (1): 76–104.

Landor, M., A. Benami, N. Segev, and B. Loberant. 2015. "Efficacy and Acceptance of a Commercial *Hoodia parviflora* Product for Support of Appetite and Weight Control in a Consumer Trial." *Journal of Medicinal Food* 18 (2): 250–58.

Li-Weber, M. 2009. "New Therapeutic Aspects of Flavones: The Anticancer Properties of Scutellaria and Its Main Active Constituents Wogonin, Baicalein and Baicalin." *Cancer Treatment Reviews*, 35(1): 57–68.

Liu, L., J. Li, and W. Kim. 2023. "The Impact of the Wellness Tourism Experience on Tourists' Well-Being: A Structural Equation Model." *Sustainability* 15 (3): 1872. https://doi.org/10.3390/su15031872

Ludwiczuk, A., K. Skalicka-Woźniak, and M.I. Georgiev. 2017. "Terpenoids." In *Pharmacognosy*, 233–66. Cambridge: Academic Press.

Mahapatra, A.D., P. Bhowmik, A. Banerjee, A. Das, D. Ojha, and D. Chattopadhyay. 2019. "Ethnomedicinal Wisdom: An Approach for Antiviral Drug Development." In *New Look to Phytomedicine*, 35–61. Cambridge: Academic Press.

Manuel-Navarrete, D. 2024. "Ayahuasca Ceremonies, Relationality, and Inner–Outer Transformations: Ethical Challenges in Global Spiritual Circuits." *Journal of Psycho-spirituality in Religion and Society* 5 (2): 101–21. https://doi.org/10.1080/26395916.2024.2339227

Manzoor, M., S. Sakhi, J.A. Bhat, and R.W. Bussmann. 2023. "The Local Medicinal Plant Knowledge in Kashmir Western Himalaya: Documentation, Threats, and Conservation Implications." *Journal of Ethnobiology and Ethnomedicine* 19: 16. https://doi.org/10.1186/s13002-023-00631-2

Marcus, O. 2022. "Ethical Tensions Among Globalized Ayahuasca Shamanisms: Between Curanderos, Therapists, and Tourists." *Frontiers in Psychology* 13: 951777. https://doi.org/10.3389/fpsyg.2022.951777

Mgbeoji, I. 2007. *Global Biopiracy: Patents, Plants, and Indigenous Knowledge*. Vancouver: UBC Press.

Mgbeoji, I. 2012. "Bio-Cultural Knowledge and the Challenges of Intellectual Property Rights Regimes for African Development." *Dalhousie Law Journal* 35: 397.

Ministry of Tribal Affairs. 2021. *Statistical Profile of Scheduled Tribes in India*. India: Government of India.

Mishra, N., A. Joshi, A. Jadon, and R. Kalvankar. 2025. "Reviving Roots, Shaping Futures: Inspiring Tribal Entrepreneurship through Tourism: A Road Map to Viksit Bharat." In *Viksit Bharat 2047: Empowering a Nation Through Entrepreneurship*, 18.

Mistry, K.S., Z. Sanghvi, G. Parmar, and S. Shah. 2014. "The Antimicrobial Activity of *Azadirachta indica*, *Mimusops elengi*, *Tinospora cardifolia*, *Ocimum sanctum* and 2% Chlorhexidine Gluconate on Common Endodontic Pathogens: An *In Vitro* Study." *European Journal of Dentistry* 8 (2): 172–7.

Müller, W. E. 2003. "Current St. John's Wort Research from Mode of Action to Clinical Efficacy." *Pharmacological Research*, 47(2):101–09.

Naik, B., & S. Panda. 2023. "Empowerment of Tribal Communities Through Innovation and Entrepreneurship: A path to sustainable progress." *International Journal of Research and Review*, 10(11): 334–43.

Nasim, N., I.S. Sandeep, and S. Mohanty. 2022. "Plant-Derived Natural Products for Drug Discovery: Current Approaches and Prospects." *The Nucleus* 65 (3): 399–411.

Nedungadi, P., S. Suresh, R. Divakaran, A. Pahwa, and A Menon. 2025. "Learning Technology for Ayurveda with Pedagogically Rich Virtual Patients: AyurSIM." *Frontiers in Medicine* 12: 1537047. https://doi.org/10.3389/fmed.2025.1537047

Negi, C.S., S. Nautiyal, L. Dasila, K.S. Rao, and R.K. Maikhuri. 2002. "Ethnomedicinal Plant Uses in a Small Tribal Community in a Part of Central Himalaya, India." *Journal of Human Ecology* 14 (1): 23–31.

Nguyen, D., L. Ampofo, B. Ncube, R. Opoku, and M. Rahman. 2024. "Harnessing Digital Platforms in Advancing Traditional and Complementary Medicine for Cancer Care in Sub-Saharan Africa." *JMIR Cancer* 10: e52250. https://doi.org/10.2196/52250

Ota, A.B. 2020. *Study on Traditional Medicinal Healing Practices among Selected Scheduled Tribe Communities in Odisha*. Available on https://repository.tribal.gov.in/bitstream/123456789/74429/1/SCST_2013_research_0350.pdf (Accessed May 25, 2025).

Plowman, T. 1984. "The Ethnobotany of Coca (*Erythroxylum* spp., Erythroxylaceae)." *Advances in Economic Botany*, 1: 62–111.

Paul, S., & M. Das. 2025. "Medicinal Plants used for Traditional Healing Practices Among the Tribal Communities of Ri-Bhoi District, Meghalaya, India." *Current Botany*, 16:157–68.

Pathak, A., A.P. Gupta, and P. Pandey. 2024. "Herbal Medicine and Sustainable Development Challenges and Opportunities." In *Herbal Medicine Phytochemistry: Applications and Trends*, 1–26.

Petrovska, B.B. 2012. "Historical Review of Medicinal Plants' Usage." *Pharmacognosy Reviews* 6 (11): 1.

Pirintsos, S., A. Panagiotopoulos, M. Bariotakis, V. Daskalakis, C. Lionis, G. Sourvinos, et al. 2022. "From Traditional Ethnopharmacology to Modern Natural Drug Discovery: A Methodology Discussion and Specific Examples." *Molecules* 27 (13): 4060.

Prabhakar, P., and M. Banerjee. 2020. "Antidiabetic Phytochemicals: A Comprehensive Review on Opportunities and Challenges in Targeted Therapy for Herbal Drug Development." *International Journal of Pharmaceutical Research* 12 (2): 1673–96.

Prabhakar, P., and B. Mamoni. 2021. "Technical Problems, Regulatory and Market Challenges in Bringing Herbal Drug into Mainstream of Modern Medicinal Practices." *Research Journal of Biotechnology* 16 (3): 212–22.

Prabhakar, P., S. Mukherjee, A. Kumar, S. Kumar, D.K. Verma, S. Dhara, et al. 2023. "Optimization of Microwave-Assisted Extraction (MAE) of Key Phenolic Compounds from Pigeon Pea (*Cajanus cajan* L.), Their Characterization, and Measurement of Their Anti-Diabetic and Cytotoxic Potential." *Journal of Food Measurement and Characterization* 17 (6): 5697–720.

Prabhakar, P., S. Tripathy, D.K. Verma, M. Banerjee, and P.P. Srivastav. 2025a. "Secondary Metabolites of Pigeonpea [*Cajanus cajan* (L.) Millsp.] with Their Biological and Pharmacological Activities." In *Integrated Improvement of Food Legumes*, 33.

Prabhakar, P., Mukherjee, S., Kumar, A., Rout, R. K., Kumar, S., Verma, D. K., ... & Banerjee, M. 2024. "In Silico, In Vitro and Ex Vivo Evaluation of The Antihyperglycaemic, Antioxidant and Cytotoxic Properties of Coccinia Grandis L. Leaf Extract." Food Technology and Biotechnology, 62(2): 188–204.

Prabhakar, P., Tripathy, S., Verma, D. K., Banerjee, M., & P. P. Srivastav. 2025. "Secondary Metabolites of Pigeonpea [*Cajanus cajan* (L.) Millsp.] with Their Biological and Pharmacological Activities. In Integrated Improvement of *Food Legumes*, (pp. 33–54)." Cham: Springer Nature Switzerland.

Prabhakar, S., & A. Nimesh. 2022. "Skill Development Programs for Women in India: Current Status and Future Perspectives." *World Journal of Advanced Engineering Technology and Sciences*, 7(01):128–36.

Prabhakar, S., A. Nimesh, and P. Prabhakar. 2023a. "Assessment of Start-Up Ecosystems for Tribal Women Entrepreneurs in India." In *Diversity, Equity, and Inclusion Efforts of Businesses in Rural Areas*, 1–22. Hershey: IGI Global.

Prabhakar, S., A. Nimesh, and P. Prabhakar. 2023b. "Startup Ecosystem for Rural Development in India." In *Socio-Economic Disparities, Vulnerable Communities, and the Future of Work and Entrepreneurship*, 26–50. Hershey: IGI Global.

Prabhakar, S., P. Prabhakar, A. Nimesh, and M. Banerjee. 2025b. "Production and Development of Nutraceuticals Using Sustainable Food System." In *Food Bioactives and Nutraceuticals: Dietary Medicine for Human Health and Well-Beings*, 59–86. Singapore: Springer Nature Singapore.

Pradeep, T. 2016. "Traditional Ethnomedicinal Knowledge of Indian Tribes." *Current Science* 110 (4): 486.

Prakash, P. A. G. N., & N. Gupta. 2005. "Therapeutic Uses of Ocimum Sanctum Linn (Tulsi) with a Note on Eugenol and its Pharmacological Actions: A Short Review." *Indian Journal of Physiology and Pharmacology*, 49(2):125.

Press Information Bureau. 2025. "From the Margins to the Mainstream". Available on https://www.pib.gov.in/PressNoteDetails.aspx?NoteId=154651&ModuleId=3&reg=3&lang=2#:~:text=India%20is%20home%20to%20one,and%20deep%20connection%20with%20nature (Accessed on July 25, 2025).
Pullaiah, T. 2018. "Biodiversity in India." In Global Biodiversity (pp. 109–44). Apple Academic Press, USA.
Puthiyedath, R., P.R. Nair, and N. Jayan. 2025. "AyurCeL: A Comprehensive Ayurveda Clinical E-Learning Platform for Sharing Case Studies and Community Learning." *Journal of Ayurveda and Integrative Medicine* 16 (1): 100845. https://doi.org/10.1016/j.jaim.2025.100845
Queiroz, L.P., I.B.R. Nogueira, and A.M. Ribeiro. 2024. "Flavor Engineering: A Comprehensive Review of Biological Foundations, AI Integration, Industrial Development, and Socio-Cultural Dynamics." *Food Research International* 196: 115100.
Quinlan, M.B. 2011. "Ethnomedicine." In *A Companion to Medical Anthropology*, 379–403.
Radha, S., N. Kosuri, and B. Kumar. 2023. "Ethnobotany and Intellectual Property Rights: Balancing Access, Benefit Sharing, and Traditional Knowledge Protection." *International Journal of Food and Nutritional Sciences* 11 (2022).
Rahman, K., & G. M. Lowe. 2006. "Garlic and Cardiovascular Disease: A Critical Review." *The Journal of Nutrition*, 136(3), 736S–40S.
Raj, M.S.A., and M. Ayyanar. 2024. "Ethnopharmacological Importance of Commonly Used Folk Medicinal Plants Among the Malayali Tribal Community in Jawadhu Hills, Tamil Nadu, India: A Review." *Ethnobotany Research and Applications* 27: 1–41.
Rathor, C., V. Gupta, and R. Singh. 2024. "Artificial Intelligence in Ayurveda: Current Concepts and Prospects." *Journal of Ayurveda and Integrative Medicine* 15 (3): 200–10.
Ravishankar, B., and V.J. Shukla. 2007. "Indian Systems of Medicine: A Brief Profile." *African Journal of Traditional, Complementary and Alternative Medicines* 4 (3): 319–37.
Reddy, S., B. Subedi, and N. Guite. 2023. "Introduction: Ethnomedicine and Tribal Healing Practices in India: Challenges and Possibilities of Recognition and Integration." In *Ethnomedicine and Tribal Healing Practices in India: Challenges and Possibilities of Recognition and Integration*, 1–31. Singapore: Springer Nature Singapore.
Registrar General & Census Commissioner. 2013. *Census of India 2011: Data on Scheduled Tribes*. India: Government of India.
Reyes-García, V., Á. Fernández-Llamazares, Y. Aumeeruddy-Thomas, P. Benyei, R.W. Bussmann, S.K. Diamond, and E.S. Brondizio. 2022. "Recognizing Indigenous Peoples' and Local Communities' Rights and Agency in the Post-2020 Biodiversity Agenda." *Ambio* 51 (1): 84–92. https://doi.org/10.1007/s13280-021-01561-7
Rizvi, S.A., G.P. Einstein, O.L. Tulp, F. Sainvil, and R. Branly. 2022. "Introduction to Traditional Medicine and Their Role in Prevention and Treatment of Emerging and Re-Emerging Diseases." *Biomolecules* 12 (10): 1442.
Saini, N., V. Lather, and S.K. Gahlawat. 2022. "Exploring Phytochemicals from Himalayan Medicinal Plants as Novel Therapeutic Agents." *Anti-Cancer Agents in Medicinal Chemistry* 22 (9): 1674–98.
Sandoval, M., Charbonnet, R. M., Okuhama, N. N., Roberts, J., Krenova, Z., Trentacosti, A. M., & M. J. Miller. 2000. "Cat's Claw Inhibits TNFα Production and Scavenges Free Radicals: Role in Cytoprotection." *Free Radical Biology and Medicine*, 29(1): 71–8.
Sangho, A., S. Dembélé, F. Traoré, A. Mounkoro, and D. Diallo. 2024. "Contribution of Ethnomedical Use Data to the Marketing of Traditional Herbal Medicines in Mali." *Journal of Ethnobiology and Ethnomedicine* 20: 59. https://doi.org/10.1186/s13002-024-00702-8
Sarawagi, A., & M. Singh. 2024. "Enhancing Livelihoods and Empowering Tribal Women: Analyzing the Role of Self-Help Groups in Sonbhadra." *Sachetas*, 3(2):23–30.

Sasidharan, S.P., Z. Viji, R.P. Sankaran, S. Muraleedharan, and K. Arunachalam. 2024. "Exploring Folklore Herbal Knowledge: A Compilation from Palakkad District, Kerala, India." *Ethnobotany Research and Applications* 27: 1–28.

Shah, A., K.A. Bharati, J. Ahmad, and M.P. Sharma. 2015. "New Ethnomedicinal Claims from Gujjar and Bakerwals Tribes of Rajouri and Poonch Districts of Jammu and Kashmir, India." *Journal of Ethnopharmacology* 166: 119–28.

Shahrajabian, M.H., and W. Sun. 2023. "Survey on Medicinal Plants and Herbs in Traditional Iranian Medicine With Anti-Oxidant, Anti-Viral, Anti-Microbial, and Anti-Inflammation Properties." *Letters in Drug Design & Discovery* 20 (11): 1707–43.

Shankar, R., and M.S. Rawat. 2013. "Medicinal Plants Used in Traditional Medicine in Aizawl and Mamit Districts of Mizoram." *Journal of Biology and Life Sciences* 4 (2): 95.

Sharma, J., S. Gairola, R.D. Gaur, R.M. Painuli, and T.O. Siddiqi. 2013. "Ethnomedicinal Plants Used for Treating Epilepsy by Indigenous Communities of Sub-Himalayan Region of Uttarakhand, India." *Journal of Ethnopharmacology* 150 (1): 353–70.

Shukla, G., Bhat, J. A., & S. Chakravarty. 2022. "Species Richness and Folk Therapeutic Uses of Ethnomedicinal Plants in West Bengal, India–A Meta-Analysis." *Phytomedicine Plus*, 2(1):100158.

Si, Y., Y. Guo, and H. Zhang. 2024. "Health Destinations, Therapeutic Landscapes, and Tourists' Perceived Quality of Life." *Tourism Management Perspectives* 49: 101115. https://doi.org/10.1016/j.tmp.2024.101115.

Singh, A., M.C. Nautiyal, R.M. Kunwar, and R.W. Bussmann. 2017. "Ethnomedicinal Plants Used by Local Inhabitants of Jakholi Block, Rudraprayag District, Western Himalaya, India." *Journal of Ethnobiology and Ethnomedicine* 13 (1): 49.

Singh, D., & P. K. Chaudhuri. 2017. "Chemistry and Pharmacology of Tinospora Cordifolia." *Natural Product Communications*, 12(2): 1934578X1701200240.

Singh, N., Bhalla, M., de Jager, P., & M. Gilca. 2011. "An Overview on Ashwagandha: A Rasayana (rejuvenator) of Ayurveda." *African Journal of Traditional, Complementary, and Alternative Medicines*, 8(5 Suppl): 208.

Singh, Y.D., M.K. Panda, and K.B. Satapathy. 2020. "Ethnomedicine for Drug Discovery." In *Advances in Pharmaceutical Biotechnology: Recent Progress and Future Applications*, 15–28. Singapore: Springer Singapore.

Sofowora, A., E. Ogunbodede, and A. Onayade. 2013. "The Role and Place of Medicinal Plants in the Strategies for Disease Prevention." *African Journal of Traditional, Complementary and Alternative Medicines* 10 (5): 210–29.

Sonam, S. K., Hans, H., & J. Bisen. 2020. "Impact Assessment of Mushroom Cultivation on Livelihood of Women Mushroom Growers of Samastipur District of Bihar." *Journal of Pharmacognosy and Phytochemistry*, Sp9(2): 251–3.

Sonam, Shishir, K., & P. Prabhakar. 2021. "Constraint Analysis in Mushroom Cultivation Among Women Mushroom Cultivators in the State of Bihar, India." *Advances in Management*, 14(2):7–11.

Sridhar, V., T. Sunitha, S.H. Prasad, and B. Padmaja. 2018. "Ethnomedicinal Practices in Different Communities of Telangana for Treatment of Wounds." *American Journal of Scientific and Medical Research* 4 (1): 20–5.

Sultana, R., N. Muhammad, and A.K.M. Zakaria. 2018. "Role of Indigenous Knowledge in Sustainable Development." *International Journal of Development Research* 8 (2): 18902–6.

Suman, K.K., S.G. Abbas, K. Kumar, B.K. Singh, and P.R. Oraon. 2019. "Identification and Documentation of Ethnomedicinal Plants Used by Tribal People of Ranchi District, Jharkhand." *Bulletin of Environment, Pharmacology and Life Sciences* 8(11): 78–87.

Süntar, I. 2020. "Importance of Ethnopharmacological Studies in Drug Discovery: Role of Medicinal Plants." *Phytochemistry Reviews* 19 (5): 1199–209.

Surendran, S., P. Prasannan, Y. Jeyaram, V. Palanivel, A. Pandian, and R. Ramasubbu. 2023. "Knowledge on Ethnogynaecology of Indian Tribes—A Comprehensive Review." *Journal of Ethnopharmacology* 303: 115880.

Teall, E.K. 2014. "Medicine and Doctoring in Ancient Mesopotamia." *Grand Valley Journal of History* 3 (1): 2.

Tripathi, S. 2019. "Ethnomedicine and Future Challenges." *Global Journal of Archaeology & Anthropology* 10 (5): 87–90.

Upasani, S.V., V.G. Beldar, A.U. Tatiya, M.S. Upasani, S.J. Surana, and D.S. Patil. 2017. "Ethnomedicinal Plants Used for Snakebite in India: A Brief Overview." *Integrative Medicine Research* 6 (2): 114–30.

Wang, J., Y.K. Wong, and F. Liao. 2018. "What Has Traditional Chinese Medicine Delivered for Modern Medicine?" *Expert Reviews in Molecular Medicine* 20: e4.

Wanzala, W., and S.I. Minyoso. 2024. "Ethnomedicines in the 21st Century: Challenges and Opportunities in the Contemporary World." *Journal of Medicinal Herbs and Ethnomedicine* 10: 12–36.

Wen, Y., L. Li, X. Zhang, and Z. Gao. 2025. "Research on Operation Strategy of Multiple Channels in the Pharmaceutical E-Commerce Supply Chain." *Scientific Reports* 15: 727. https://doi.org/10.1038/s41598-025-00727-7

Yang, Y., Q. Zhang, J. Li, and H. Sun. 2025. "A Blockchain and IPFS-Based System for Monitoring the Geo-Authentic Quality of Medicinal Herbs." *Sustainable Cities and Society* 110: 105123. https://doi.org/10.1016/j.scs.2025.105123

Yogeesha, A., and G. Krishnakumar. 2023. "Ethnobotanical Study of Medicinal Plants Used in the Treatment of Neurological Disorders in the Western Ghats Region of Dakshina Kannada District, Karnataka, India." *Ethnobotany Research and Applications* 26: 1–30.

2 Herbal Medicine in Chinese, Indian, and Western Medical System

P. L. Bharati, Devanjal Bora, Manajit Bora, and Puneet Kumar Singh

2.1 INTRODUCTION

One of the oldest medical systems in the world is found in China. Although Chinese herbal treatments and acupuncture have been used for at least 2,200 years, the Huangdi Neijing, or The Yellow Emperor's Inner Classic, which was composed in the third century BCE, is the oldest known written account of Chinese medicine. The goal of traditional Chinese medicine (TCM), a medical system that has been practiced for at least 23 centuries, is to prevent or treat illness by preserving or re-establishing yin and yang equilibrium. The theoretical underpinnings of TCM that still serve as the foundation for its practice today were supplied by that work. Traditional Chinese healers basically aim to bring the two complimentary forces—yin (passive) and yang (active), which permeate the human body as well as the world overall—back into dynamic equilibrium. TCM states that a person is healthy when these two forces are in harmony; sickness, on the other hand, arises when the balance between yin and yang is upset.

Western medicine diagnoses and treats illnesses and diseases using an evidence-based approach. Healthcare professionals in Western medicine treat and prevent illness and enhance your general health using scientifically validated techniques. To give you the greatest care possible, they employ contemporary screening technologies, cancer therapies, and surgical techniques.

Numerous herbal medicines have been traditionally used in India for a very long period of time as a safe and effective alternative medicine or traditional systems of Medicine, namely AYUSH (Ayurveda, Yoga and Naturopathy, Unani, Siddha, Homoeopathy). The most popular herbal medical system, Ayurveda, places a strong emphasis on holistic medicine that considers the body, mind, and spirit as one. The underlying principle is that the human beings achieve physical, mental, and emotional health through harmonious coexistence with nature (Liu et al. 2005). Ayurveda is "science of life" which is focus on prevention of health of healthy person and cure of the illness.

Prayojanam chasya swasthasya swasthya rakshanamaturasya vicarprashamanam ch.

(Carak Samhita)

DOI: 10.1201/9781003623748-2

The concept of health according to Ayurveda is well-balanced human components physically and mentally (Acharya 2014, Tripathi 2006).

Samadoshah, samagnischa samadhatumalakriyah
Prasannaatmenindriyamanah swasthyaityabhidheeyate

(Susrut Samhita)

Samdosha means balanced *Vata* (principle of movement), *Pitta* (principle of transformation), and *Kapha* (principle of structure); *Samagni* (balanced agni) means proper Metabolism, *Samdhatu* (balanced tissues) means adequate quantity of *Ras* (nutritive fluid), *Rakta* (blood), *Mansa* (muscles), *Meda* (fat), *Asthi* (bone), *Majja* (bone marrow), and *Shukra* (ovum and sperm); proper elimination of *Mutra* (urine), *Pureesh* (stool), and *Sweda* (sweats); and satisfied or content *Atma* (soul), *Indriya* (sensory and motor organ), and *Mana* (mind). Above definition of health in Ayurveda is showing physical, mental, and spiritual wellbeing.

Herbal Medicines has a potency and play major role to maintain the health of healthy persons. Herbal drugs are mainly used in Ayurveda system of medicine besides of this animals, minerals, and marine drugs also used for the treatment in this system as single as well as compound form. The characteristics of the medications promote the body's inherent healing ability, enhance the action of medicinal herbs, and enhance the functioning of specialized organs. Numerous institutes and universities have conducted extensive basic and clinical research using cutting-edge techniques on medicinal plants and their formulations. As a rich source for antioxidants Indian medicinal plants are well-known to prevent/delay different diseased states (Vaidya et al. 2007).

2.2 HERBAL MEDICINE IN CHINESE MEDICAL SYSTEM

Similar to the Ayurvedic medical system, TCM has been utilized by the Chinese since long ago. The Chinese theory of the connection between nature and humans is the foundation of TCM, a healing system. The Five Elements, zang-fu, channels and collaterals, qi, blood, bodily fluids, diagnosis techniques, and the differentiation of symptom-complexes are all mentioned in its doctrines, etc. (Matos et al. 2021). Products made from plants are only used after undergoing some sort of processing, such as soaking in wine or vinegar or stir-frying. In clinical practice, a complex and frequently customized therapy may be prescribed after a typical diagnosis. Over 50% of people in China frequently utilize traditional remedies, with rural areas having the highest incidence of use. About 5,000 traditional remedies are available in China; they account for approximately one fifth of the entire Chinese pharmaceutical market (Li 2000). The regulatory framework for Chinese herbal medicine categorizes it into three distinct groups: raw herbal materials, sliced herbal preparations, and patent medicines. The production landscape includes 561 resource centres for raw herbs, complemented by more than 1,500 manufacturers of sliced herbs and 684 manufacturers of patent formulas. Crucially, herbal extracts fall under a separate classification as biomedicine, not

TCM, as they are derived from isolated active ingredients and are not formulated according to the tenets of Chinese herbology

2.3 HERBAL MEDICINE IN WESTERN MEDICAL SYSTEM

Western herbal medicine (WHM), also termed as phytotherapy or botanical medicine, is a complex system of medicine, wherein medical professionals use traditional evidence-based methods to manage and treat illness. WHM is based on received wisdom or tradition knowledge that has been acquired throughout time by observation and practical experience (Niemeyer, Bell, and Koithan 2013). WHM is a clinical practice that employs natural plant materials—such as roots, bark, and flowers—with minimal industrial processing. Practitioners use these botanicals in complex formulations to address disease and dysfunction while also promoting overall health. The name "WHM" is a recent designation to set apart herbalism based on Anglo-American traditions from other major herbal systems like Ayurveda and TCM. This field, which encompasses a variety of practitioners, has several alternate names, including traditional Western herbalism, medical herbalism, and phytotherapy, and is commonly practised across the Western world, including in North America, Europe, and Australasia (Snow 2016). The foundation of WHM lies in traditional knowledge accrued from empirical observation and experience. A key manifestation of this is the practice of creating individualized, multi-herb formulations from whole plant extracts. WHM treatment is holistic, seeking to address the underlying cause of illness by strengthening the patient's overall constitution and correcting systemic patterns of dysfunction. This approach is grounded in a holistic paradigm, which posits that both people and plants are integrated, multidimensional entities whose totality cannot be understood by their constituent parts alone. This contrasts with a reductionist, mechanistic model that focuses narrowly on isolated symptoms or disease pathways.

2.4 INDIAN MEDICINAL PLANTS AND THEIR PHARMACOLOGICAL ACTIONS

Since Indian medicinal plants include advantageous components, such as ingredients for functional foods, they have been used all over the world. Dietary fibres, vitamins, minerals, antioxidants, oligosaccharides, essential fatty acids, and lignins are a few components that contribute to the functional nature of food. There are plenty of these substances in medicinal plants. Indian medical systems hold that, in contrast to Western medicine, which uses single medications to treat complex illnesses, complex plant combinations can be used. Therefore, in India, whole foods are employed as functional foods instead of supplements. Spices such as onion, garlic, mustard, red chilli, turmeric, clove, cinnamon, saffron, curry leaf, fenugreek, and ginger are examples of medical plants and food ingredients with useful properties. Some herbs have shown antitumor effects such as Bixa orellana and vegetables like amla, wheat grass, soybean, and Garcinia cambogia (Dixit, Devasagayam, and Ghaskadbi 2005;

Kulkarni et al. 2006). Notable medicinal plants with similar functional properties include *Aegle marmelos*, *Allium cepa*, *Aloe vera*, *Andrographis paniculata*, *Azadirachta indica*, and *Brassica juncea* (Tilak and Devasagayam 2006a; Tilak-Jain and Devasagayam 2006b). Numerous historical sources contain multiple references to the commercial activity in spices and Indian medicinal herbs. For example, Indian aloe is utilized extensively in medicine, cosmetics, and nutraceuticals. In a similar vein, the active ingredients in the plant *Adhatoda vasica* have been identified through significant research on cough. Here, we are discussing about pharmacological actions of seven important Indian medicinal plants, namely *Curcuma longa* L., *Piper nigrum* L., *Glycyrrhiza glabra* L., *Asparagus racemosus* Willd., *Zingiber officinale* Roscoe, *Withania somnifera (L.) Dunal*, and *Tinospora cordifolia (Willd.) Hook. f. & Thomson.*

2.4.1 *Curcuma Longa* L. (*Haridra*)

In Ayurveda, the Indian System of Medicine, *Curcuma longa* L. (Zingiberaceae) has been used as an ingredient in the preparation of Ayurvedic formulations. It is a rhizomatous geophyte native to India and introduced in many parts of the world in the seasonally dry tropical biome (POWO 2025). Ayurvedic properties of this plant are *Ras* (taste)-*Katu* (pungent or acrid) and *tikta* (bitter), *Guna* (properties)-*Laghu* (easily digestible and feeling lightness), and *Ruksha* (dry), *Veerya* (potency or active components)-*Ushna* (hot), *Vipak* (effect)-*Katu* (pungent or acrid) (Sharma et al. 2000a). The rhizome shown to have anti-microbial, carminative, anti-inflammatory, and antioxidant activities (Vankar 2008). Curcumin is regarded as the most biologically active constituent of *Curcuma longa* L. (Sharma, Steward, and Gescher 2007). Curcumin, bisdemethoxycurcumin, and demethoxycurcumin isolated from turmeric found to possess *in vitro* antimicrobial activity against medically important bacteria such as *B. subtilis, K. pneumonia, E. coli, Enterobacter aerogenes, Pseudomonas aeruginosa, S. aureus, P. Mirabilis, Aspergillus niger*, and *Candida albicans* (Singh and Jain 2012). Crude extracts of curcuminoids and essential oil of *Curcuma longa* showed potent *in vitro* antibacterial activity against *Bacillus subtilis, Bacillus macerans, Bacillus licheniformis*, and *Azotobacter* (Naz et al. 2010). Aqueous extract of *Curcuma longa* L. shown to inhibit hepatitis B virus replication in liver cells (Kim et al. 2009). Zingiberene, turmerone, and ar-turmerone, 1-phellandrene, and α-terpinolene present in turmeric oil showed strong free radical scavenging activities in the DPPH free radical scavenging assay. Turmeric oils also showed high inhibitory potency against cancer cell lines (Yu 2006). Turmerin, a protein from turmeric, the powdered rhizome of the herb *Curcuma longa* L., showed potent antioxidant activity (Mukunda 2010). In HIV-infected patients, turmerin shown to inhibit HIV-infected T-cell proliferation. Curcumin specifically inhibits cyclooxygenase-2 of cancer cells of colon and breast in the human. Curcumin shown to have efficacious in treating oral submucous fibrosis by fibrinolytic properties as well as by inhibiting the inflammatory markers, lipid peroxidation, and cellular proliferation. Curcumin demonstrates anti-cancer activities by elevating levels of vitamins C and E and prohibiting lipid peroxidation and DNA damage (Alok et al. 2015). Curcumin showed chemopreventive effects by inhibiting the growth of *Helicobacter pylori*, associated with the gastric and colon cancer (Mahady et al. 2002). Curcuminoids have anti-wrinkle activity through the inhibition of UV-B-induced upregulation of nuclear factor-kappa B (NF-jB) activity in keratinocytes and TNF-α induction in fibroblast Jang et al. (2012). Curcumin shown to inhibit TRPV1-mediated pain hypersensitivity by blocking the transient receptor potential vanilloid 1 (TRPV1) activation by capsaicin in a competitive manner (Alok et al. 2015). Aqueous extract of *Curcuma longa* and its polysaccharide fraction showed anti-inflammatory activity by inhibiting the secretion of IL-12 and PGE2 *in vitro* (Chandrasekaran et al. 2013). Curcumin has been shown to decrease serum total cholesterol, triglycerides, and LDL-C levels in methionine-treated rats (Kapoor, Ansari, and Bhandari 2008). Pre-treatment of rats with curcumin at the doses of 100mg and 200mg/kg body wt. significantly prevented lindane-induced oxidative damage in rat liver (Singh and Sharma 2011). Rhizome of *Haridra* is useful for the medicine. *Haridra churna* (Powder) can be used as spices and medicine by internal and externally like *lepa* (Paste). Classical Ayurvedic compound of *Haridra* is *Chandraprabha vati*, *Mahatikta ghrita*, *Haridra Khanda*, *Dashmoolarist*, etc.

2.4.2 *Piper nigrum* L. (*Marich*)

Piper nigrum L. (Piperaceae), commonly known as black pepper, is used in Ayurveda and other traditional system medicine for the treatment of gastro-intestinal disease, respiratory disease, skin cancer, pain, and other diseases. It is a climber native to India and introduced in many parts of the world in the wet tropical biome (POWO 2025). Ayurvedic properties of this plant are *Ras* (taste)-*Katu* (pungent or acrid), *Guna* (properties)-*Laghu* (easily digestible and feeling lightness), *Teekshna* (sharp) and *Ruksha* (dry), *Veerya* (potency or active components)-*Ushna* (hot), and *Vipak* (effect)-*Katu* (pungent) (Lavekar et al. 2008a).The plant contains active compound piperine, shown to have anti-inflammatory, antimicrobial, antioxidant, anti-cancer, anxiolytic, hepatoprotective, and larvicidal activities (Rai et al. 2018). Piperine showed anti-mycobacterial activity against *M. tuberculosis* H37Rv strains at a concentration of 100μg/ml (Deepthi et al. 2012). Piperine was found to be cytotoxic towards DLA and EAC cells at a concentration of 250μg/ml (Sunila and Kultan 2004). Piperine possesses various neuro-protective activities such as anticonvulsant, anti-depressant, and cognitive enhancing effect. Piperine shown to inhibit monoamine oxidase activity and increases the level of noradrenaline and serotonin in mouse brain. Piperine at a dose between 5 and 20mg/kg body weight attenuated the increase in lipid peroxidation and acetylcholinesterase activity in cholinergic-deficient rats (Priprem et al. 2011). Piperine has been reported to act on various cell cycle proteins such as cyclin D, cyclin T, CDK2, and CDK4. Piperine was observed to attach with cell cycle proteins via hydrogen bonding and affect the cell cycle progression (Tripathi, Ray, and Mishra 2022). Piperine has been shown to act as a modulating agent of both innate and acquired immune responses. Inhibition of the expression of inflammatory mediators like tumour necrosis factor-α, inducible NO synthase, and LPS-induced COX-2 is observed to be enhanced in the presence of piperine, especially in macrophages and suppresses release of IL-1β in LPS-primed bone marrow-derived macrophages (Ghavami, Froushani, and Tehrani 2024). Fruit of *Marich* is useful for the medicine. *Marich churna* (powder) can be used single as spices and medicine by internal. Classical Ayurvedic compound of *Marich* is *Trikatu choorna* (powder), *Agnikumar ras, Tribhuan kirti ras, Sutshekhar ras, Swaskuthar ras, Vidangarist*, etc.

2.4.3 *Glycyrrhiza glabra* L. (*Madhuyasthi*)

Glycyrrhiza glabra L. (Fabaceae) is an important medicinal plant of Chinese, Greek, Egyptian, Indian, and Roman civilizations. It is a perennial or subshrub native to Central Mediterranean to Mongolia and Pakistan including India and introduced in many parts of the world in the temperate biome (POWO 2025). The plant has been used in traditional medicinal systems for treating diabetes, bronchopulmonary diseases, gastro-intestinal diseases, urinary diseases, heart disease, skin disease, wounds, and cancer (Bnouham et al. 2002; Nair, Divyasree, and Gopakumar 2010; Pieroni and Quave 2005; Said et al. 2002). Ayurvedic properties of this plant are *Ras* (taste)-*Madhur* (sweet), *Guna* (properties)-*Guru* (delayed digestible and feeling heavyness) and *Snigdha* (unctuous or oily), *Veerya* (potency or active components)-*Sheeta* (cold), *Vipak* (effect)-*Madhur* (sweet) (Sharma et al. 2001a). Glycyrrhizin, a triterpenoid compound, is a major constituent of *Glycyrrhiza glabra* L. (Sharma, Katiyar, and Agrawal 2017). Roots of the plant are rich source of flavonoids such as isoflavones, glabridin, and glabrene (Simons et al. 2011). Glabridin was found to be active against both yeast and fungi *in vitro*. Glabridin has shown resistance modifying activity against drug resistant mutants of *Candida albicans* (Fatima et al. 2009). Glabridin treatment decreased lipid peroxidation and ameliorated the severity of degenerative changes in streptozotocin-induced diabetes on small intestine and pancreatic β-cells in rats (Abo Gazia and Hasan 2012). Glabridin, hispaglabridin A, hispaglabridin B, 4′-O-methylglabridin, and 3′-hydroxy-4′-O-methylglabridin were shown to be effective in protecting mitochondrial function against oxidative stresses (Haraguchi et al. 2000). The constituents of *Glycyrrhiza glabra* L. found to exhibit similar activity, that is, anti-inflammatory of hydrocortisone. Glycyrrhizin stabilizes lysosomes via inhibition of phospholipase A2 activity in the lysosomal membrane (Shiki et al. 1986). Glycyrrhizin shown to inhibit collagen-induced platelet aggregation in a concentration-dependent manner (Okimasu, Moromizato, and Watanabe 1983). Licochalcone C, a flavonoid from *Glycyrrhiza glabra* L., shows to attenuate LPS-IFN-γ-induced inflammatory response significantly by modulating the antioxidant network activity of superoxide dismutase, catalase, and glutathione peroxidase activity (Franceschelli et al. 2011). Glycyrrhizin and other components of *Glycyrrhiza glabra* L. possess anticarcinogenic properties by inhibiting abnormal cell proliferation as well as tumour formation in breast, liver, and skin. Glabridin stimulates creatine kinase activity in rat uterus, epiphyseal cartilage, diaphyseal bone, aorta, and left ventricle of the heart via binding with the human oestrogen receptor. Glabridin inhibits the growth of breast cancer cells independently of oestrogen receptors Tamir et al. (2000). *Glycyrrhiza glabra* L. stimulates anti-pepsin activity and accelerates mucin secretion through increased synthesis of glycoprotein and prolonged life of the epithelial cells, thereby exerting its antiulcer activity (Wynn and Fougere 2007). Root of *Yastimadhu* is useful for the medicine. *Madhuyasthi churna* (Powder) can be used single as medicine by internal. Classical Ayurvedic compound of *Madhuyasthi* is *Eladi gutika (*Tablet*)*, *Saptamrit Loh*, *Swadist virechan churna*, *Pushyanug churna*, *Kankasav*, etc.

2.4.4 *Asparagus racemosus* Willd. (*Shatavari*)

A*sparagus racemosus* Willd. (Asparagaceae) is used as Ayurvedic rasayana (rejuvenative tonic), which helps prevent ageing, increase longevity, impart immunity, improve mental function and vigour, and add to the vitality of the body. It is also used to treat nervous disorders, dyspepsia, tumours, inflammation, neuropathy, and hepatopathy (Sharma 2000c). It is a climber native to tropical Africa to N. Australia including India and grows primarily in the seasonally dry tropical biome (POWO 2025). Ayurvedic properties of this plant are *Ras* (taste)-*Madhur* (sweet) and *tikta* (bitter), *Guna* (properties)-*Guru* (delayed digestible and feeling heavyness), *Snigdha* (unctuous or oily), *Veerya* (potency or active components)-*Sheeta* (cold), and *Vipak* (effect)-*Madhur* (Sweet) (Sharma et al. 2000b). In Ayurveda, A*sparagus racemosus* is considered a rejuvenative tonic for females. It increases folliculogenesis and ovulation and prepares the womb for implantation. The plant acts as post-partum tonic by increasing lactation and normalizing the uterus and the various hormones. It has also been used as the traditional medicine for treating leucorrhoea and menorrhagia (Sharma and Bhatnagar 2011). Apart from these, the plant has been used as anthelmintic, febrifuge, as an aphrodisiac, and in the treatment of rheumatism, epistaxis, haematuria, cough, diarrhoea, dysentery, tongue ulcers, and wound healing (Jain 1991).The therapeutic activity of *Asparagus racemosus* is due to the presence of potent phytocomponents, namely phytosterols, saponins, polyphenols, flavonoids, and ascorbic acid (Visavadiya and Narasimhacharya 2009). Steroidal saponins (Shatavarins I–IV) are the main active constituents of *Asparagus racemosus* primarily present in the roots. It has been found that Shatavarin IV has exhibited immuno-modulation activity against specific T-dependent antigens in immune-compromised animals (Kamat et al. 2000). Racemoside A, a water-soluble steroidal saponin purified from the fruits of *Asparagus racemosus*, is a potent anti-leishmanial molecule effective against *Leishmania donovani* (Dutta et al. 2007). The phytoestrogenic activity of *Asparagus racemosus*, which exerts hormone-like actions in the body, partly due to the isoflavones with mild estrogenic activity that help to balance the oestrogen levels, is attributed to the presence of steroidal saponins (Asolkar, Kakker, and Charkre 1992). Tuberous root of *shatavari* is used to prepare *Shatavari churna* (powder), which can be used as a medicine. Classical Ayurvedic compound of *Shatavari* is *Dhatupoushtik churna*, *Nardeeya laxmivilas ras*, *Saraswatarista*, *Chandra kala ras*, etc.

2.4.5 *Zingiber Officinale* Roscoe (*Shunthi*)

Zingiber officinale Roscoe (Zingiberaceae) is known as Ginger (En) and *Shunthi* (Sa), and used in Ayurveda and other traditional medicinal systems since ancient times for treating variety diseases such as dyspepsia, diarrhoea, arthritis, gingivitis, asthma, nausea, coryza, sciatica, headache, haemorrhoids, common colds, and coughs Imtiyaz et al. (2013). It is a perennial or rhizomatous geophyte native to India to S. Central China and introduced in many parts of the world in the seasonally dry tropical biome (POWO 2025). Ayurvedic properties of this plant are *Ras* (taste)-*Katu* (pungent or acrid), *Guna* (properties)-*Laghu* (easily digestible and feeling lightness), and *Snigdha* (unctuous or oily), *Veerya* (potency or active components)-*Ushna* (hot), *Vipak* (effect)-*Madhur* (sweet) (Lavekar et al. 2008b). The plant is used as an antioxidant, anti-inflammatory, analgesic, anti-arthritic, anti-microbial, digestive, cardiotonic, sialagogue, anti-hyperglycaemic, hypo-cholesteremic, and hypo-lipidaemic activities (Byadgi et al. 2021). The major bioactive compounds in ginger are gingerols, shogaols, quercetin, zingerone, gingerenone-A, and zingiberene (Mao et al. 2019). The anti-inflammatory activity of Shunthi is mediated by inhibiting activation of macrophage and neutrophils, as well as inhibiting migration of monocyte and leukocyte (Ezzat et al. 2018). Essential oils from Shunthi showed antibacterial activities against both Gram-positive and Gram-negative bacteria (Wibowo et al. 2020). Shunthi and its bioactive compounds such as gingerols, shogaols, paradol, and zingerone possess potent antioxidant and anti-inflammatory properties. The antioxidant mechanism of Shunthi is linked to activation of Nrf2 signalling pathway (Ayustaningwarno et al. 2024). 6-Shagaol present in Shunthi showed potent antioxidant and anti-inflammatory properties due to its alpha, beta unsaturated Ketone moiety (Ekka et al. 2017). The anticancer activity of Shunthi is due to the presence of active bioingredients such as gingerols, shogaol, and paradols (Mashhadi et al. 2013). Rhizome of *Shunthi* is useful for the preparation of *Shunthi churna* (powder) and *kalka* (paste), and both can be used as spices and medicine. Classical Ayurvedic compound of *Shunthi* is *Trikatu churna, Anand bhairav ras, Sootshekhar ras, Sootikabharan ras, Kaishor guggulu, Lohasav, Jirakadyarist*, *Punarnava mandoor*, etc.

2.4.6 *Withania Somnifera* (L.) Dunal (*Ashwagandha*)

Withania somnifera (L.) Dunal (Solanaceae) is commonly known as *Ashwagandha* or Indian ginseng and used in Ayurveda and other traditional medicinal systems since ancient times for treating various diseases such as anxiety, stress, and insomnia (Majeed et al. 2023). It is a shrub, and its native range extends from southern Europe through Central China and Africa to Myanmar, including India, and it grows primarily in the subtropical biome (POWO 2025). Ayurvedic properties of this plant are *Ras*(taste)–*Tikta (bitter), Katu* (pungent or acrid), Madhur (sweet), *Guna* (properties)-*Laghu* (easily digestible and feeling lightness), *Snigdha* (unctuous or oily), *Veerya* (potency or active components)-*Ushna* (hot), and *Vipak* (effect)-*Madhur* (sweet) (Sharma et al. 2001b). Alkaloids and steroidal lactones, two bioactive substances found in the plant's leaves, have a variety of pharmacological properties, including anti-inflammatory, antioxidant, antibacterial, anti-arthritic, anti-stress, immune-modulatory, and anticarcinogenic effects. Withanine is the major constituent, and the other alkaloids include somniferine, somnine, somniferinine, and withananine, among others. The leaves of *Withania somnifera* contain steroidal lactones called as withanolides (Kaur, Junaid, and Raman 2013). Roots, leaves, and seeds of *Ashwagandha* are known to have medicinal properties. *Ashwagandha churna* (powder) is prepared by using these parts of the plant. Classical Ayurvedic compound of *Ashwagandha* is *dhatupaustik choorna* (powder), *Rasraj ras*, *Lakshadi Guggulu*, *Ashwagandharist*, etc.

2.4.7 *Tinospora Cordifolia* (Willd.) (*Guduchi*)

Tinospora cordifolia (Willd.) Hook. f. & Thomson (Menispermaceae) is also known as *Giloy, Amrita*. It is a liana or woody climber native to Indian subcontinent to Indo-China and grows primarily in the wet tropical biome (POWO 2025). Ayurvedic properties of this plant are *Ras* (taste)-*Tikta (bitter), Kashay* (astringent), *Guna* (properties)-*Laghu* (easily digestible and feeling lightness), *Snigdha* (unctuous or oily), *Veerya* (potency or active components)-*Ushna* (hot), and *Vipak* (effect)-*Madhur* (sweet) (Sharma et al. 2001c). *Tinospora cordifolia* is used to treat a variety of conditions, including rheumatoid arthritis (RA), hepatitis, peptic ulcer disease (PUD), diabetes, high cholesterol, hay fever, gout, upset stomach, gonorrhoea, and syphilis, and to strengthen the immune system. Various scientific studies reported antidiabetic, antimicrobial, osteoprotective, hepatoprotective, immunomodulatory, wound healing, and antioxidant activities of the plant (Upadhyay et al. 2018). The pharmacological activities of the plant are attributed to the presence of various bio-active phytochemicals such as alkaloids, steroids, glycosides, aliphatic compounds, phenolics, and polysaccharides (Sharma et al. 2019). Roots, stems, and leaves of *Guduchi* are of medicinal value. Classical Ayurvedic compound of *Guduchi* is *Giloyadi churna* (powder), *Guduchyadi taila*, *Dashmoolarist*, *Sanjivani vati*, *chyavanprash*, *Panchnimba choorna*, *Guduchyadi ghrita*, *Amritadi guggulu*, *Guduchi satva*, *Amritarist*, *Sarvajwarhar lauha*, etc. The images of some well-known medicinal plants and the most common phytochemical compounds present are given below (Figures 2.1–2.3) (Sharma et al. 2001d).

2.5 CONCLUSION

Traditional medicine systems have long constituted a cornerstone of healthcare and cultural identity in numerous countries, with historical roots extending back millennia. In India, for instance, the use of herbal medicines is not merely a historical practice but a vibrant and deeply ingrained component of a rich, diversified cultural heritage. This profound knowledge of medicinal plants, their properties, and their applications has been successfully preserved and passed down through generations via oral traditions and ancient texts like the Ayurvedic scriptures. In the contemporary era, this traditional wisdom is being rigorously re-examined through the lens of modern science. With the aid of advanced scientific methodologies—such as *in vitro* and *in vivo* assays, chromatography, and mass spectrometry—the purported bioactivities of these herbal drugs are being systematically isolated, characterized, and

FIGURE 2.1 (a) *Zingiber officinale* Roscoe, (b) *Curcuma longa* L., (c) *Asparagus racemosus* Wild., (d) *Glycyrrhiza glabra* L., (e) *Withania somnifera* (L.) Dunal, (f) *Piper longum* L., (g) *Tinospora cordifolia* (Willd.) (From Hook.f. & Thomson.)

validated. This pharmacological analysis is crucial, as it does more than just confirm ancient knowledge; it paves the pathway for novel drug discoveries by identifying active lead compounds and offers insights for optimizing the efficacy and bioavailability of existing herbal formulations. Consequently, the strategic amalgamation of empirically-grounded traditional medicine with cutting-edge scientific techniques represents a powerful, synergistic approach. This collaboration holds immense potential to accelerate and improve the drug discovery process, ultimately serving the betterment of global health and mankind by providing new, effective, and potentially more natural therapeutic options.

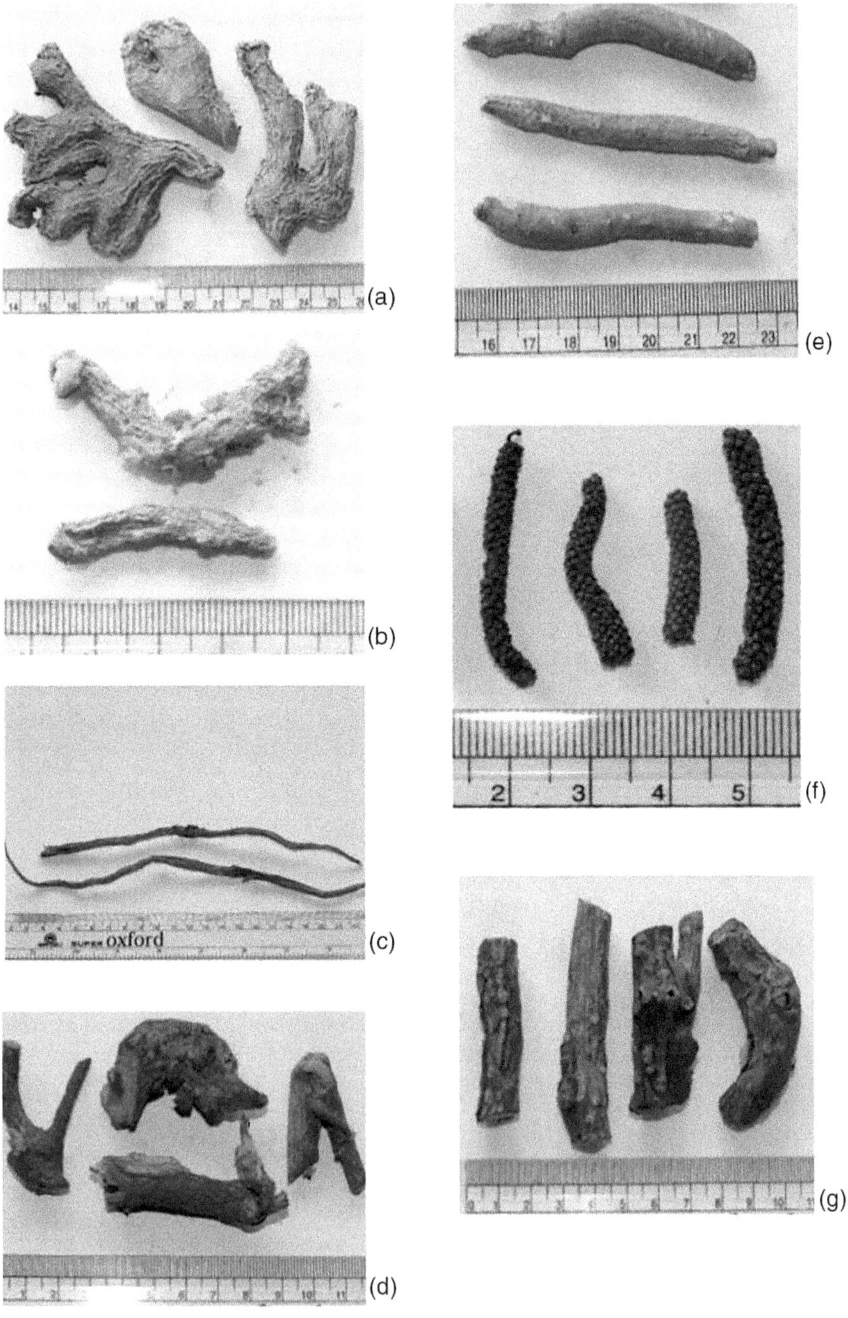

FIGURE 2.2 (a) *Zingiber officinale* (Rhizome), (b) *Curcuma longa* (Rhizome), (c) *Asparagus racemosus* (Roots), (d) *Glycyrrhiza glabra* (Roots), (e) *Withania somnifera* (Roots), (f) *Piper longum* (Fruits), (g) *Tinospora cordifolia* (Stem parts).

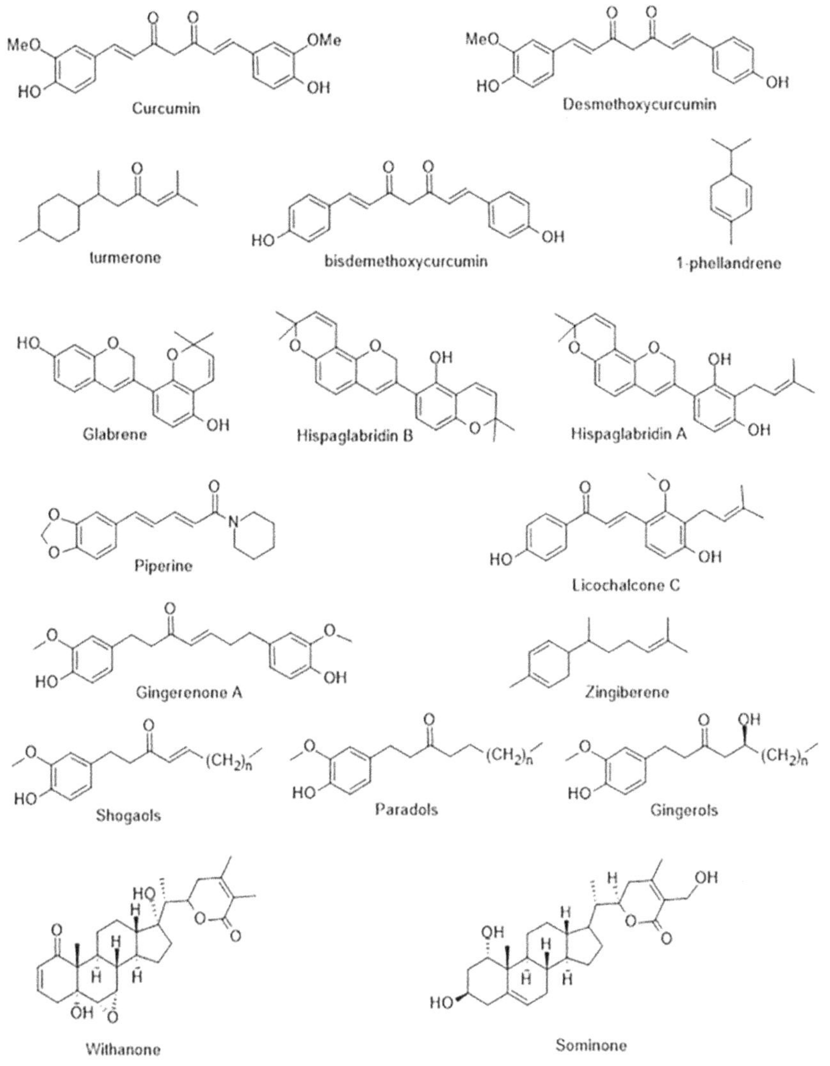

FIGURE 2.3 Important phytochemicals.

REFERENCES

Abo Gazia, M.M., and N.M. Hasan. 2012. "Effect of Glabridin on the Structure of Ileum and Pancreas in Diabetic Rats: A Histological, Immunohistochemical and Ultrastructural Study." *Nature and Science* 10: 78–90.

Acharya, J.T. 2014. *Sushrut Samhita Nibandh sangrah Commentary sutra sthan 15/41*. Varanasi: Chaukhambha Sanskrit Sansthan.

Alok, A., I.D. Singh, S. Singh, M. Kishore, and P.C. Jha 2015. "Curcumin - Pharmacological Actions And its Role in Oral Submucous Fibrosis: A Review." *Journal of Clinical and Diagnostic Research* 9(10): ZE01–3. doi: 10.7860/JCDR/2015/13857.6552.

Asolkar, L.V., K.K. Kakker, and O.J. Charkre 1992. *Second Supplement to Glossary of Indian Medicinal Plants with Active Principles Part-I*, 100–101. New Delhi: Publication and Information Directorate (CSIR).

Ayustaningwarno, F., G. Anjani, M.A. Ayu, and V. Fogliano 2024. "A Critical Review of Ginger's (*Zingiber officinale*) Antioxidant, Anti-Inflammatory, and Immunomodulatory Activities." *Frontiers in Nutrition*. doi: 10.3389/fnut.2024.1364836.

Bnouham, M., H. Mekhfi, A. Legssyer, and A. Ziyyat. 2002. "Ethnopharmacology Forum Medicinal Plants Used in the Treatment of Diabetes in Morocco." *International Journal of Diabetes & Metabolism* 10: 33–50.

Byadgi, P.S., D.S. Kanashetti, R. Tiwari, B.N. Maurya, M. Rana, V.K. Dwivedi, R.D. Suresh, and K.N. Dwivedi 2021. "Shunthi (*Zingiber officinale* Rosc.): A Miraculous Medicinal Plant." *International Journal of Advanced Research in Medicinal Chemistry* 3(1): 8–13.

Chandrasekaran, C.V., K. Sundarajan, J.R. Edwin, G.M. Gururaja, D. Mundkinajeddu, and A. Agarwal. 2013. "Immune-Stimulatory and Anti-Inflammatory Activities of Curcuma Longa Extract and Its Polysaccharide Fraction." *Pharmacognosy Research* 5: 71–9.

Deepthi, S.P.R., V. Junise, P. Shibin, S. Senthila, and R.S. Rajesh. 2012. "Isolation, Identification and Antimycobacterial Evaluation of Piperine from *Piper longum*." *Der Pharmacia Lettre* 4: 863–8. www.scholarsresearchlibrary.com

Dixit, P., T.P.A. Devasagayam, and S.S. Ghaskadbi. 2005. "Antioxidant Effects of Germinated Fenugreek." *Phytotherapy Research* 19: 977–83.

Dutta, A., A. Ghoshal, D. Mandal, N.B. Mondal, S. Banerjee, N.P. Sahu, and C Mandal. 2007. "Racemoside A, An Anti-Leishmanial, Water-Soluble, Natural Steroidal Saponin, Induces Programmed Cell Death in *Leishmaniadonovani*." *Journal of Medical Microbiology* 56: 1196–204.

Ekka, D., S. Dubey, S.D. Khichariya, D.S. Dhruw, and S.M. Parhate. 2017. "Antioxidant and Anti-Inflammatory Effect of Sunthi in *Pranvaha Srotas*." *Ayushdhara* 4 (1): 1056–9.

Ezzat, S.M., MI. Ezzat, M.M. Okba, E.T. Menze, and A.B. Abdel-Naim 2018. "The Hidden Mechanism Beyond Ginger (ZingiberofficinaleRosc.) Potent In Vivo and In Vitro Anti-Inflammatory Activity." *Journal of Ethnopharmacology* 214: 113–23. doi: 10.1016/j.jep.2017.12.019.

Fatima, A., V.K. Gupta, S. Luqman, A.S. Negi, J.K. Kumar, K. Shanker, D. Saikia, S. Srivastava, M.P. Darokar, and S.P. Khanuja. 2009. "Antifungal Activity of Glycyrrhiza glabra Extracts and Its Active Constituent Glabridin." *Phytotherapy Research* 23: 1190–3.

Franceschelli, S., M. Pesce, I. Vinciguerra, A. Ferrone, G. Riccioni, A. Patruno, A. Grilli, M. Felaco, and L. Speranza. 2011. "Licocalchone-C Extracted from *Glycyrrhiza glabra* Inhibits Lipopolysaccharide-Interferon-γ Inflammation by Improving Antioxidant Conditions and Regulating Inducible Nitric Oxide Synthase Expression." *Molecules* 16: 5720–34.

Ghavami, A., S.M.A. Froushani, and A. Tehrani. 2024. "Immunomodulatory Potential of Piperine in Rats." *Turkish Journal of Immunology* 12 (1): 1–8. doi: 10.4274/tji.galenos.2024.20082.

Haraguchi, H., N. Yoshida, H. Ishikawa, Y. Tamura, K. Mizutani, and T. Kinoshita. 2000. "Protection of Mitochondrial Functions Against Oxidative Stresses by Isoflavans from *Glycyrrhiza glabra*." *Journal of Pharmacy and Pharmacology* 52: 219–23.

Imtiyaz, S., K. Rahman, A Sultana, M Tariq, and SS Chaudhary. 2013. "Zingiber Officinale Rosc.: A Traditional Herb with Medicinal Properties." *TANG* 3 (4): e26. https://doi.org/10.5667/tang.2013.0009

Jain, S.K. 1991. *Dictionary of Indian Folk Medicine and Ethnobotany*, 29–30. New Delhi: Deep Publications.

Jang, S., J. Chun, E.M. Shin, H. Kim, and Y.S. Kim 2012. "Inhibitory Effects of Curcuminoids from Curcuma Longa on Matrix Metalloproteinase-1 Expression in Keratinocytes and Fibroblasts." *Journal of Pharmaceutical Investigation*. doi: 10.1007/s40005-012-0005-8.

Kamat, J.P., K.K. Boloor, T.P. Devasagayam, and S.R. Venkatachalam. 2000. "Antioxdant Properties of *Asparagus racemosus* Against Damaged Induced by Gamma Radiation on Rat Liver Mitochondria." *Journal of Ethanopharmacology* 71: 425–35.

Kapoor, P., M.N. Ansari, and U. Bhandari. 2008. "Modulatory Effect of Curcumin on Methionine- Induced Hyperlipidemia and Hyperhomocysteinemia in Albino Rats." *Indian Journal of Experimental Biology* 46: 534–40.

Kaur, N., N. Junaid, and B. Raman. 2013. "A Review on Pharmacological Profile of Withania somnifera (Ashwagandha). Research and Reviews." *Journal of Botanical Sciences* 2: 6–14.

Kim, H.J., H.S. Yoo, J.C. Kim, C.S. Park, M.S. Choi, M. Kim, H. Choi, J.S. Min, Y.S. Kim, S.W. Yoon, and J.K. Ahn. 2009. "Antiviral Effect of Curcuma Longa Linn Extract Against Hepatitis B Virus Replication." *Journal of Ethnopharmacology* 124: 189–96.

Kulkarni, S.D., J.C. Tilak, R. Acharya, N.S. Rajurkar, T.P.A. Devasagayam, and A.V.R. Reddy. 2006. "Evaluation of the Antioxidant Activity of Wheatgrass (Triticum aestivum L.) as a Function of Growth and Under Different Conditions." *Phytotherapy Research* 20: 218–27.

Lavekar, G.S., et al. 2008a. *Database on Medicinal Plants Used in Ayurveda& Siddha: Marich*. Vol. 5, 187. New Delhi: CCRAS.

Lavekar, G.S., et al. 2008b. *Database on Medicinal Plants Used in Ayurveda & Siddha: Shunthi*. Vol. 5, 315. New Delhi: CCRAS.

Li, X. 2000. "History of Use of Traditional Herbal Medicines." *IARC Monographs* 82, 43–4.

Liu, X.M., J.Q. Zou, Z.X. Shen, and G.Q. Su. 2005. "Overview of Traditional Indian Medicine." *Modern Traditional Chinese Medicine Materia Medica World Science Technology* 7 (6): 86–8.

Mahady, G.B., S.L. Pendland, G. Yun, and Z.Z. Lu. 2002. "Turmeric (*Curcuma longa*) and Curcumin Inhibit the Growth of Helicobacter pylori, a Group 1 Carcinogen." *Anticancer Research* 22 (6C): 4179–81.

Majeed, M., et al. 2023. "A Standardized *Ashwagandha* Root Extract Alleviates Stress, Anxiety, and Improves Quality of Life in Healthy Adults by Modulating Stress Hormones: Results from a Randomized, Double-Blind, Placebo-Controlled Study." *Medicine (Baltimore)* 102: e35521.

Mao, Q.Q., X.Y. Xu, S.Y. Cao, R.Y. Gan, H. Corke, T Beta, and H.B Li. 2019. "Bioactive Compounds and Bioactivities of Ginger (*Zingiber officinale* Roscoe)." *Foods* 8(6):185. doi: 10.3390/foods8060185.

Mashhadi, N.S., R. Ghiasvand, G. Askari, M. Hariri, L. Darvishi, and M.R. Mofid 2013. "Anti-Oxidative and Anti-Inflammatory Effects of Ginger in Health and Physical Activity: Review of Current Evidence." *International Journal of Preventive Medicine* 4(Suppl 1): S36–42.

Matos, L.C., et al. 2021. "Understanding Traditional Chinese Medicine Therapeutics" An Overview of the Basics and Clinical Applications." *Healthcare (Basel)* 9 (3): 257. https://doi.org/10.3390/healthcare9030257

Mukunda, C. 2010. "Turmerin, a Protein from *Curcuma longa* L. Prevent Oxidative Organ Damage Against *Najanaja*venom Phospholipase A2 in Experimental Animal." *Journal of Current Pharmaceutical Research* 3: 29–34.

Nair, C.K.K., P. Divyasree, and G Gopakumar. 2010. *Ethnomedicinal Plants to Fight Neoplastic Diseases in Chattopadhyay, D., Ethnomedicine: A Source of Complementary Therapeutics*, 203–26. Trivandrum: Research Signpost.

Naz, S., S. Jabeen, S. Ilyas, F. Manzoor, F. Aslam, and A. Ali. 2010. "Antibacterial Activity of *Curcuma longa* Varieties Against Different Strains of Bacteria." *Pakistan Journal of Botany* 42: 455–62.

Niemeyer, K., I.R. Bell, and M. Koithan 2013. "Traditional Knowledge of Western Herbal Medicine and Complex Systems Science." *Journal of Herbal Medicine* 3(3):112–9. doi: 10.1016/j.hermed.2013.03.001.

Okimasu, E., Y. Moromizato, and S. Watanabe, et al. 1983. "Inhibition of Phospholipase A2 and Platelet Aggregation by Glycyrrhizin, an Antiinflammation Drug." *Acta Medica Okayama* 37: 385–91.

Pieroni, A., and C.L. Quave. 2005. "Traditional Pharmacopoeias and Medicines Among Albanians and Italians in Southern Italy: A Comparison." *Journal of Ethnopharmacology* 101: 258–70.

POWO. (2025). "Plants of the World Online. Facilitated by the Royal Botanic Gardens, Kew." Published on the Internet. https://powo.science.kew.org/. Retrieved 02 January 2025.

Priprem, A., P. Chonpathompikunlert, S. Sutthiparinyanont, and J. Wattanathorn. 2011. "Antidepressant and Cognitive Activities of Intranasal Piperine-Encapsulated Liposomes." *Advances in Bioscience and Biotechnology* 2: 108–16.

Rai, N, M Singhai, AK Singhai, and R. Shukla. 2018. "Amagical Medicinal Fruit of Piper Nigrum." *World Journal of Pharmaceutical Research* 7: 418–25. doi: 10.20959/wjpr20188-11748.

Said, O., S. Khalil, S. Fulder, and H. Azaizeh. 2002. "Ethnopharmacological Survey of Medicinal Herbs in Israel, the Golan Heights and the West Bank Region." *Journal of Ethnopharmacology* 83: 251–65.

Sharma, K., and M. Bhatnagar. 2011. "*Asparagus racemosus*(Shatavari): A Versatile Female Tonic." *International Journal of Pharmaceutical & Biological Archives* 2: 855–63.

Sharma, P., B.P. Dwivedee, D. Bisht, A.K. Dash, and D. Kumar. 2019. "The Chemical Constituents and Diverse Pharmacological Importance of *Tinospora cordifolia*." *Heliyon* 5 (9): e02437. doi: 10.1016/j.heliyon.2019.e02437.

Sharma, P.C., et al 2000a. *Database on Medicinal Plants Used in Ayurveda: Haridra*. Vol. 1, 153. New Delhi: CCRAS.

Sharma, P.C., et al. 2000b. *Database on Medicinal Plants Used in Ayurveda: Shatavari*. Vol. 1, 419. New Delhi: CCRAS.

Sharma, P.C., et al. 2000c. *Database on Medicinal Plants Used in Ayurveda: Guduchi*. Vol. 3, 256. New Delhi: CCRAS.

Sharma, P.C., et al. 2001a. *Database on Medicinal Plants Used in Ayurveda: Yastimadhu*. Vol. 3, 561. New Delhi: CCRAS.

Sharma, P.C., et al. 2001b. *Database on Medicinal Plants Used in Ayurveda: Ashwagandha*. Vol. 3, 88. New Delhi: CCRAS.

Sharma, P.V. 2001d. *Charaka Samhita*. Vol. 2, 7–14. Varanasi: Chaukhambha Orientalis.

Sharma, R. A., W.P. Steward, and A.J. Gescher. 2007. "Pharmacokinetics and Pharmacodynamics of Curcumin." *Advances in Experimental Medicine and Biology* 595: 453–70.

Sharma, V., A. Katiyar, and R.C. Agrawal. 2017. "Glycyrrhiza glabra: Chemistry and Pharmacological Activity." In *Sweeteners*, 87–100. doi: 10.1007/978-3-319-27027-2_21.

Shiki, Y., Y. Ishikawa, K. Shirai, Y. Saito, and S. Yoshida. 1986. "Effect of Glycyrrhizin on Lysosomes Labilization by Phospholipase A2." *American Journal of Chinese Medicine* 14: 131–7.

Simons, R., J.P. Vincken, L.A.M. Mol, A.M. Susan, T.F.H. Bovee, T.J.C. Luijendijk, M.A. Verbruggen, and H. Gruppen. 2011. "Agonistic and Antagonistic Estrogens in Licorice Root (*Glycyrrhiza glabra*)." *Analytical and Bioanalytical Chemistry* 4: 305–13.

Singh, R., and P. Sharma. 2011. "Hepatoprotective Effect of Curcumin on Lindane-Induced Oxidativestress in Male Wistar Rats." *Toxicology International* 18: 124–9.

Singh, R.P., and D.A. Jain. 2012. "Evaluation of Antimicrobial Activity of Curcuminoids Isolated from Turmeric." *International Journal of Pharmacy & Life Sciences* 3: 1368–76.

Snow, J. 2016. "Context Effects in Western Herbal Medicine: Fundamental to Effectiveness." *Explore* 12 (1): 55–62. https://doi.org/10.1016/j.explore.2015.10.004

Sunila, E.S., and G. Kultan. 2004. "Immunomodulatory and Antitumor Activity of *Piper longum* Linn. and Piperine." *Journal of Ethnopharmacology* 90: 339–46.

Tamir, S., M. Eizenberg, D. Somjen, N. Stern, R. Shelach, A. Kaye, and J. Vaya. 2000. "Estrogenic and Antiproliferative Properties of Glabridin from Licorice in Human Breast Cancer Cells." *Cancer Research* 60: 5704–9.

Tilak, J.C., and T.P.A. Devasagayam. 2006a. "Indian Medicinal Plants: A Potential Reservoir in Health and Disease." In *Contemporary Perspectives on Clinical Pharmacotherapeutics*, edited by K. Kohli, M. Gupta, and S. Tejwani, 29–43. New Delhi: Elsevier.

Tilak-Jain, J.A., and T.P.A. Devasagayam. 2006b. "Cardioprotective and Other Beneficial Effects of Some Indian Medicinal Plants." *Journal of Clinical Biochemistry and Nutrition* 38: 9–18.

Tripathi, A.K., A.K. Ray, and S.K. Mishra. 2022. "Molecular and Pharmacological Aspects of Piperine as a Potential Molecule for Disease Prevention and Management: Evidence from Clinical Trials." *Beni-Suef University Journal of Basic and Applied Sciences* 11: 16. https://doi.org/10.1186/s43088-022-00196-1

Tripathi, B. 2006. *Caraka Samhita, Sutra sthan 30/26*. Varanasi: Chokhamba Surbharati Prakashan.

Upadhyay, S.N., M. Bora, L. Kawlni, K. Mukherjee, and J. Hazra. 2018. "Comprehensive Pharmacology Review of Guduchi [Tinospora cordifolia (Willd.) Miers]." *Journal of Drug Research in Ayurvedic Sciences* 3 (1): 48–52.

Vaidya, A.D.B., et al. 2007. "Current Status of Herbal Drugs in India: An Overview." *Journal of Clinical Biochemistry and Nutrition* 41: 1–11. DOI: https://doi.org/10.3164/jcbn.2007001 https://www.jstage.jst.go.jp/article/jcbn/41/1/41_1_1/_article/-char/en

Vankar, P.S. 2008. "Effectiveness of Antioxidant Properties of Fresh and Dry Rhizomes of *Curcuma longa* (Long and Short Varieties) With Dry Turmeric Spice." *International Journal of Food Engineering* 4: 1–8.

Visavadiya, N.P., and A.V. Narasimhacharya. 2009. "*Asparagus* Root Regulates Cholesterol Metabolism and Improves Antioxidant Status in Hypercholesteremic Rats. Evidence-Based." *Complementary and Alternative Medicine* 6: 219–26.

Wibowo, D.P., R. Mariani, S.U. Hasanah, and D.L. Aulifa. 2020. "Chemical Constituents, Antibacterial Activity and Mode of Action of Elephant Ginger (Zingiber officinale var. officinale) and Emprit Ginger Rhizome (*Zingiberofficinale* var. amarum) Essential Oils." *Pharmacognosy Journal* 12 (2): 404–9.

Wynn, S.G., and B.J. Fougere 2007. *Veterinary Herbal Medicine*, 592–4. St. Louis: MOSBY Elsevier.

Yu, Y. 2006. Comparison of bioactivities and composition of curcumin-free turmeric (*Curcuma longa* l.) oils from different sources. All Thesis, paper 29. http://tigerprints.clemson.edu/all_theses/29

Sharma, P.C. et al. 2000c- "*Database on Medicinal Plants Used in Ayurveda*" Vol. 1, 418. New Delhi CCRAS

3 Taxonomic and Therapeutic Approach of High Value 25 Medicinal Plants of India

Swapnil G. Jaiswal and Satyanarayan Naik

3.1 INTRODUCTION

India is the hub of medicinal plants since ancient time and referred as medicinal garden to the world. Indian medicine system, which includes ayurveda, siddha, unani, homeopathy, yoga and naturopathy, has the strong base of medicinal plants against several diseases and disorders. Similar to India, China, Egypt and South America also utilize these medicinal plants as per their medicinal system to treat people in various medical situations (Sharma et al. 2008). Tribal communities of India living in forest areas are closely associated with medicinal plants. They have the applied knowledge about each and every plant part (root, leaves, bark, seeds, flowers) through experience, and they are using that plant part to prepare herbal medicine to cure various ailments. Ethnobotanical study of these underutilized plants on the basis of gathered botanical information from tribal communities may give more knowledge about their phytochemical profile and simplify their classification in a certain group known as plant family and species (Kshirsagar 2018).

Due to shifting trends from natural medicines to allopathic one due to speedy recovery, people are ignoring the long-lasting effect of natural medicines in health promotion. Moreover, urbanization of agricultural land, most of the important species of medicinal plants shifted in danger zone or disappeared. Conservation of these medicinal plants and their timely promotion is important today to maintain sustainability. Apart from this, lack of supply towards increasing demand for medicinal plants in international market facing critical situation. To overcome this condition and increase export trade, Indian government started promotional schemes to attract farmers, processors and suppliers to get involved. Statistical data for the year show that India's medicinal plants sector is growing, with the market valued at ₹4.2 billion (US$56.6 million) in 2019 and projected to surge at a 38.5% annual growth rate, reaching ₹14 billion (US$188.6 million) by 2026. Globally, the herbal trade is a massive US$120 billion industry, but India still captures only a small share of this market. Past statistical data showed that in 2017–18, India exported US$330 million worth of raw herbs, growing at 14.22% from the past year. Moreover, extracts and value-added products exported with a growth rate of 12.23% and hit the market worth of US$456 million over a last year (IBEF 2020).

 DOI: 10.1201/9781003623748-3

In the present chapter, an information related to 25 most potent medicinal plants that are found everywhere in India are focused taxonomically, therapeutically and commercially. List of those 25 medicinal plants includes tulsi, amla, aloe vera, neem, ginger, turmeric, ashwagandha, stevia, mulethi, guggal, isabgol, harida, baheda, brahmi, nuxvomica, shatavari, senna, giloy, rosemary, sage, thyme, mint, parsley, basil, curry leaves and mandookparni.

3.2 PLANTS FROM LAMIACEAE FAMILY

Plants from lamiaceae family are known by different names such as labiatae family or mint family or sage family or family of flowering plants as per the interest of botanists, chemist and biologist. Plants of this family are aromatic in nature and found in different formats like shrub, herb, trees and vines. The wide interest of botanist, chemist, pharmacologist and foodies is in their flavouring and aromatic part which is due to the specific group of compounds present in it and responsible for various biological activities. It is rich in antioxidant and antimicrobial compounds and act against inflammatory, cancerous, depressive, bacterial and viral diseases (Cocan et al. 2018; Raja 2012). In this article, five plants (Figure 3.1) from lamiaceae family, namely tulsi, rosemary, sage, mint and thyme, are selected for compiling information related to taxonomy, therapeutic and bioactive compounds present in it responsible for biological activities.

3.2.1 Taxonomic Information

Taxonomy gives systematic naming science where any biological group (plant, animals and microorganisms) defined in more scientific way. In the present article, all selected 25 medicinal plants when classified as per the taxonomy, all are coming under the kingdom plantae. Right after that it is further classified as order, family, genus and species. These parameters provide further distinct features responsible for uniqueness nomenclature in each and every botanical plant. Below Table 3.1 gives taxonomic information on the selected 25 medicinal plants.

Tulsi, rosemary, sage, mint and thyme as mentioned in Table 3.1 belong to the family *Lamiaceae* and are cultivated throughout the different regions of world. Mainly all these five herbs known for their aromatic phytoconstituents present in

FIGURE 3.1 Pictures of medicinal plants from natural habitat. *Lamiaceae* Family Plants—(a) Rosemary, (b) Tulsi, (c) Thyme, (d) Sage, and (e) Mint.

TABLE 3.1
Taxonomic Profile of 25 Medicinal Plants

Sr. No.	Common Name	Order	Family	Genus	Species	Scientific Name	Reference
1	Tulsi	Lamiales	Lamiaceae	*Ocimum*	tenuiflorum	*Ocimum tenuiflorum L.*	Pattanayak et al. (2010)
2	Rosemary			*Salvia*	*rosmarinus*	*Saliva rosmarinus*	Datiles and Acevedo-Rodríguez (2022)
3	Sage			*Salvia*	*officinalis*	*Saliva officinalis*	Sharma et al. (2019)
4	Mint			*Mentha*	*piperita*	*Mentha piperita*	Vining et al. (2020)
5	Thyme			*Thymus*	*vulgaris*	*Thymus vulgaris*	Prasanth Reddy et al. (2014)
6	Brahmi	Lamiales	Plantaginaceae	*Bacopa*	*monnieri*	*Bacopa monnieri*	Gohil and Patel (2010)
7	Isabgol			*Plantago*	*ovata*	*Plantago ovata*	Fatma et al. (2023)
8	Haritaki	Myrtales	Combretaceae	*Terminalia*	*chebula*	*Terminalia chebula*	Chattopadhyay and Bhattacharyya (2007)
9	Baheda			*Terminalia*	*bellirica*	*Terminalia bellirica*	Sharma et al. (2021)
10	Mulethi	Fabales	Fabaceae	*Glycyrrhiza*	*glabra*	*Glycyrrhiza glabra*	Vibha et al. (2009)
11	Senna			*Senna*	*alexandrin*	*Senna alexandrina*	Savulescu et al. (2018)
12	Mandukparni	Apiales	Apiaceae	*Centella*	*asiatica*	*Centella asiatica*	Sudhakaran (2017)
13	Parsley			*Petroselinum*	*crispum*	*Petroselinum crispum*	Milka et al. (2021)

14	Ginger	Zingiberales	Zingiberaceae	*Zingiber*	*officinale*	*Zingiber officinale*	Moghaddasi and Kashani (2012)
15	Turmeric			*Curcuma*	*longa*	*Curcuma longa*	Ayati et al. (2019)
16	Aloe Vera	Asparagales	Asphodelaceae	*Aloe*	*barbadensis*	*Aloe barbadensis*	Kaur and Bains (2024)
17	Stevia	Asterales	Asteraceae	*Stevia*	*rebaudiana*	*Stevia rebaudiana*	Yadav et al. (2011)
18	Shatavari	Asparagales	Asparagaceae	*Asparagus*	*racemosus*	*Asparagus racemosus*	Singla and Jaitik (2014)
19	Guggal	Sapindales	Burseraceae	*Commiphora*	*wightii*	*Commiphora wightii*	Bhardwaj and Alia (2019)
20	Neem	Sapindales	Meliaceae	*Azadirachta*	*indica*	*Azadirachta indica*	Tiwari et al. (2014)
21	Giloy	Ranunculales	Menispermaceae	*Tinospora*	*cordifolia*	*Tinospora cordifolia*	Kattupalli et al. (2019)
22	Amla	Malpighiales	Phyllanthaceae	*Phyllanthus*	*emblica*	*Phyllanthus emblica*	Singh et al. (2019)
23	Curry Leaves	Sapindales	Rutaceae	*Murraya*	*koenigii*	*Murraya koenigii*	Gahlawat et al. (2014)
24	Ashwagandha	Solanales	Solanaceae	*Withania*	*somnifera*	*Withania somnifera*	Bano et al. (2015)
25	Nux Vomica	Gentianales	Loganiaceae	*Strychnos*	*nux-vomica*	*Strychnos nux-vomica*	Behera et al. (2017)

it which gives unique aroma and extracted in the form of essential oil. Botanically, these herbs are coming under the category of shrub and can be cultivated in the field as well potted in the kitchen gardens. They are not grown more than one metre height and available easily throughout the year for their use as culinary and medicinal herb against routine infectious diseases (Datiles and Acevedo-Rodríguez 2022; Pattanayak et al. 2010; Prasanth Reddy et al. 2014; Sharma, Fagan and Schaefer 2019; Vining et al. 2020).

3.2.2 Potent Bioactive Compounds

Phytochemical profile of rosemary categorized in volatile and non-volatile part. Non-volatile part is rich in phenolic acids, flavonoids and terpenoids (diterpenes and triterpenes). Volatile part contains essential oils (monoterpenoids) responsible for specific aroma and biological activities. Volatile components, namely 1,8-cineole, alpha pinene, camphor, camphene, beta pinene, borneol and limonene, contribute broadly in rosemary oil composition. Rosemary extract (non-volatile part) contains mainly diterpenes (carnosic acid, carnosol and rosmanol) and triterpenes (ursolic acid, betulinic acid and micromeric acid) followed by flavonoids (apigenin, diosmetin, genkwanin and hispidulin) and phenolic acids, namely Rosmarinic acid, caffeic acid (Chan, Wong, and Chan 2022; Mersin and Işcan 2022).

Sage plant parts (leaves, stem and flowers) exhibited highest amount of biologically active compounds responsible for unique aroma and flavour in the category of Lamiaceae family herbs. Likewise, rosemary, sage is rich in polyphenols, flavonoids and terpenoids along with polysaccharide compounds. Volatile oil and non-volatile extracts of sage divide the category of compound present in it. 1,8 cineole, borneol, camphor, pinene, thujone, humulene and caryophyllene are abundant in concentrations of sage oil phytochemical profile. Out of these concentrations of these volatile fractions varies according to geographic location, growing conditions and plant parts. Flowers of sage are rich in alpha pinene and 1,8-cinole, stem contains linalool in highest concentrations and leaves are abundant source of camphene, camphor, limonene, thujone, humulene and borneol acetate. Non-volatile extracts of sage contain polyphenols who contributed significantly are caffeic acid, rosmarinic acid, salvianolic acids, sagecoumarin, lithospermic acid, sagerinic acid and yunnaneic acids. In view of flavonoids in sage gives major contribution by luteolin, apigenin, hispidulin, kaempferol and quercetin. Glucose, galactose, rhamnose mannose and xylose are found predominant polysaccharide in sage plant (Ghorbani and Esmaeilizadeh 2017; Lopresti 2017).

Thyme is known due to its commercial application as thyme oil, which is rich in thymol and its isomer carvacrol followed by p-cymene and alpha terpineol. Chemical composition of thyme oil contains thymol individually more than 50% and other are in less than 20% concentration (Prasanth Reddy et al. 2014; Singletary 2016). Mint exhibits menthol (40–60%) as major bioactive compound in volatile fraction of oil. Apart from this, others are menthone (15–30%), methyl acetate (1–23%), 1,8-cineole (1–13%), menthofuran (1–10%), isomenthone (2–8%), limonene (1–7%) and beta caryophyllene (2–4%). Phenolic compounds responsible for antioxidant effect are rosmarinic acid, chlorogenic acid, caffeic acid, gallic acid, coumaric acid and protocatechuic acid (Chakraborty, Chakravarti, and Bhattacharjee 2022; Wani et al. 2022).

Tulsi or Holy Basil used from ancient times to treat several ailments and diseases due to their wide phytochemical profile. Nutritionally, the whole tulsi plant fulfil almost all requirement of body like carbohydrate (polysaccharides), protein, lipid (fatty acid derivatives), vitamins (Vit. A and C) and minerals (calcium, phosphorus, zinc, iron). Besides, it is also rich in antioxidant compounds, namely polyphenols and flavonoids. Tannins, saponins, steroids (β-sitosterol, β-sitosterol-3-O-β-D-glucopyranoside, stigmasterol and campesterol in leaves and stem) and alkaloids are also present which are responsible for other biological activities. A polysaccharide extracted from the leaves part contains monosaccharides such as glucose, galactose, xylose, rhamnose and arabinose), while fatty acid derivatives isolated from the roots and leaves of tulsi are rich in palmityl glucoside, sanctuomic acid (mosquitocidal activity) and cerebrocides (antistress activity). Volatile oil and non-volatile oil were also separated from the plant parts leaves, stems, seeds and roots, respectively. Eugenol or methyl eugenol, beta caryophyllene, linalool, 1,8-cineole, isocaryophyllene and caryophyllene oxide are the major volatile oil bioactive compounds, while linoleic acid and linolenic acid are the major non-volatile bioactive compounds isolated from tulsi plant (Kumari et al. 2024; Pattanayak et al. 2010; Singh and Chaudhuri 2018).

3.2.3 Therapeutic Applications

Medicinal plants through traditional medicine system are in use since centuries due to their long-lasting beneficial effect on health. They are used in different formats like in powder form, liquid form, or semisolid paste form as per the need and applicability. There are certain unique phytochemicals present in each and every medicinal plant who gives their special effect in reliving from diseased condition. In the present article, 25 medicinal plants of commercial and medicinal importance are selected, and therapeutic properties of each plant are mentioned in this section.

Phytochemicals present in rosemary extract make it more potent for pharmacological applications. Instead of using a group of compounds (diterpenes) from rosemary against cancerous diseases (colon, prostate, lung, liver, breast, kidney, skin), whole rosemary extract has found to be more effective as an anticancer agent due to their cytotoxic effect of polyphenols, flavonoids and triterpenes (Chan, Wong, and Chan 2022). In another study, neuropharmacological properties of rosemary were highlighted by mentioning their role as memory booster, antidepressant, neuroprotective and effective against Alzheimer disease, addiction, neuropathic pain and inflammation (Ghasemzadeh Rahbardar and Hosseinzadeh 2020). Polyphenols and flavonoids present in rosemary act as antioxidant, antibacterial, antimicrobial, antifungal, anti-inflammatory, antidiabetic, antiobesity and antiulcer agent (Aziz, Batool, and Akhtar 2022).

In view of food applications, rosemary used as flavouring agent, preservatives by keeping away food from oxidative changes and microbial actions. In a study, rosemary oil in encapsulated form showed antimicrobial effect against mesophilic bacteria in Minas Frescal cheese and Mozzarella cheese by extending shelf life of cheese.

Rosemary oil has been observed to be effective in preserving a wide range of non-vegetarian food items and food industries are utilizing this oil in meat, poultry and fish products. Rosemary oil can also be used as active packaging material to life span on animal-based food products to some extents. Previous research studies showed strong evidence to act rosemary oil as a preservative in active packaging material by enhancing shelf life of lamb cutlets, fish fillets and keeping freshness of packaged shrimp. Use of rosemary oil is not limited to food applications only, and it is also useful in the cosmetic industry due to its wide range of applicability in products like hair toner, bath essence, hair shampoos, cologne water and serums (Aziz et al. 2022; Mersin and Işcan 2022).

Sage all the way gives similar kind of health effect as described for rosemary due to the somewhat similar kind of phytochemicals present in plants parts. Antioxidant, antimicrobial, antibacterial, anti-inflammatory, anticancer, antidepressant and neuroprotective are the pharmacological effects of sage bioactive compounds as discussed in earlier studies (Ghorbani and Esmaeilizadeh 2017; Lopresti 2017). Pharmacological properties of thyme showed that it is used as antimicrobial, antiviral, antioxidant, anti-inflammatory and insecticidal. Medicinally, it can be used against throat, skin, intestine, liver, respiratory tract and urinary tract-related infections. In aromatherapy, thyme oil has special importance due to their pain reliving property and soothing effect. Pharmaceutical, cosmeceutical and the food industry have wide applications of thyme oil due to their speciality aroma and flavouring compound thymol. It is used as aroma agent in perfumes, soaps, toothpaste, mouth washes and chewing gums. It is used as a flavouring agent in culinary recipes such as baked foods, vegetable salads, stewed meat, egg dishes, sea foods and marinades of chicken and fish (Prasanth Reddy et al. 2014; Singletary 2016).

Mint is used in stomach or intestine-related problems, vomiting and nausea-related symptoms, diabetes-related issues, cough and cold, muscle pain and irregular menstrual cycle. Pharmacological properties of mint include antimicrobial, antiviral, antibacterial, anti-inflammatory, antiallergic, antifungal and hepatoprotective. Food applications cover mint as aroma and flavouring effect in food like chewing gum, candies, herbal teas, vegetable salads, ice cream and frozen desserts, roasted lambs, pork and chicken. Pharma applications involve mint as pain-reliving creams, inhalers, digestion disorder-related syrups or tablets, cough and cold-related syrups, mouth washes and teeth-related pain relievers (Chakraborty, Chakravarti, and Bhattacharjee 2022; Wani et al. 2022).

Pharmacological potential of tulsi already described and proved in ayurveda and siddha. It was used traditionally as a medicine for fever, cough and cold, headache, boosting immunity, tiredness, digestive disorders (dysentery, colic, piles, stomach pain), bronchitis, blood purifier, skin-related infections, inflammations, eye diseases and wound cut. Rather than this, it can also be used for antioxidant, antimicrobial, anticancer, antidiabetic, antistress, antiobesity, antifertility, mosquitocidal, Lieshmanicidal, radioprotective and immunomodulatory effect. Tulsi leaves can be used as functional food ingredient to get unique nutritional benefits to an individual. In a study, tulsi leaves were incorporated in bakery products (bread and biscuits), dairy products (milk and cottage cheese) and beverages like tea, wine, juice and RTS (Kumar et al. 2022; Kumari et al. 2024; Pattanayak et al. 2010; Singh and Chaudhuri 2018).

3.3 PLANTS FROM PLANTAGINACEAE, COMBRETACEAE, FABACEAE, APIACEAE AND ZINGIBERACEAE FAMILY

3.3.1 Taxonomic Information

Ten plant species, each two from single family, namely Plantgenaceae (Brahmi and isabghol), Combretaceae (Haritaki and Baheda), Fabaceae (Mulethi and Senna), Apiaceae (Mandukparni and Pasley) and Zingiberaceae (Ginger and turmeric), were selected for the present article (Figure 3.2).

Isabgol (*Plantago ovata*) and Brahmi (*Bacopa monnieri* L.) both are herbs belongs from the family *Plantaginaceae.* Both are native to Indian subcontinent and found in northwestern and eastern parts (Iran, Japan, Korea and Egypt, China, Taiwan and Vietnam). In context of plant part used, seeds and husk of isabgol used for medicinal formulation while whole plant of brahmi used in formulation of medicinal drug. Isabgol is a rich source of phytoconstituent hemicellulose (Effective against constipation) while brahmi is a rich in alkaloids, saponins and sterols responsible for enhancing brain-related activity. Botanically, both the plants not grown above the one metre height and having soft herb with leaves cylindrical to ovoid shape grown in wet land and dry land areas near to water resources (Fatma et al. 2023; Gohil and Patel 2010).

Harida (*Terminalia chebula*) and Baheda (*Terminalia belorica*) commonly known as Hartaki and Bhibhitaki in Sanskrit, respectively. Botanically, it comes from the genus Terminalia and family *Combretaceae*. Fruits of both the plants are utilized against different ailments due to the presence bioactive compounds alkaloids and tannins in abundant concentrations. Geographically, both the plants are

FIGURE 3.2 Pictures of medicinal plants from natural habitat. *Plantaginaceae, Combretaceae, Fabaceae, Apiaceae, Zingiberaceae* Family Plants—(1a) Brahmi, (1b) Isabgol, (2a) Haritaki, (2b) Baheda, (3a) Mulethi, (3b) Senna, (4a) Mandukparni, (4b) Parsley, (5a) Ginger, (5b) Turmeric.

native to Indian subcontinent including India, Bangladesh, Nepal and Sri Lanka. In India, it is found in tropical and subtropical regions (Assam, Uttar Pradesh, Gujarat, Maharashtra, Andhra Pradesh, Tamil Nadu, Karnataka and Kerla). Botanically, description of both the plants shows that it grows 25- to 30-meter height and found in deciduous forest of Indian states. Blooming season of flowers starts from the month of October and ends in the month of December with fruiting. Flowers and fruits are grown at the end of branches. Flowers are unpleasant in smell, and fruits are oblong to ovoid drupes with yellowish to dark brown colour (Chattopadhyay and Bhattacharyya 2007; Sharma et al. 2021).

Mulethi (*Glycyrrhiza glabra*) is a name in Hindi also called as licorice a common name in English. Botanically, it is known as *Glycyrrhiza glabra* belongs to the family Fabaceae (Leguminoceae). It is native to southwest and central Asia (Pakistan, China, Egypt). In India, it is cultivated in Jammu and Kasmir, Delhi, Haryana, Uttar Pradesh and Gujarat. It is a perennial herb found in subtropical regions and grows in upward direction three to five feet from soil. Only root or rhizome part is utilized for medicinal purpose. Glycyrrhizin compound present gives 50 times sweetness than the sugar; hence, it is also called as sweet wood (Kaur et al. 2021; Vibha et al. 2009). Senna is another medicinal herb known by various synonyms in scientific way such as Cassia acutifolia Delile, Cassia angustifolia Vahl., Cassia lanceolata Forssk. and Cassia senna L. Again, it belongs to the family Fabaceae and grows two to three feet erect shrub. Warm and dry weather condition mostly suited for their growth. It is native to tropical and subtropical regions of the world, namely India, Africa, Saudi Arabia and Pakistan. In India, Gujarat, Maharashtra, Rajasthan and Tamil Nadu are the major senna producing states. Economically leaves and pods are utilized because it contains sennosoids compound in high concentration which has laxative in nature and used in medicinal formulation (Savulescu et al. 2018).

Parsley and Mandukparni a plant of 'Apiaceae' family comes under the category of aromatic plants. Parsley, *Petroselinum crispum*, is native to south-western Europe and western Asia subtropical regions. Generally, parsley is not cultivated, and it grows naturally in the stony soil and temperate zones, but still there are three cultivated varieties reported in some literature. These are *P. crispum var. crispum*—curly leaf parsley; *P. crispum var. neapolitanum*—flat leaf parsley; and *P. crispum var. tuberosum*—rooted parsley. Sowing to cultivation period for parsley is around seven to eight months which starts from spring season (March) to mid of winter (November). All the plant parts like leaves, roots and fruits are utilized in formulation of herbal drugs and flavouring agent in the food industry (Marthe 2020; Milka et al. 2021).

Madukparni (Brahmamanduki), *Centella asiatica* is native to Asia (India, China, Pakistan, Sri Lanka, Indonesia, Malaysia) and not cultivated. It is grown naturally in warmer to wet sandy areas like floodplains, besides river and other water resources. As per the ancient ayurveda, the whole plant has the medicinal value as cardiotonic, brain food (memory enhancer), diuretic and astringent (Sudhakaran 2017).

Ginger and turmeric both are from the same family zingiberaceae mostly known to its local name in India as 'Adrak' and 'Haldi', respectively. According to the literature, the Greek word zingiberis comes from the Sanskrit word 'Shringavera' means branched shaped structure or deers antlers. Both ginger and turmeric are rhizomes with antler-like structures that grow underground in the soil. Both are monocot perennial herbs grown in tropical and subtropical regions with humid conditions. Ginger and turmeric originated and cultivated mostly in India, China and Southeast Asia and exported to other regions of the world as medicinal and culinary herbs. In the North East Part of India, Maharashtra, West Bengal and Tamil Nadu are the major cultivars of ginger and turmeric. Both are around eight- to nine-month crops and usually harvested in the month of January to April (Kumari, Kumar, and Solankey 2020; Yadav and Tarun 2017).

3.3.2 Potent Bioactive Compounds

Brahmi is a good source of phytoconstituents, namely triterpenoids, saponins, alkaloids, sterols, flavonoids and phenolics related to cognitive and other health-related effects. The major bioactive compounds are triterpenoid saponins bacosides, which are a group of compounds, namely bacoside A, B, A3, bacosaponins and bacopasides, associated with all brain-related functioning. Alkaloids present in brahmi are brahmine, herpestine and nicotine, while the sterol group includes bioactive compounds beta-sitosterol and stigmasterol. Phenolic and flavonoids majorly present in brahmi are caffeic acid, chlorogenic acid, betulinic acid and luteolin, apigenin and quercetin, respectively. Additionally, another major glyocidic compound seen in brahmi is pseudojujubogenin (Alzobaidi et al. 2021; Gohil and Patel 2010).

Besides, another medicinal plant mandukparni (*Centella asiatica*) of Apiaceae family possesses a similar kind of phytochemical group as found in brahmi, but the type of active constituents present in it is different. Triterpenoid saponin, phytosterol, glycoside, flavonoid andpolyphenols are the group of active phytochemicals in mandukparni. Out of which triterpenoid saponins asiaticoside, centelloside, madekossida and asiatic acid majorly contribute to all pharmacological functions (Bhuyar et al. 2021; Irham et al. 2019).

Isabghol husk or husk along with seeds is a rich source of mixed polysaccharides (soluble, 35% and insoluble, 65%). Xylose, arabinose, galacturonic acid, galactose, mannose, glucose and rhamnose are the sources of sugars in husk and seeds. Xylose, arabinose, rhamnose and galacturonic acid are the sources of carbohydrates in isabghol. Seeds contain fatty acids (Linoleic acid, oleic acid, stearic acid, palmitic acid, lignoceric acid, linolenic acid). Mineral content in isabghol covers calcium, magnesium, phosphorus, potassium, sodium, sulphur, zinc and iron. Moreover, polyphenols and mucilaginous compounds (Pentosans and aldobionic acid) make it more potent for antimicrobial and laxative effects (Fatma et al. 2023; Sutradhar, Tahsin, and Ashrafi 2022). Phytochemical analysis of senna plant revealed that extracts prepared from senna leaves is a rich source of alkaloids, tannins, saponins, sterols, carbohydrates, glycosides (Sennosides A, B, C, D and anthraquinones), phenols and flavonoids (Elansary et al. 2018; El-hassan et al. 2012; Viswanathan and Nallamuthu 2012). Phenolic compounds present in senna fruits are gallic acid, gentisic acid, caffeic acid, quercetin, benzoic acid, vanillic acid, syringic acid, protocatechuic acid, neochlorogenic acid, isoquercetin, kaempferol, rhamnetin, luteolin and psoralene (Elansary et al. 2018). Another medicinally important plant from the Fabaceae family showed their potential is mulethi, which is 30–50 times sweeter than sugar with long-lasting sweet taste in the mouth. This sweet taste is due to the triterpene saponin compound glycyrrhizin (a mixture of calcium, magnesium and potassium salt of glycyrrhizic acid with a percentage range from 2% to 25%. Additionally, it also contains flavonoids (liquiritin, liquiritigenin, isoliquiritin, neoliquiritin and rhamnoliquirilin), Isoflavones (glabridin, galbrene, glabrone, shinpterocarpin, licoisoflavones A and B, formononetin, glyzarin and kumatakenin), polysaccharides, steroids (oestrogen), pectins, simple sugars, amino acids, vitamins (C, E and B complex) and mineral salts (silicon, potassium, iron, zinc, calcium, copper, magnesium, selenium, manganese, phosphorus and sodium). Yellow colour of mulethi powder is due to the flavonoid content of liquiritin and isoliquiritin (Kumar et al. 2023; Shakyawal and Mahobiya, 2023; Vibha et al. 2009).

Haridra and Baheda both are the rich sources of biologically active compounds hydrolysable tannins of 30–40% concentration (Gallic acid, ellagic acid, chebulic acid, chebulegic acid, chebulinic acid, neochebulinic acid and corilagin). Followed by tannins, it contains terpenoids, saponins, alkaloids and glycosides, namely glucose, fructose, sucrose, galactose and mannose (Chattopadhyay and Bhattacharyya 2007; Sharma et al. 2021).

Parsley, a medicinal herb of the Apiaceae family, is a rich source of vitamins, minerals, pigments and polyphenols. Overall, the whole plant is divided into three parts: leaves, roots and fruits, which is famous due to the presence of aromatic essential oil in spread throughout the plant's parts in varying ratio. Vitamin C is found in higher amounts in whole plants as compared to other vitamins (B-Complex, vitamin D, E, K and A). Carotenoids are plant pigments with unique colouring characteristics. Minerals, namely zinc, iron, calcium, phosphorus, sodium, magnesium and potassium, are present in the leaves of parsley. Fatty acids (linolenic and palmitic acid) along with phenols and flavonoids, namely luteolin, coumarins and myristicin, are responsible for antioxidant activity (Chauhan and Aishwarya 2018; Punosevac et al. 2021). Ginger and turmeric, both spices, contain unique aroma, flavour and colour due to the presence of notable bioactive compounds gingerols, shogaols and curcuminoids, respectively. Moreover, the essential oils of both ginger and turmeric are rich in terpenoids (mono and sesquiterpenes). Bisabolene, zingiberene, zingiberol, sesquiphellandrene and curcurmene are observed majorly in ginger oil, while turmerone, ar-turmerone, curcumene, germacrone and ar-curcumene are found in turmeric oil (Chanda and Ramchandra 2019; Moghaddasi and Kashani 2012).

3.3.3 Therapeutic Applications

Brahmi is known for their brain-related pharmacological action like a brain cell enhancer, intellectual function and overall mental health restoration effect. Past studies showed that oral intake of brahmi extract as per the prescribed doses significantly enhances audiovisual functioning associated with the brain and also sharpens memory (Calabrese et al. 2008; Kumar et al. 2016). Triterpenoid saponins are useful for the listed brain-related activities and in conditions like Parkinson's and Alzheimer's disease. Alkaloids listed in brahmi are contributing towards sedative and anxiolytic effects in treating anxiety, epilepsy and insomnia-related diseases. Sterols take part in cardio-protective functions, immunomodulatory functions and antidiabetic effects. Saponin and glycosides give wound healing effect and act as an antistress as well as an antifertility agent in raised situations. Phenolic compounds and flavonoids are responsible for hepatoprotective, antidiabetic, antioxidant, anti-inflammatory and anticancer effects (Alzobaidi et al. 2021; Gohil and Patel 2010; Meena et al. 2022; Thorat, Bagkar, and Patil 2018).

In view of food applications, Meena et al. 2022 compiled information about its usage in various food products like herbal products (whey beverages, health drink, ice cream, tea, buffalo ghee, tablets and oil), cereal bars, chicken nuggets, soup mix, Nutri-balls and dried powder.

Similarly, *Centella asiatica* (Mandukparni) is also responsible for protecting mental health. Triterpenoids, asiaticoside, asiatic acid, madecassoside and madecassic acid are a group of compounds responsible for the pharmacological effect. All the therapeutic properties are the same as mentioned for Brahmi, but would healing effect is an add-on property in the context of mandukparni (Sun et al. 2020). Triterpenoid saponin compounds individually highlighted their role as asiatocoside to improve memory and the immune system. It also rejuvenates blood circulation to the brain to the skin cells by repairing skin and blood cells. Madecosside synthesizes collagen, which is associated with the improvement of cell repair. Asiatic acid has the potential to act as an antiseptic agent and protect the body from the effects of free radicals. Apart from this, it also has anti-inflammatory, antiapoptotic, antinociceptive and memory boosting properties (Bhuyar et al. 2021; Irham et al. 2019). In a different study, the role of mandukparni extract was described as an active ingredient in food packaging. The presence of terpenoid and polyphenol compounds can make it usable in biopolymer film and modify packaging film characteristics (Pillai, Bhosale, and Roy 2024).

Pharmacological properties of isabghol showed its potential mostly in intestinal, ulcer and constipation-related disorders. Apart from this, it also has anticancer, antidiabetic, anti-inflammatory, wound healing activity, hypoglycemic activity and hypocholestrolemic activity (Sarfraz et al. 2017; Sutradhar, Tahsin, and Ashrafi 2022). In a different study, an extract of isabghol obtained through the supercritical fluid extraction method, which was rich in polyphenols, possesses antiobesity, antimicrobial and cytotoxic properties (Qanash et al. 2025). Isabgol husk was incorporated along with aloe vera juice in the formulation of vermicelli, which was found to be effective in the shelf-life enhancement of vermicelli up to six months by reducing microbial activity (Vaghashiya et al. 2018). A similar kind of pharmacological properties seen in isabghol are also observed in senna. The Senna plant is mostly used in herbal drug formulation, which is related to constipation and weight loss-related effect. It has a laxative effect, is a blood cleanser and is effective in skin-related queries due to the presence of bioactive compounds sennosides. It can give antidiabetic, antiparasitic, anti-inflammatory, antioxidant, antibacterial and antifungal effects (Elansary et al. 2018; Ikram et al. 2023). In some studies, senna extract incorporated with bread has been found to be effective in inhibiting mould growth in bread by enhancing their shelf life. It was found to be a promising alternative to synthetic antioxidants from preservation point of view (Muhamad Mahir, Mohamed Rehan, and Mohd Fauzi 2023).

Mulethi, generally called sweet wood, is useful against general viral, fungal and bacterial infection as well as sour throat, cough, ulcers, wound healing, skin-related infections, allergic symptoms. Apart from this, it acts as antioxidant, anti-inflammatory, anticancer, antidiabetic and hepatoprotective (Kaur et al. 2021; Shakyawal and Mahobiya, 2023). From a food application point of view, mulethi powder was utilized along with amla powder in the preparation of herbal ladoo (Bishnoi, Gehlot, and Siddiqui 2018).

Haridra contains tannins (Chebulic acid, chebulagic acid, corilagin and gallic acid), a major constituent responsible for medicinal property. It gives pharmacological effect like antioxidant, antimicrobial, antiviral, antibacterial, antimutagenic, radioprotective, hypolipidemic, wound healing, chemo-preventive and immunomodulatory activity (Chattopadhyay and Bhattacharyya 2007). Baheda contains alkaloid, coumarin, flavones, saponin, tannins and glycosides. It gives effects like antimicrobial, antioxidant, antimutagenic, antidiabetic, antiulcer, analgesic, antidiarrheal and hepatoprotective (Sharma et al. 2021).

Haridra and Baheda also used as functional ingredients in food and pharma applications. Since ancient time haridra-based jam, a sweet dish is used in South Asian countries in their food. Trifala churna a combination of haridra, baheda and amla is used in ayurvedic medicine as an oral drug to treat as antioxidant, antimicrobial, antimutagenic, antidiabetic, anticancer, immunomodulator, blood purifier, rejuvenator and antiaging (Jha and Sit 2023). In another study, it is suggested that oral administration (prescribed doses) of haridra-based decoction could be useful in preventing cardiovascular, cancerous and metabolic diseases (Pellati et al. 2013).

As the bioactive potential of plant parts of parsley, its pharmacological property is also changed. Phenols and flavonoids present in leaves and stems showed antioxidant effect. Flavonoid myristicin of parsley responsible for oestrogen production during menopause. Leaves and stems also responsible to regularize menstruation cycle in women, intestinal cramps, kidney stone, indigestion, anorexia, high blood pressure control, control nose bleeding, diarrhoea, anticancer, anti-inflammatory, immune-suppressant, anticoagulant, antihyperlipidemic and antihypertensive (Chauhan and Aishwarya 2018; Punosevac et al. 2021).

Ginger and turmeric are both spice category rhizomes generally used for culinary purpose, but this kitchen spice has an enormous health beneficial effect. Gingerols and shogaols (ginger) as well as curcuminoids (turmeric) are the group of biologically active compounds along with terpenoids (essential oil) responsible for pharmacological effect (Jaiswal and Naik 2018, 2021). Both the spices give more or less similar pharmacological properties like antioxidant, antimicrobial, antiulcer activity, anti-inflammatory, anticancer (turmeric), cardiac protecting, wound healing (turmeric) and protecting digestive health. Active constituents in ginger give relief in cough and cold, fever-like symptoms, sore throat and migraine-like symptoms. Both the spices are used as culinary herbs in the Indian kitchen since ancient times. Several studies have been carried out on its application as an antioxidant in preserving lipid-based food. Unique aroma and flavour of ginger and turmeric make them usable in spice tea and turmeric used as colouring compound in food preparations. Candies, cookies, breads, cakes, ales, beer, seasonal beverages and pickles are some of the products made from ginger (Guneri 2021; Kim and Chun 2005; Laelago Ersedo et al. 2023).

3.4 PLANTS FROM ASPHODELACEAE, ASTERACEAE, ASPARAGACEAE, BURSERACEAE, MELIACEAE, MENISPERMACEAE, PHYLLANTHACEAE, RUTACEAE, SOLANACEAE AND LOGANIACEAE FAMILY

3.4.1 Taxonomic Information

Ten plants, namely aloe vera (Asphodelaceae), stevia (Asteraceae), shatavari (Asparagaceae), guggal (Burseraceae), neem (Meliaceae), ashwagandha (Solanaceae), giloy (Menispermaceae), curry leaves (Rutaceae), amla (Phyllanthaceae) and nux vomica (Loganiaceae), were selected for the present article (Figure 3.3).

Ashwagandha (*Withania somnifera* L.), the Indian ginseng, is name after its power reflects as it gives a power like a horse, after its consumption. Roots are the main ingredient in medical formulations and smells like horse. Previous review reveals that ashwagandha used in maximum medical formulations and in almost all medicinal systems (Ayurveda, Unani and Siddha). Alkaloids are the main focusing bioactive constituents that gives health promoting effects and act as blood purifier. Geographically, this plant is found in Southern Mediterranean regions to the Canary Island and from South to East Africa; from Palestine to North India, covering Israel, Jordan, Egypt, Sudan, Iran, Afghanistan, Baluchistan and Pakistan. Botanically, this plant comes from the *Solanaceae* family gros around 1 metre height and sunny climate is preferable for their growth because cloudy and humid weather attracts pests and declined the growth of plants (Bano et al. 2015). In the same range of pharmacological properties, another plant Shatavari (*Asparagus racemosus* L) belongs from different family

FIGURE 3.3 Pictures of medicinal plants from natural habitat. Family: Asphodelaceae (1. Aloe Vera), Asteraceae (2. Stevia), Asparagaceae (3. Shatavari), Burseraceae (4. Guggal), Meliaceae (5. Neem), Menispermaceae (6. Giloy), Phyllanthaceae (7. Amla), Rutaceae (8. Curry leaves), Solanaceae (9. Ashwagandha), Loganiaceae (10. Nux Vomica).

Asparagaceae followed by subfamily *Asparagoideae.* It is again rejuvenating overall health of human beings, especially females. Steroidal saponins followed by alkaloids and flavonoids are the compounds of interest responsible for health-promoting effects. Geographically shatavari plant distributed all over the world and cultivated mainly in India, Southern part of China, Sri Lanka and Africa. Botanically, Shatavari plant requires mostly sunny climate and grows in all types of soils. Sowing time for this plant is generally in July and August. Once sowing completed, with proper care taken for one to two year a crop is ready to harvest (Singla and Jaitik 2014).

Tinospora cordifolia commonly known as gudchi or giloe in Hindi is a climbing shrub or herb widely distributed in Asian countries (India, China, Sri Lanka, Myanmar, Thailand, Philippines, Indonesia, Malaysia) and African countries. It grows almost all types of soil conditions and sunny weather best suited for their constant growth. Alkaloids and terpenoids are the special group of targeting compounds of the family *cordifolia* which have numerous health benefits (Kattupalli et al. 2019). Nux vomica (*Strychnos nux-vomica* L.) is poisonous alkaloids belongs from the family *Loganiaceae.* Due to the presence of toxic constituent alkaloid in the whole plant, it gains wide applicability throughout all medicinal systems (Chinese, Tibetan, Unani, Ayurveda and homeopathy). Geographically this plant is distributed evergreen forest of India, Pakistan and Burma. Botanical description provides wide knowledge about their morphometric characteristics like medium tall (12–15 metre) slow growing deciduous tree with smooth black barks, blooming starts from mid-February to till end of April. Fruits ripened over ten months period with orange red colour (Behra et al. 2017).

Aloe vera (*Aloe barbadensis*) is a very popular and worldwide cultivated succulent plant. Origin of this *Liliaceae* family plant is confusing but historical literature related to this plant shows that it was originated from Arabian Peninsula and after that spread all over the world. Aloe vera grows in tropical and subtropical areas and generally loamy soil having water drainage capacity suited best for the growth. Moreover, sunny climate is preferable. Shade practices also give a suitable environment for the growth. This plant can tolerate in rainier as well as extreme dry environment. Life span of aloe vera is generally 25 years and stem grows height up to 30 cm on an average. Each single leaf grows 40–50 cm long and 5–7 cm width. Pharmacologically, aloe vera can be used mostly in skin-related remedies (Kaur and Bains 2024).

Neem (*Azadirachta indica*) is a holy tree found in tropical and subtropical regions. It is native to India and Asian subcontinents and belongs to the family *Meliaceae.* Neem plant can grow height up to 15 to 20 metres in some cases 35 to 40 metres. Nimbin is the main active compound having bitter in taste as well as alkaloids, saponins and steroids are other secondary metabolites found in the neem plants (Tiwari et al. 2014). Guggal (*Commiphora wightii*) is a shrub of 2–3 metre height and grown in arid zone. It is basically native to India but found in Northern Africa to Northern India (Gujarat, Rajasthan and Maharashtra) (Bhardwaj and Alia 2019). Curry leaves (*Murraya koenigii*) basically used in culinary purpose. Apart from this whole plant parts have been utilized for medicinal purpose. The plant grows around the height of 6–8 metre height and native to India, Sri Lanka and other south Asian countries. In India it is distributed almost all the states. Essential oil contains alkaloids, flavonoids and steroids have enormous medicinal properties (Gahlawat, Jakhar, and

Dahiya 2014). Aonla (*Embelica officinalis*) is also called as Indian gooseberry. It is cultivated in subtropical areas mostly southern and northern part of India. In Ayurveda, aonla considered as rich source of rejuvenating rasayana. It is rich source of phytoconstituents like minerals, vitamins, amino acids, fixed oils, tannins, phosphatides and essential oils (Singh et al. 2019).

Stevia (*Stevia rebaudiana*) is a perennial herb or shrub native to South America and then widely popular and cultivated in Japan, China, Taiwan, Thailand, Korea, Brazil and Malaysia. In some countries, it is directly consumed along with food. But in United States, its use is banned at commercial level or direct involvement in food. Steroids are the major group of compounds found in stevia responsible for extreme sweet effect than sucrose. Leaves and stems are utilized for medicinal use. Botanically it belongs to the family '*Asteraceae*' and subfamily '*Asteroideae*'. Stevia plant can grow in hilly and subtropical regions with a climatic condition of 20–25°C. Plant height up to the range of 45 cm and floral appearance is small white colour (15–17 mm). Flowering season could start from January to March in Southern Hemisphere and September to December in Northern hemisphere (Yadav et al. 2011). Figure 3.4 represents the structure of major biologically active compounds found in medicinal plants.

3.4.2 Potent Bioactive Compounds

Guggal (*Commiphora wightii*) is rich in volatile oils (Mono, di and triterpenoids), steroids, lignans, ferulates and variety of inorganic ions. Monoterpenoid component in oil are myrcene, dimyrcene and polymyrcene. Sesquiterpenoids covers cadinene while diterpenoids includes alpha-camphorene and cembrene. Triterpenoids present in oil are myrrhanol (A,B and C) and myrrhanone (A and B), Commipherol, commipherin. Besides guggal also contain steroids (E and Z guggalsterone, I-VI guggalsterol), flavonoids (Muscanone, naringenin and quercetin), lignans (Sasamin and diayangambin), sugars (Arabinose, galactose and fructose) and amino acids (Bhardwaj and Alia 2019).

Aloe vera (*Aloe barbadensis*) has the same pharmacological properties as mentioned for ginger and turmeric. Aloe vera has the more than 200 bioactives that includes polyphenols, tannins, saponins, mucilage, flavonoids, indoles, alkanes, pyrimidines, alkaloids, organic acids, aldehydes, dicarboxylic acids, ketones, alcohols, fatty acids, phytosterols, terpenoids, cardiac glycosides, anthracene and quinone derivatives (Kaur and Bains 2024; Moghaddasi and Kashani 2012; Yadav and Tarun 2017).

Shatavari root is an abundant source of steroidal saponins (Shatavari I-V) along with phenolics, flavonoids ((kaempferol, quercetin, rutin), alkaloids (Phytoestrogens), carbohydrates (Glucose, galactose, arabinose, rhamnose and xylose), vitamins (A, E, C, B1 and B2) and minerals (Calcium, magnesium, zinc, iron, potassium, copper). Besides, shatavari volatile oil has some essential fatty acids, namely gamma linolenic acid, eugenol and thymol in their composition (Singh et al. 2018).

In ashwagandha, steroidal lactone are the primary bioactive constituents (Withanolides) responsible for pharmacological activities. Secondly, roots of ashwagandha contain alkaloids (Anahygine, berberine, cuscohygrine) while in rest of the plant alkaloids, namely caffeine, choline and withanamide A-I are found. After

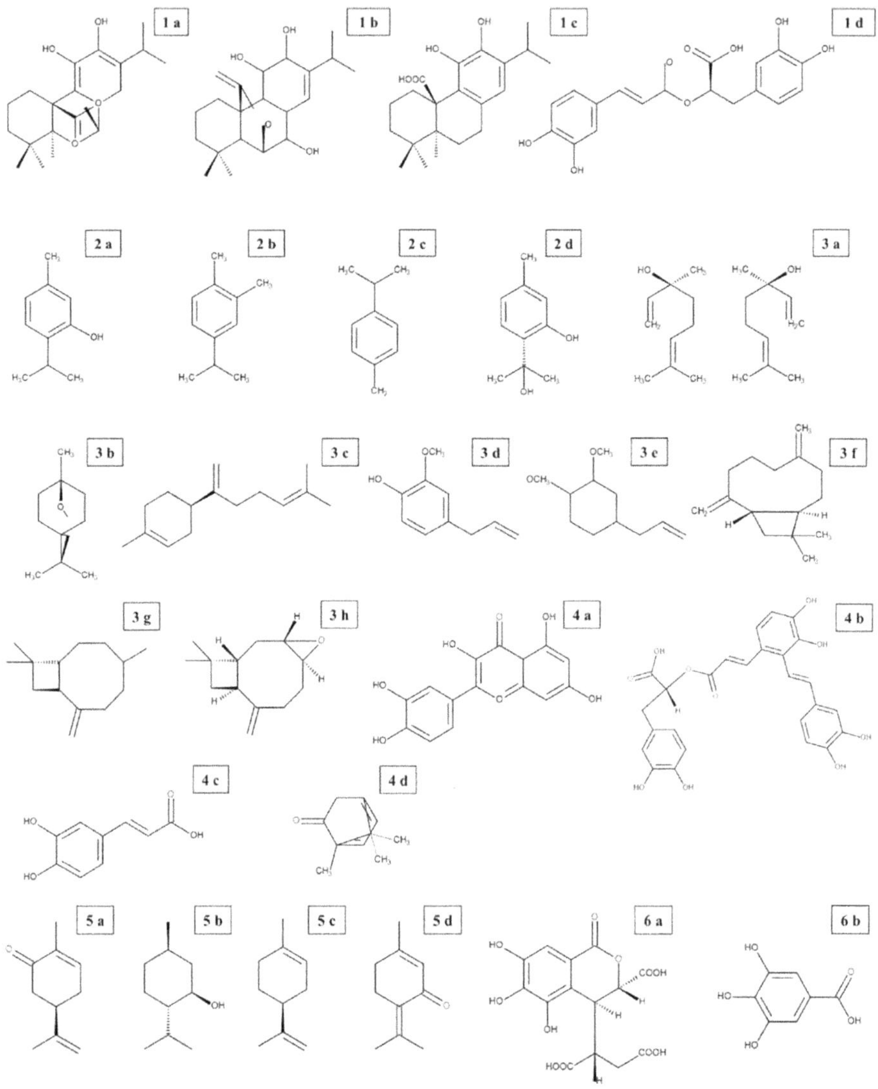

FIGURE 3.4 Major biologically active compounds in medicinal plants. (1a) Carnosol, (1b) Rosmanol, (1c) Carnosic acid, (1d) Rosemarinic acid. (2a) Thymol, (2b) Carvacrol, (2c) *p*-cyemene, (2d) alpha terpineol. (3a) Linalool, (3b) 1,8-cineole, (3c) Beta bisabolene, (3d) Eugenol, (3e) Methyl eugenol, (3f) Beta caryophyllene, (3g) Isocaryophyllene, (3h) Caryophyllene. (4a) Quercetin, (4b) Salivanolic acid, (4c) Caffeic acid, (4d) Camphor. (5a) Carvone, (5b) Menthol, (5c) Limonene, (5d) Piperitone. (6a) Chebulic acid, (6b) Gallic acid.

alkaloids, flavonoids (Kaempferol and quercetin), coumarines (Fruit: Aesculetin and scopoletin), phytosterols (Campesterol, stigmasterol and cycloartenol) and fatty acids (Linoleic and oleic) are observed in root as well as other parts of the ashwagandha plant (Shinde et al. 2023).

Curry leaves, stems, bark, roots and seeds are the major sources of alkaloids. Carbazole-type alkaloids are found in all parts of *Murraya koenigii*. Alkaloids, namely murrayastine, murrayline, pypayafolinecarbazole, trterpenoid, cyclonabanimbine, tetrahydromahanimbine, coumarine (murrayone imperatoxin) along with vitamin A, C and nicotinic acid with some minerals (calcium) are present in whole plant part of *Murraya koenigii* (Gahlawat, Jakhar, and Dahiya 2014). Giloy (Gudchi) composition varied slightly as per the plant or plant part on which they grow. Otherwise, basic active constituents remain same in which alkaloids (Protoberberine), terpenoids and polysaccharides are majorly found followed by glycosides, steroids, sesquiterpenoids and phenols (Choudhary et al. 2013).

Amla or Indian gooseberry is a rich source of tannins (Emblicanin A and B, punigluconin, pedunculagin, ellagitannin, dehydroellagitannin and benzopyran tannin) followed by alkaloids (phyllantine, phyllembein and phyllantidine). Other active constituents include carbohydrates (Pectin), Vitamin (Ascorbic acid), Flavonoids (Kaempferol and quercetin) and amino acids, namely glutamic acid, aspartic acid, proline, alanine, cystine, lysine (Diwan et al. 2018). In context of stevia, glycosides are the major group of compounds present in the chemical composition followed by phenolic and flavonoids and terpenoids. Eight diterpene glycosides are present in stevia out of which four are known due to their sweetening properties, namely stevioside, rebaudioside-A, rebaudioside-C and dulcoside-A. These four glycosides are 210, 242, 30 and 30 times sweeter than sucrose (Yadav et al. 2011).

Nux vomica, the whole plant part contains phytoconstituents in the form of alkaloids, flavonoids, tannins and triterpenoids, glycosides, lignins and steroids. strychnine and brucine are the principal toxic alkaloids that can be found in the marked range 50–60% of total alkaloid content. Therefore, purification and detoxification treatment are the only way to use it further (Behera 2019).

Nimbin is the principle active compound in neem plant. Apart from this, other compounds extracted from neem categorized in two classes such as isoprenoids and non-isoprenoids. Terpenoids and steroids including azadirone, limonoids, protomelacins, gedunin, vilasinin, azadirachtin, nimbin and salanin falls under isoprenoids class. Protein, polysaccharides, polyphenolics, tannins and glycosides fall in non-isoprenoid class (Tiwari et al. 2014).

3.4.3 Therapeutic Applications

Pharmacological properties of guggal includes hypolipidemic and hypoglycemic activity, cardiac protective and thyroid stimulant activity, antimicrobial, anti-inflammatory, antiarthritic, antiobesity, antifertility and cytotoxic activity (Bhardwaj and Alia 2019; Singh, Dhyani, and Kaur 2015).

Medicinally, aloe vera can act as antioxidant, anticancer, antimicrobial, antifungal, antibacterial, anti-inflammatory, antiulcer, antiaging, hepatoprotective, wound healing, skin-related allergies and improve gut health (Kaur and Bains 2024). In view of food applications, aloe vera can be used in dairy products (yoghurt and ice creams), snack foods, famine foods, as a cooked vegetable, food products (confectionaries and beverages), as a preservative. As per the previous findings it was observed that not all the species of aloe vera used in food and other cosmeceutical applications due to the presence of high concentration of toxic alkaloids and their reactive nature (Grace 2011).

Shatavari (*Asparagus racemosus*) and Ashwagandha (*Withania somnifera*) in ayurveda called as 'rasayana'. due to the presence steroidal saponins in abundant quantity along with alkaloids, flavonoids, dihydrophenanthrene derivatives, furan derivatives shatavari is responsible for healing and cleansing activity in the body. Shatavari kalp in pregnancy proved to antiabortion agent and its consumption in prescribed format helps to increase foetal weight. As per the reported literature shatavari and ashwagandha helps to build body immune system. Along with this, it acts as antidepressant, antioxidant, antibacterial, antidiarrheal, anticancer, antiulcer, anti-aflatoxigenic and cardioprotective (Bano et al. 2015; Singla and Jaitik 2014). Likewise, curry leaves, stevia, aonla and giloy have the more or less same pharmacological effect as reported for the shatavari and ashwagandha (Diwan et al. 2018; Kattupalli et al. 2019; Marcinek and Krejpcio 2016).

Commercially, shatavari and ashwagandha root powder has high demand in market due to their biological activity like immune boosting, cancer preventing, rejuvenating nervous system and solving stomach-related issues. In processed foods, it also used as functional ingredients in the form of additives in biscuits, bakery, confectionary, dairy products, low glycaemic food product (chapati, ladoo, naan and thepla) and feed supplement for dairy animals (Shinde et al. 2023; Singh et al. 2018).

Nux vomica after ensuring their detoxification can be used in treatment of several disorders and ailments. Dut to the toxic residues in plant, different medicinal system prescribed some ways to purify it before use it against specific disease or disorder. Generally, it can be used as antioxidant, antimicrobial, anticancer, immunomodulatory, anti-alcoholic, antiallergic, antamnesic, antitumour, anticonvulsant, insecticidal, rodenticidal and nematocidal (Behra et al. 2017).

Neem tree leaves have the enormous health benefits as well as used in agriculture as pesticide or herbicide. It can also be used in the preparation of floor cleaner, strain removal in textiles, dish washer formulation, insect repellent, cosmeceutical application (face wash, facial scrub, skin conditioner). Pharmacological properties are due to the nimbin active derivative present in neem extract and gives anticancer, antimicrobial and immunomodulatory effect (Tiwari et al. 2014).

3.5 EMPLOYMENT POTENTIAL

3.5.1 Status of Medicinal Plants

India has great heritage of medicinal plants due to their geographical location and climatic conditions on worlds map. Apart from this ancient medicinal system of India confirms the applicability of medicinal plants for betterment of human life. This gold mine of medicinal plants in India has been categorized in 15 agroclimatic zones right from Himalayan range to Kanyakumari and Western ghats to north-eastern part of India. As per the information shared by Indian Brand Equity Foundation (IBEF), around 18,000 medicinal plants species were available of which 7000 species explored for therapeutic applications. Ayurveda, Unani, Siddha, Homeopathy and allopathy are medicinal systems of India continuously working on it to get better formulations for rejuvenating overall human health. Commercially around 1000 species are traded of which nearly 175–180 medicinal plants have demand more than 100 metric tons annually. From these facts and figures, around 70–80% medicinal plants collected from wild forests and remaining procured from destructive farming practices. But as per the observation made by All India Trade Survey of Prioritised Medicinal Plants in 2019, requirement of high value medicinal plants in the global market has increased by twice while supply declines nearly half. This could be the alarming situation for all the cultivars, pharmaceutical industries and supply chain to do the root cause analysis for declining supply and find a solution to overcome from it. In a study related to threatened medicinal plants of India, state-wise and species-wise information is provided on medicinal plants which are in extinct category, endangered category, endangered, vulnerable and threatened zone (Gowthami et al. 2021; IBEF 2020).

3.5.2 Potential Areas of Employment Generation

3.5.2.1 Cultivation and Farm Practices

Medicinal plants of India which is the second largest market for world after China need to be cultivated in proper way. It could be profitable business for Indian farmers by optimizing some parameters like upgradation of old farm practices to implementation of new farm technologies. Certain criteria's work strategies need to be followed to become successful cultivar:

Gathered information related to demand of medicinal plants in the global market and plan accordingly to make a selection of specific medicinal plant variety for cultivation. At the same time, other parameters like soil quality and health need to be check, required environmental conditions for proper growth and yield need to be planned by adopting controlled farm practices. In addition to this, skilled labourers must require who have the knowledge of all the necessary tools required for the growth, type of pest attacked and how to control it, care need to be taken during adverse weather condition, knowledge of fertilizer and pest control required during growing phase of crop and most importantly knowledge of harvesting at certain stage of growth of medicinal plants. Last but not the least to become successful cultivar and profit maker farmer, networking with other successful cultivars must be done, permits and licencing must be done in the case of contract farming, proper storage facility and preliminary processing structures need to be developed for a medicinal plant if required, approach to research organization in order to analyse quality of cultivated crop in the market, search a market to get appropriate cost against demand (Bosak and Schroeder 2004; Mushtaq and Manjul 2024).

3.5.2.2 Processing and Value Addition

In this area volatile essential oil extraction units, drying and dehydration equipment's are involved. One can set up business in this area because raw herbs or plant parts need to be processed for its applicability in any food, pharmaceutical and cosmeceutical sector. Any kind of disease or disorder can be cured by herbal drug treatment which is a formulation of certain group compounds extracted from herbs. Therefore, the extraction of biologically active compounds from raw materials is necessary.

In other cases, drying, dehydration and crushing equipment's helpful for powdered herbal drug formulation (Kar 2023; Panda and Giri 2024).

3.5.2.3 Research and Development, Marketing and Export

Research and development activities are the key aspects through which medicinal plants can be explored in a deeper way. Therefore, setting up certified research laboratories for medicinal plants and their products could be one of the important steps not only from marketing and export point of view but also from skilled employment generation point of view. Research and development are not limited to plant profile analysis, but it starts from the identification of new unexplored plant species, development of new hybrid varieties in order to get more yield of target-oriented compounds in the crop. Research and development are also useful for study of pharmacological properties, new herbal drug formulation. Research activities also certify medicinal plant seed or products in more qualitative way and certified products after quality check have more demand in market. Marketing right product to right end user or farmer builds a trust on brand. Once trust build up by end user on brand it could be easy to sale the product in the domestic or international market through export (Sharma et al. 2008).

3.5.3 Entrepreneurship and Startup

Very few names of medicinal plant processing industries are known in market and supply of raw materials mostly from forest reservoirs in a unorganized way. In order to make availability of quality raw materials for processing and getting high value and profitable product, there is a need of group of trained people who is having idea about proper raw material collection from forest reservoirs. In connection to this an agency should be set up who trained the people and provide consultancy. In another way a small-scale industry or start up can be set up who provide semi processed raw materials to pharmaceutical and food industry. Consultancy project for providing training to famers in view good agricultural practices, seed analysis for better germination, available modern techniques for crop profile analysis through AI and data analysis, modern tools and techniques available for cultivation, awareness about supply chain facility, storage facility and export facility can be done. Interested people can set up their business in the above listed areas and generate revenue as well as provide employment to other needy people. Setting ayurvedic medicine and wellness centres, providing herbal spa and naturopathy treatment and manufacturing herbal cosmetic products through co-operative societies are also other ways to start a career in the area of medicinal plants and their products (Caplins and Halvorson 2017; Kumar et al. 2011; Salamon 2022).

3.5.4 Government Strategy to Boost Employment

The Government of India through certain government agencies has started several measures to boost the medicinal plant sector of India. The National Medicinal Plant Board (NMPB) is offering 75% subsidy to farmers for cultivating medicinal plants, developing some schemes and guidelines for financial assistance in various zones of medicinal plant divisions, securing some commercial and promotional plans for government and nongovernment associations to achieve yield-oriented targets for medicinal crops. In 2017, NMPB launched a concept, namely 'Voluntary Certification Scheme for Medicinal Plants Produce (VCSMPP)' for developing good agricultural practices (GAPs) and good field collection practices (GFCPs) in order to encourage Indian farmer towards trading. AYUSH Ministry of India provide financial assistance to exporters at the international level for organizing trade fairs, business summits and conferences to boost the medicinal market at the international level.

The Merchandise Exports from India Scheme (MEIS) provides incentives to the exporting community for specified goods in order to counteract infrastructural inefficiencies and the associated costs of exporting products manufactured in India; thereby, giving special emphasis to these products which are of India's export interest and have the capability to generate employment and enhance the country's competitiveness in the world market.

Export Promotion Council (EPC), an entity of Commerce Department, Government of India, is monitoring all export related activities through SHEFEXIL (Shellac and Forest Products Export Promotion Council) and PHARMEXCIL (Pharmaceutical Export Promotion Council). EPC also facilitating financial assistance in all types of promotional activities which is related to exports of medicinal plants and their products (IBEF 2020).

3.6 CONCLUSION

India is the country blessed with all categories of medicinal plants right from north to south and east to west. With this, our Indian medicinal system also practised in the application of these herbs in curing diseases and disorders since ancient time. Due to the fastest changing economy, infrastructure and livelihood, natural resources and their end users are declining day by day. Due to urbanization, agricultural land size already gets affected. Overexploitation of these natural medicinal plants without adopting conservation practices and good agricultural practices is an alarming situation for present and future generations. In the present article, an attempt has been made to represent all time important 25 medicinal plant species botanically, therapeutically and economically. Conservation of these species through their systematic farming under government schemes and guidelines may change the scenario from alarming situation to profitable earnings. Besides, it may help to increase the overall export of the medicinal plants and their products in the international market, as well as create employment opportunities for today's youth. From a future point of view and a kind of situation created in the near future in health aspects, more commercial unexplored medicinal plants need to be explored therapeutically and commercially as a source of income and wealth for the needy.

ACKNOWLEDGEMENT

First author was thankful to his family for providing all possible support to complete the present work.

REFERENCES

Alzobaidi, N., H. Quasimi, N. A. Emad, A. Alhalmi, and M. Naqvi. 2021. "Bioactive Compounds and Traditional Herbal Medicine: Promising Approaches for the Treatment of Dementia." *Degenerative Neurological and Neuromuscular Disease* 11: 1–14.

Ayati, Z., M. Ramezani, M. S. Amiri, A. T. Moghadam, H. Rahimi, A. Abdollahzade, A. Sahebkar, and S.A. Emami. 2019. "Ethnobotany, Phytochemistry and Traditional Uses of Curcuma spp. and Pharmacological Profile of Two Important Species (C. longa and C. zedoaria): A Review." *Current Pharmaceutical Design* 25: 871–935.

Aziz, E., Batool, R., Akhtar, W., Shahzad, T., Malik, A., Shah, M.A., Iqbal, S., Rauf, A., Zengin, G., Bouyahya, A., Rebezov, M.B., Dutta, N., Khan, M.U., Khayrullin, M., Babaeva, M.K., Goncharov, A., Shariati, M.A. and M. Thiruvengadam. 2022. Rosemary Species: A Review of Phytochemicals, Bioactivities and Industrial Applications. *South African Journal of Botany* 151: 3–18. https://doi.org/10.1016/j.sajb.2021.09.026

Bano, A., N. Sharma, H. S. Dhaliwal, and V. Sharma. 2015. "A Systematic and Comprehensive Review on Withania somnifera (L.) Dunal- An Indian Ginseng." *Journal of Pharmaceutical Research International* 7: 63–75.

Behera, M. C. 2019. "Strychnos nux-vomica Linn -Strychnine Tree." In *Medicinal Plants in India Importance and Cultivation*, Vol. II, 390–404. New Delhi: Jaya Publishing House.

Behera, M. C., T. L. Mohanty, and B. K. Paramanik. 2017. "Silvics, Phytochemistry and Ethnopharmacy of Endangered Poison Nut Tree (*Strychnos nux-vomica* L.): A Review." *Journal of Pharmacognosy and Phytochemistry* 6: 1207–16.

Bhardwaj, M., and A. Alia. 2019. "*Commiphora wightii* (Arn.) Bhandari. Review of Its Botany, Medicinal Uses, Pharmacological Activities and Phytochemistry." *Journal of Drug Delivery and Therapeutics* 9: 613–21.

Bhuyar, P., M. H. Rahim, G. P. Maniam, and N. Govindan. 2021. "Isolation and Characterization of Bioactive Compounds in Medicinal Plant Centella asiatica and Study the Effects on Fungal Activities." *Journal of Microbiology, Biotechnology and Food Sciences* 10 (4): 631–5.

Bishnoi, J. P., R. Gehlot, and S. Siddiqui. 2018. "Utilization of Glycyrrhiza glabra for Preparation of Herbal Aonla Ladoo." *Journal of Pharmacognosy and Phytochemistry* 7 (2): 446–50.

Bosak, K., and K. Schroeder. 2004. "Biodiversity Conservation and the Struggle for the Nanda Devi Biosphere Reserve." *The American Geographical Society's Focus on Geography* 48: 1–6. https://doi.org/10.1111/j.1949-8535.2004.tb00129.x

Calabrese, C., W. L. Gregory, M. Leo, D. Kraemer, K. Bone, and B. Oken. 2008. "Effects of a Standardized Bacopa monnieri Extract on Cognitive Performance, Anxiety, and Depression in the Elderly: A Randomized, Double-Blind, Placebo-Controlled Trial." *Journal of Alternative and Complementary Medicine* 14 (6): 707–13.

Caplins, L., and S. J. Halvorson. 2017. "Collecting *Ophiocordyceps sinensis*: an Emerging Livelihood Strategy in the Garhwal, Indian Himalaya." *Journal of Mountain Science* 14: 390–402. https://doi.org/10.1007/s11629-016-3892-8

Chakraborty, K., A. R. Chakravarti, and S. Bhattacharjee. 2022. "Bioactive Components of Peppermint (*Mentha piperita* L.), Their Pharmacological and Ameliorative Potential and Ethnomedicinal Benefits: A Review." *Journal of Pharmacognosy and Phytochemistry* 11 (1): 109–14.

Chan, E. W. C., S. K. Wong, and H. T. Chan. 2022. "An Overview of the Chemistry and Anticancer Properties of Rosemary Extract and Its Diterpenes." *Journal of Herbmed Pharmacology* 11 (1): 10–9.

Chanda, S., and T. V. Ramchandra. 2019. "Phytochemical and Pharmacological Importance of Turmeric (*Curcuma longa*): A Review." *Research & Reviews: A Journal of Pharmacology* 9 (1): 16–23.

Chattopadhyay, R. R., and S. K. Bhattacharyya. 2007. "PHCOG REV.: Plant Review Terminalia chebula: An Update." *Pharmacognosy Reviews* 1: 151–6.

Chauhan, E. S., and J. Aishwarya. 2018. "Nutraceuticals Potential of Petroselinum Crispum: A Review." *Journal of Complementary Medicine and Alternative Healthcare* 7 (2). https://doi.org/10.19080/JCMAH.2018.07.555707

Choudhary, N., M. B. Siddiqui, S. Azmat, and S. Khatoon. 2013. "*Tinospora cordifolia*: Ethnobotany, Phytopharmacology and Phytochemistry Aspects." *International Journal of Pharmaceutical Sciences and Research* 4 (3): 891–9.

Cocan, I., E. Alexa, C. Danciu, I. Radulov, A. Galuscan, D. Obistioiu, and C. A. Dehelean. 2018. "Phytochemical Screening and Biological Activity of Lamiaceae Family Plant Extracts." *Experimental and Therapeutic Medicine* 15: 1863–70.

Datiles, M. J., and P. Acevedo-Rodríguez. 2022. cabicompendium.47678, CABI Compendium. CABI International, *Rosmarinus officinalis* (rosemary). https://doi.org/10.1079/cabicompendium.47678

Diwan, G., K. Sinha, N. Lal, and N. R. Rangare. 2018. "Tradition and Medicinal Value of Indian Gooseberry: A Review." *Journal of Pharmacognosy and Phytochemistry* 7: 2326–33.

Elansary, H. O., A. Szopa, P. Kubica, H. Ekiert, H. M. Ali, M. S. Elshikh, E. M. Abdel-Salam, M. El-Esawi, and D. O. El-Ansary, 2018. "Bioactivities of Traditional Medicinal Plants in Alexandria." *Evidence-Based Complementary and Alternative Medicine*: 1463579. https://doi.org/10.1155/2018/1463579

El-hassan, A. M., M. E. Shayoub, M. A. Abdalkreem, H. M. Osman, and K. Khalifa. 2012. "Design, Formulation, and Evaluation of Senna Effervescent Tablets." *Journal of Forest Products and Industries* 1 (2): 21–5.

Fatema, S., M. Hashir, A. Perveen, and S. Shamim. 2023. "Efficacy of *Plantago ovata* (aspaghol): A Review." *Indian Journal of Unani Medicine* 16: 143–50.

Gahlawat, D. K., S. Jakhar, and P. Dahiya. 2014. "*Murraya koenigii* (L.) Spreng: An Ethnobotanical, Phytochemical and Pharmacological Review." *Journal of Pharmacogncy and Phytochemistry* 3: 109–19.

Ghasemzadeh Rahbardar, M., and H Hosseinzadeh. 2020. "Therapeutic Effects of Rosemary (*Rosmarinus officinalis* L.) and Its Active Constituents on Nervous System Disorders." *Iran Journal of Basic Medical Sciences* 23: 1100–12.

Ghorbani, A., and M. Esmaeilizadeh. 2017. "Pharmacological Properties of *Salvia officinalis* and Its Components." *Journal of Traditional and Complementary Medicine* 7 (4): 433–40.

Gohil, K. J., and J. A. Patel. 2010. "A Review on *Bacopa monniera*: Current Research and Future Prospects." *International Journal of Green Pharmacy* 4: 1–9.

Gowthami, R., N. Sharma, R. Pandey, and A. Agrawal. 2021. "Status and Consolidated List of Threatened Medicinal Plants of India." *Genetic Resources and Crop Evolution* 68: 2235–63. https://doi.org/10.1007/s10722-021-01199-0

Grace, O. M. 2011. "Current Perspectives on the Economic Botany of the Genus Aloe L. (Xanthorrhoeaceae)." *South African Journal of Botany* 77: 980–7.

Guneri, N. 2021. "A Review on Turmeric (*Curcuma longa* L.) and Usage in Seafood." *Marine Science and Technology Bulletin* 10 (1): 71–84.

IBEF. 2020. High demand for medicinal plants in India. https://www.ibef.org/blogs/high-demand-for-medicinal-plants-in-india

Ikram, A., W. Khalid, F. Saeed, M. S. Arshad, M. Afzaal, and M. U. Arshad. 2023. "Senna: As Immunity Boosting Herb Against Covid-19 and Several Other Diseases." *Journal of Herbal Medicine* 37: 100626. https://doi.org/10.1016/j.hermed.2023.100626

Irham, W. H., Tamrin, Marpaung, L. and Marpongahtun. 2019. "Bioactive Compounds In Pegagan Leaf (Centella asiatica L. Urban) for Wound Healing." *IOP Conference Series: Journal of Physics*, 1232 (2019): 012019. https://doi.org/10.1088/1742-6596/1232/1/012019

Jaiswal, S., and S. Naik. 2018. "Contribution of Agricultural Produce Spice Zingiber Officinale to a Sustainable Food System: Green Extraction and Stability Study of Antioxidant Compounds." *Open Agriculture* 3: 326–38. https://doi.org/10.1515/opag-2018-0036

Jaiswal, S. G., and S. N. Naik. 2021. "Turmeric Oil: Composition, Extraction, Potential Health Benefits and Other Useful Applications." *Avicenna Journal of Medical Biochemistry* 9: 93–106.

Jha, A. K., and N. Sit. 2023. "Methods of Extraction of Bioactive Compounds from *Terminalia chebula* (Haritaki) and Their Applications." *Food Bioengineering* 2: 139–50.

Kar, N. R. 2023. "To Improve Employability by Cultivation of Medicinal Plants: A Current Viewpoint." *Journal of Pharmaceutical Research* 6: 1872–82.

Kattupalli, S., V. Vesta, S. Vangara, and U. Spandana. 2019. "The Multi-Activity Herbaceous Vine - *Tinospora cordifolia*." *Asian Journal of Pharmaceutical and Clinical Research* 12: 23–6.

Kaur, L., R. Kaur, A. Singh, and N. Kaur. 2021. "A Brief Review on Ethnobotanical, Pharmaceutical and Therapeutical Uses of *Glycyrrhiza glabra*." *Pharma Innovation* 10: 997–1006.

Kaur, S., and K. Bains. 2024. "Aloe Barbadensis Miller (Aloe Vera): Pharmacological." *International Journal for Vitamin and Nutrition Research* 94: 308–21.

Kim, M., and J. Chun. 2005. "Bacterial Community Structure in Kimchi, a Korean Fermented Vegetable Food, as Revealed by 16S rRNA Gene Analysis." *International Journal of Food Microbiology* 103 (1): 91–6.

Kshirsagar, S. R. 2018. "Taxonomy and Occurrence of Some Medicinal Plants Used in Traditional Medicines." *Archives of Pharmacy and Pharmacology Research* 1: 1–8.

Kumar, M., R. W. Bussmann, M. Joshi, and P. Kumar. 2011. "Ethnomedicinal Uses of Plants Close to Rural Habitation in Garhwal Himalayas, India." *Journal of Medicinal Plants Research* 5: 2252–60.

Kumar, N., L. G. Abichandani, V. Thawani, K. J. Gharpure, M. U. R. Naidu, and G. Venkat Ramana. 2016. Efficacy of Standardized Extract of Bacopa monnieri (Bacognize®) on Cognitive Functions of Med Ical Students: A Six-Week, Randomized Placebo-Controlled Trial. *Evidence Based Complimentary and Alternative Medicine*: 4103423. https://doi.org/10.1155/2016/4103423

Kumar, R., P. Saha, P. Lokare, K. Datta, P. Selvakumar, and A. Chourasia. 2022. "A Systemic Review of *Ocimum sanctum* (Tulsi): Morphological Characteristics, Phytoconstituents and Therapeutic Applications." *International Journal for Research in Applied Sciences and Biotechnology* 9 (2): 221–6.

Kumar, S., H. Dhankhar, N. Dinodia, and S. Kumari. 2023. "A Review on Phytochemicals and Pharmacological Effects Reported in *Glycyrrhiza glabra* (Mulethi)." *Ecology, Environment and Conservation Journal* 29: 423–30.

Kumari, M., M. Kumar, and S. S. Solankey. 2020. "Zingiber officinale Roscoe: Ginger." In *Medicinal, Aromatic and Stimulant Plants. Handbook of Plant Breeding*, edited by Novak, Johannes, Blüthner, Wolf-Dieter, Vol. 12. Cham: Springer. https://doi.org/10.1007/978-3-030-38792-1_20

Kumari, S., P. A. Singh, S. Hazra, R. Sindhwani, and S. Singh. 2024. "*Ocimum sanctum*: The Journey from Sacred Herb to Functional Food." *Recent Advances in Food, Nutrition & Agriculture* 15 (2): 83–102.

Laelago Ersedo, T., T. A. Teka, S. Fikreyesus Forsido, E. Dessalegn, J. A. Adebo, M. Tamiru, and T. Astatkie. 2023. "Food Flavor Enhancement, Preservation, and Bio-Functionality of Ginger (*Zingiber officinale*): A Review." *International Journal of Food Properties* 26: 928–51.

Lopresti, A. L. 2017. "Salvia (Sage): A Review of Its Potential Cognitive-Enhancing and Protective Effects." *Drugs in R&D* 17 (1): 53–64.

Marcinek, K., and Z. Krejpcio. 2016. "*Stevia rebaudiana* Bertoni: Health Promoting Properties and Therapeutic Applications." *Journal of Consumer Protection and Food Safety* 11: 3–8. https://doi.org/10.1007/s00003-015-0968-2

Marthe, F. 2020. "*Petroselinum crispum* (Mill.) Nyman (Parsley)." In *Medicinal, Aromatic and Stimulant Plants. Handbook of Plant Breeding*, edited by Novak, Johannes, Blüthner, Wolf-Dieter, Vol. 12. Cham: Springer. https://doi.org/10.1007/978-3-030-38792-1_13

Meena, S. S., D. C. Rai, P. Panda, and S. Kumar. 2022. "A Comprehensive Review on Bacopa monnieri (L). Pennell (Brahmi): Utilization as a Functional Food Ingredient and Health-Promoting Attributes." *Annals of Phytomedicine* 11: 142–50. https://doi.org/10.54085/ap.2022.11.1.14

Mersin, B., and G. S. Işcan. 2022. "Rosmarinus officinalis L." In *Novel Drug Targets With Traditional Herbal Medicines: Scientific and Clinical Evidence*, 525–41. Cham: Springer International Publishing.

Milka, P., R. Jelena, L. Aleksandra, and K. Tatjana. 2021. "A Review of Botanical Characteristics, Chemical Composition, Pharmacological Activity and Use of Parsley." *Archives of Pharmacy* 71: 177–96.

Moghaddasi, M. S., and H. H. Kashani. 2012. "Ginger (Zingiber officinale): A Review." *Journal of Medicinal Plants Research* 6: 4255–8.

Muhamad Mahir, M. N., A. Mohamed Rehan, and N. A. Mohd Fauzi. 2023. "The Determination of Antioxidant Activity of Senna Alexandrina Leaves and Phoenix Dactylifera Fruits for Food Preservation." *Progress in Engineering Application and Technology* 4 (2): 984–95.

Mushtaq, A., and D. Manjul. 2024. "Nucleus Model of Sacred Groves (Sacred Groves: the Nuclei of Biodiversity Cells) Traditional Beliefs, Myths, Associated Anthropogenic Threats and Possible Measures of Conservation in Western Most Regions of Lesser Himalayas, India." *Ecological Frontiers* 44: 217–33. https://doi.org/10.1016/j.chnaes.2023.05.002

Panda, D. S., and R. K. Giri. 2024. "Medicinal Plant Cultivation: Current Scenario, Challenges, and Opportunities from Indian Farmer's Perspective." *Journal of Health Science and Medical Research* 42: e20231009. https://www.jhsmr.org/index.php/jhsmr/article/view/1009

Pattanayak, P., P. Behera, D. Das, and S. K. Panda. 2010. "Ocimum sanctum Linn. A Reservoir Plant for Therapeutic Applications: An Overview." *Pharmacognosy Reviews* 4: 95–105.

Pellati, F., R. Bruni, D. Righi, A. Grandini, M. Tognolini, F. Pio Prencipe, F. Poli, S. Benvenuti, D. Del Rio, and D. Rossi. 2013. "Metabolite Profiling of Polyphenols in a Terminalia chebula Retzius Ayurvedic Decoction and Evaluation of Its Chemopreventive Activity." *Journal of Ethnopharmacology* 147 (2): 277–85.

Pillai, A. R. S., Y. K. Bhosale, and S. Roy. 2024. Extraction of Bioactive Compounds from *Centella asiatica* and Enlightenment of Its Utilization into Food Packaging: A Review. *International Journal of Food Science*, 1249553. https://doi.org/10.1155/2024/1249553

Prasanth Reddy, V., K. Ravi Vital, P. V. Varsha, and S. Satyam. 2014. "Review on Thymus vulgaris Traditional Uses and Pharmacological Properties." *Medicinal and Aromatic Plants* 3: 167.

Punosevac, M., J. Radovic, A. Lekovic, and T. Kundakovic-Vasovic. 2021. "A Review of Botanical Characteristics, Chemical Composition, Pharmacological Activity and Use of Parsley." *Archives of Pharmacy* 71: 177–96.

Qanash, H., A. S. Bazaid, N. K. Binsaleh, A. S. Alshammari, and R. Eltayeb. 2025. "Therapeutic Potential of *Plantago ovata* Bioactive Extracts Obtained by Supercritical Fluid Extraction as Influenced by Temperature on Anti-Obesity, Anticancer, and Antimicrobial Activities." *Plants (Basel, Switzerland)* 14 (12): 1813.

Raja, R. R. 2012. "Medicinally Potential Plants of Labiatae (Lamiaceae) Family: An Overview." *Research Journal of Medicinal Plants* 6 (3): 203–13.

Salamon, I. 2022. "The Importance of Medicinal and Aromatic Plant Cultivation in Sustainable Development of the Agricultural Landscape." *Macedonian Pharmaceutical Bulletin* 68: 203–4. https://doi.org/10.33320/maced.pharm.bull.2022.68.04.093

Sarfraz, R. M., H. Khan, and S. Maheen, et al. 2017. "Plantago ovata: A Comprehensive Review on Cultivation, Biochemical, Pharmaceutical and Pharmacological Aspects." *Acta Poloniae Pharmaceutica* 74 (3): 739–46.

Savulescu, E., M. I. Georgescu, V. I. Popa, and V. Luchian. 2018. "Morphological and Anatomical Properties of the *Senna Alexandrina* Mill. (*Cassia Angustifolia* Vahl.)." *Agriculture for Life, Life for Agriculture" Conference Proceedings* 1: 305–10.

Shakyawal, S., and P. Mahobiya. 2023. "Glycyrrhiza glabra (Mulethi) A Boon in Ayurveda: A Review." *Journal of Natural and Ayurvedic Medicine* 7 (2). https://doi.org/10.23880/jonam-16000405

Sharma, A., C. Shanker, L. K. Tyagi, M. P. Singh, and R. P. Marg. 2008. "Herbal Medicine for Market Potential in India: An Overview." *Academic Journal of Plant Sciences* 1: 26–36.

Sharma, P., K. K. Verma, H. Raj, and N. Thakur. 2021. "A Review on Ethnobotany, Phytochemistry and Pharmacology on Terminalia Belerica (Bibhitaki)." *Journal of Drug Delivery and Therapeutics* 11: 173–81.

Sharma, Y., J. Fagan, and J. Schaefer. 2019. "Ethnobotany, Phytochemistry, Cultivation and Medicinal Properties of Garden Sage (*Salvia officinalis* L.)." *Journal of Pharmacogncy and Phytochemistry* 8: 3139–48.

Shinde, S., A. K. Balasubramaniam, V. Mulay, G. Saste, A. Girme, and L. Hingorani. 2023. "Recent Advancements in Extraction Techniques of Ashwagandha (*Withania somnifera*) With Insights on Phytochemicals, Structural Significance, Pharmacology, and Current Trends in Food Applications." *ACS Omega* 8 (44): 40982–1003.

Singh, A. K., S. Singh, P. L. Saroj, D. S. Mishra, P. P. Singh, and R. K. Singh. 2019. "Aonla (Emblica officinalis) in India: A Review of Its Improvement, Production and Diversified Uses." *Indian Journal of Agricultural Sciences* 89: 1773–81.

Singh, A. K., A. Srivastava, V. Kumar, and K. Singh. 2018. "Phytochemicals, Medicinal and Food Applications of Shatavari (Asparagus racemosus): An Updated Review." *The Natural Products Journal* 8: 32–44.

Singh, D., and P. K. Chaudhuri. 2018. "A Review on Phytochemical and Pharmacological Properties of Holy Basil (*Ocimum sanctum* L.)." *Industrial Crops and Products* 118: 367–82.

Singh, D., S. Dhyani, and G. Kaur. 2015. "A Critical Review on Guggulu (*Commiphora wightii*)." *International Journal of Ayurveda and Pharma Research* 3 (1): 1–9.

Singla, R., and V. Jaitik. 2014. "Shatavari (Asparagus racemosus Wild): A Review on Its Cultivation, Morphology, Phytochemistry and Pharmacological Importance." *International Journal of Pharmaceutical Sciences and Research* 5: 742–57.

Singletary, K. W. 2016. "Thyme: History, Applications, and Overview of Potential Health Benefits." *Nutrition Today* 51: 40–9.

Sudhakaran, M. V. 2017. "Botanical Pharmacognosy of *Centella asiatica* (Linn.) Urban." *Pharmacognosy Journal* 9: 546–58.

Sun, B., L. Wu, Y. Wu, C. Zhang, L. Qin, M. Hayashi, M. Kudo, M. Gao, and T. Liu. 2020. "Therapeutic Potential of *Centella asiatica* and Its Triterpenes: A Review." *Frontiers in Pharmacology* 11: 568032. https://doi.org/10.3389/fphar.2020.568032

Sutradhar, K., M. R. Tahsin, and S. Ashrafi, et al. 2022. "Phytochemical, Biochemical and Pharmacological Properties of Plantago ovata (Ispaghula Husk) – A Review." *Dhaka University Journal of Pharmaceutical Sciences* 21 (2): 231–43.

Thorat, B. S., T. A. Bagkar, and R. P. Patil. 2018. "Brahmi" The Memory Booster Medicinal Herb." *Journal of Medicinal Plant Studies* 6 (1): 185–7.

Tiwari, R., A. K. Verma, S. Chakraborty, K. Dhama, and S. V. Singh. 2014. "Neem (*Azadirachta indica*) and Its Potential for Safeguarding Health of Animals and Humans: A Review." *Journal of Biological Sciences* 14: 110–23.

Vaghashiya, J. M., D. Raj, H. G. Suthar, P. B. Zinzala, and R. B. Bharai. 2018. "Impact of Aloe vera Juice and Isabgol Husk on Microbial Quality of Wheat Flour Vermicelli." *International Journal of Chemical Studies* 6 (4): 393–7.

Vibha, J. B., K. Choudhary, M. Singh, M. S. Rathore, and N.S. Shekhawat. 2009. "A Study on Pharmacokinetics and Therapeutic Efficacy of Glycyrrhiza glabra: A Miracle Medicinal Herb." *Botany Research International* 2: 157–63.

Vining, K. J., K. E. Hummer, N. V. Bassil, B. M. Lange, C. K. Khoury, and D. Carver. 2020. "Crop Wild Relatives as Germplasm Resource for Cultivar Improvement in Mint (Mentha L.)." *Frontiers in Plant Science* 11: 1217.

Viswanathan, S., and T. Nallamuthu. 2012. "Phytochemical Screening and Antimicrobial Activity of Leaf Extracts of *Senna alexandrina* Mill. Against Human Pathogens." *International Journal of Current Science* 2: 51–6.

Wani, S. A., H. Naik, J. I. Wagay, N. A. Ganie, M. Z. Mulla, and B. N. Dar. 2022. "Mentha: A Review on Its Bioactive Compounds and Potential Health Benefits." *Quality Assurance and Safety of Crops & Foods* 14 (4): 154–68.

Yadav, A. K., S. Singh, D. Dhyani, and P. S. Ahuja. 2011. "A Review on the Improvement of Stevia [*Stevia rebaudiana* (Bertoni)]." *Canadian Journal of Plant Science* 91: 1–27. https://doi.org/10.4141/cjps10086

Yadav, R. P., and G. Tarun. 2017. "Versatility of Turmeric: A Review the Golden Spice of Life." *Journal of Pharmacognosy and Phytochemistry* 6: 41–6.

4 Phytocompounds and Their Therapeutic Effects

Naveen Chandrakar, Pawan Prabhakar, and Mamoni Banerjee

4.1 INTRODUCTION

Plants naturally contain bioactive substances called phytocompounds having capacity to alter important biological processes and affect the course of disease, and they are well known for their therapeutic potential (Riaz et al. 2023). Currently, the world is increasingly adopting efficacious herbal remedies as the synthetic medications possess adverse effects, are remain insufficient, and are prohibitively expensive in developing countries (Ahmed et al. 2023; Saggar et al. 2022). In recent decades, scientists worldwide have studied the natural product and conducted ethnomedicinal survey of plants that establishes the foundation for the discovery of novel therapeutic substances. Now they have focused on natural medicinal substances as a valuable source of novel therapeutic agents, driven by the urgent need to address challenges such as antibiotic resistance, the side effects of synthetic drugs, and the rising prevalence of chronic and lifestyle-related diseases (Abdallah et al. 2023; Atanasov et al. 2015). Among these natural resources, plants hold a prominent position due to their vast chemical diversity and long-standing role in traditional medicine systems across cultures.

Ethnomedicinal surveys systematic studies documenting the use of plants by indigenous and local communities have become an essential foundation for drug discovery, as they provide critical insights into bioactive species that have already been "pre-screened" by centuries of human experience. Such surveys not only catalogues plant species and their traditional therapeutic applications but also offer valuable ethnobotanical, ecological, and cultural context that can guide targeted pharmacological research (Sharma, Singh, and Sharma 2022). The process of translating this traditional knowledge into new medicines involves isolating active phytochemicals, characterising their structures, and evaluating their pharmacological properties through *in vitro, in vivo,* and *in silico* methods (Prabhakar, Kumar, and Banerjee 2024b). This approach not only accelerates the search for new drug candidates but also supports biodiversity conservation and the recognition of indigenous intellectual property rights, ensuring fair benefit-sharing under frameworks like the Nagoya Protocol. As a result, ethnomedicinal surveys are not merely academic exercises they represent a bridge between ancient wisdom and modern science,

 DOI: 10.1201/9781003623748-4

offering sustainable and culturally respectful pathways for the development of innovative, safe, and effective therapeutics for global healthcare needs (Heinrich et al. 2020; Smith et al. 2017). Over the time, much evidence has been accumulated that illustrates the application of plants in pharmacology. Herbal medicines are therapeutic agents obtained from plants and plant-derived substances, utilised for the prevention, treatment, and management of several health disorders. They include the use of leaves, roots, bark, flowers, seeds, and other plant components, either in its unrefined state or transformed into extracts, powders, capsules, teas, oils, and ointments (Msomi and Simelane 2019). Herbal remedies lie in ancient therapeutic traditions, such as Ayurveda, traditional Chinese medicine (TCM), Unani, and Indigenous medical systems, have been essential to human healthcare for millennia. These herbal medicines have garnered increased global interest due to an increasing consumer preference for natural, holistic, and sustainable healthcare solutions, alongside escalating worries regarding the adverse consequences of synthetic pharmaceuticals (Yuan et al. 2016). Plants such as turmeric (*Curcuma longa*) are esteemed for their anti-inflammatory and antioxidant attributes, Ashwagandha (*Withania somnifera*) for their efficacy in stress alleviation and immunological enhancement, and Neem (*Azadirachta indica*) for its antibacterial and antifungal characteristics. Contemporary research and clinical trials have begun to substantiate numerous traditional assertions, resulting in the integration of plant-derived substances into conventional medications (Pebam et al. 2022).

Phytochemicals are a varied group of naturally occurring bioactive molecules present in plants, essential for plant development, defence, and adaptation, while also providing considerable health advantages to humans. These compounds, while not categorised as necessary nutrients such as vitamins and minerals, exhibit significant therapeutic promise owing to their antioxidant, anti-inflammatory, antibacterial, anticancer, and cardioprotective attributes (Prabhakar and Banerjee 2021). Phytochemicals are classified into major categories including alkaloids, flavonoids, phenolic acids, tannins, terpenoids, saponins, lignans, and glucosinolates, each possessing unique chemical structures and biological functions (Kumar et al. 2023). The therapeutic effects of phytochemicals frequently stem from their capacity to influence critical biochemical processes, including the neutralisation of free radicals, inhibition of disease-related enzymes, regulation of gene expression, and augmentation of immunological responses. They play a critical role in the intervening mechanisms of action against various diseases by interacting with multiple molecular targets and biochemical pathways within the human body (Gupta et al. 2024; Kumar et al. 2023). These chemicals are predominantly synthesised by various plant species, primarily blooming plants, as well as by certain animals. Plants synthesise and accumulate various organic substances, including amino acids, proteins, carbohydrates, lipids, and alkaloids, typically classified as secondary metabolites. They are present in many parts of the plant leaves, stems, roots, and fruits in differing quantities. Initially proposed as mere waste products of plants, recent research indicates that they serve significant biological functions within plant systems. The modern research on therapeutic efficacy has increasingly concentrated on isolating, characterising, and assessing these chemicals for pharmacological and nutraceutical purposes, with numerous plant-derived molecules now functioning as lead compounds in therapeutic development (Prabhakar et al. 2025b). Moreover, phytochemicals are crucial in preventive healthcare, since diets abundant in plant-based foods are significantly linked to reduced occurrences of chronic diseases such as diabetes, hypertension, and obesity (Aware et al. 2022; Manisha et al. 2025). Their efficacy is contingent upon parameters like bioavailability, concentration, synergistic interactions with other drugs, and extraction and processing processes. With the increasing desire for natural and sustainable health solutions, phytochemicals are leading global initiatives to merge traditional botanical knowledge with advanced biomedical research, presenting great opportunities for therapeutic innovation and preventative medicine (Aquil et al. 2013; Prabhakar et al. 2024a). Furthermore, the therapeutic potential of phytocompounds is well acknowledged; nonetheless, the safety of phytocompounds and the control of regulatory agencies continue to be significant concerns for the public's health and the efficient use of medicinal substances (Gamil et al. 2025). Therefore, the chapter deals with an extensive study of therapeutic effect, mechanism of action, its pharmacokinetics, and safety concerns of phytocompounds to ensure the way to a safe and effective therapeutics.

4.2 MAJOR CLASSES OF PHYTOCHEMICALS

Because of their numerous forms and structures, phytochemicals have not yet been precisely classified. Generally, the phytochemicals have been divided into primary and secondary metabolites based on how they function in the metabolism of plants. Common sugars, proteins, amino acids, purines and pyrimidines of nucleic acids, chlorophylls, and other substances are examples of primary metabolites. Polyphenols, alkaloids, terpenoids, the major group of the secondary metabolites.

4.2.1 Polyphenols

Polyphenols constitute a class of natural chemicals characterised by phenolic structures. This family comprises four primary subclasses: flavonoids, stilbenes, phenolic acids, and lignans. Flavonoids are categorised into flavanones, flavones, flavonols, and anthocyanidins. are a varied class of naturally occurring phytochemicals prevalent in plants, noted for their strong antioxidant, anti-inflammatory, and health-enhancing attributes (Tijjani et al. 2020). By scavenging free radicals, modifying enzyme activity, and affecting cell signalling pathways, their biological activities help prevent and treat chronic illnesses like diabetes mellitus, cardiovascular disorders, neurodegenerative diseases, and different types of cancer (Maniglia et al. 2021). Their capacity to alter the composition of the gut microbiota, improve metabolic health, and guard against oxidative stress-induced cellular damage has been demonstrated in recent studies. Furthermore, certain polyphenols can inhibit α-glucosidase, lipase, and matrix metalloproteinases enzymes that are associated with the course of disease, which may have therapeutic implications for diseases like cancer, diabetes, and obesity (Aryal et al. 2024; Ciupei et al. 2024). Quercetin and naringenin are two extensively studied and powerful flavonoids, recognised for their diverse health advantages. Quercetin, present in foods such as apples, onions, and berries, is renowned for its potent antioxidant and anti-inflammatory effects, which contribute to lowering the risk of chronic diseases, including heart disease and diabetes. Naringenin, predominantly located in citrus fruits like grapefruit, possesses comparable antioxidant properties and is recognised for enhancing metabolic health through improved insulin sensitivity and less inflammation. The choice of quercetin and naringenin for research or health interventions is based on their significant therapeutic potential, prevalence in commonly consumed foods, and proven ability to enhance cardiovascular health, regulate glycaemic levels, and reduce oxidative stress. However, despite their encouraging pharmacological benefits, polyphenols frequently struggle with low stability and bioavailability, which makes novel formulation techniques like conjugation with carrier molecules or nanoencapsulation necessary to increase their therapeutic efficiency (Ballard and Junior 2019; Ullah et al. 2020).

4.2.2 Terpenoids

Terpenoids, or isoprenoids, constitute a broad and varied category of naturally occurring organic compounds originating from five-carbon isoprene units organised in distinct structural configurations. Terpenoids, prevalent in plants, fungi, marine organisms, and certain mammals, are essential for ecological interactions, plant defence, and human health (Câmara et al. 2024). They account for the distinctive scents, flavours, and colours of numerous plants, such the fragrance of lavender, the hue of carotenoids in carrots, and the bitterness of hops in beer. Terpenoids demonstrate a diverse array of pharmacological activity, encompassing anti-inflammatory, anticancer, antibacterial, antiviral, antioxidant, and hepatoprotective properties (Nabavi et al., 2019). Many prominent pharmaceuticals, including the antimalarial artemisinin, the anticancer compound paclitaxel (Taxol), and cardiac glycosides, originate from terpenoid structures. In plants, terpenoids function as signalling molecules, attract pollinators, and provide defence against herbivores and pathogens (Chaachouay and Zidane 2024; Yeshi et al. 2022). Their structural diversity, encompassing simple monoterpenes to intricate triterpenes and tetraterpenes, enhances their extensive functional adaptability. Nonetheless, despite their significant therapeutic potential, obstacles such as low solubility, inadequate bioavailability, and challenges in large-scale extraction restrict their direct pharmaceutical use, necessitating continued research into synthetic biology and nanotechnology-based delivery systems to optimise their advantages (Tufail et al. 2025).

4.2.3 Alkaloids

Alkaloids are nitrogenous chemicals synthesised by marine and terrestrial organisms, frequently exhibiting potent biological effects. They are one of the major categories of natural products, predominantly occurring in plants. The name alkaloid pertains to their basic characteristics (alkali-like), resulting from the presence of one or more nitrogen atoms, as primary, secondary, or tertiary amines, often within heterocycles. Quaternary ammonium alkaloids are additionally present (Kukula-Koch and Widelski 2017).

Alkaloids shown a broad spectrum of therapeutic capabilities. Many of them exhibit local anaesthetic effects; nevertheless, their clinical application is constrained. Morphine is one of the most recognised alkaloids that has been utilised and continues to be employed for medical purposes. This alkaloid is a potent narcotic employed for pain treatment, although its efficacy is constrained due to its addictive characteristics. Quinine is a potent antimalarial agent, although it is increasingly supplanted by synthetic medicines that are more efficacious and less dangerous (Heinrich, Mah, and Amirkia 2021). Quinidine, another alkaloid derived from Cinchona species, is utilised medically for the treatment of arrhythmias or abnormal cardiac rhythms. The methyl ether derivative of morphine, codeine, is a naturally occurring compound next to morphine in the opium poppy, exhibiting significant analgesic properties and demonstrating a comparatively low potential for addiction. These alkaloids function as stimulants for respiration or heart activity (Bribi 2018). The alkaloid utilised as a medicine in numerous clinical applications is atropine whose administration is used to manage bradycardia (reduced heart rate) (Aldossary 2022). Many alkaloids are constituents of the human diet, present in both food and beverages. Plants in the human diet containing alkaloids include not only coffee seeds (caffeine), cacao seeds (theobromine and caffeine), and tea leaves (theophylline, caffeine) but also tomatoes (tomatine) and potatoes (solanine). Caffeine, the most prevalent alkaloid, is utilised as an ingredient in soft drinks to enhance flavour and in beverages designed for active individuals engaged in sports (Akinboye et al. 2023).

4.2.4 Carotenoids

Carotenoids are a diverse group of naturally occurring pigments that give many fruits, vegetables, and other plant-based foods their vibrant yellow, orange, and red colours, while also being present in green leafy vegetables where their colour is masked by chlorophyll. These compounds are synthesised by plants, algae, and certain photosynthetic bacteria, playing an essential role in photosynthesis by absorbing light energy for use in the process and protecting plant tissues from photooxidative damage through their antioxidant properties (Sun et al. 2022). Structurally, carotenoids are lipophilic molecules and are broadly classified into two main groups: carotenes, which are purely hydrocarbons (such as β-carotene and lycopene), and xanthophylls, which contain oxygen atoms (such as lutein and zeaxanthin) (Zia-Ul-Haq 2021). Carotenoids have garnered significant attention due to their potent antioxidant activity, helping to neutralise free radicals and reduce oxidative stress, which is linked to chronic diseases such as cardiovascular disorders, certain cancers, and age-related macular degeneration. Some carotenoids, notably β-carotene, α-carotene, and β-cryptoxanthin, are provitamin A compounds that the body can convert into vitamin A, essential for vision, immune function, reproduction, and skin integrity (Bohn 2019). Non-provitamin A carotenoids like lutein, zeaxanthin, and lycopene also have specific health-promoting roles, including eye protection, skin photoprotection, and potential anti-inflammatory effects. Dietary intake of carotenoids is best achieved through a diet rich in colourful fruits and vegetables such as carrots, sweet potatoes, pumpkins, tomatoes, red peppers, spinach, kale, and corn, and their absorption is enhanced when consumed with healthy fats due to their fat-soluble nature. Beyond nutrition, carotenoids are widely used as natural colorants in the food, cosmetic, and pharmaceutical industries, making them valuable both biologically and commercially (Toti et al. 2018).

4.3 THERAPEUTIC EFFECT OF PHYTOCOMPOUNDS

4.3.1 Antimicrobial Activity

Phytocompounds, the diverse array of bioactive secondary metabolites produced by plants, possess remarkable antimicrobial activity that enables them to combat a wide spectrum of pathogenic microorganisms, including bacteria, fungi, viruses, and protozoa. These compounds such as alkaloids, flavonoids, phenolics, terpenoids, tannins, glycosides, and saponins employ multiple mechanisms to inhibit or destroy microbial invaders (Angelini 2024; Verma et al. 2024). Many phytocompounds target and disrupt microbial cell walls and membranes, altering their permeability and causing leakage of vital cellular components, while others interfere with key enzymatic systems, inhibit protein or nucleic acid synthesis, or block metabolic pathways critical for microbial survival and reproduction (Hashempur et al. 2025). For example, flavonoids can form complexes with bacterial cell walls and inactivate microbial enzymes, tannins can precipitate proteins and deprive microbes of essential nutrients, and saponins can bind to sterols in fungal membranes, leading to cell lysis. The antimicrobial action of phytocompounds is often enhanced by their synergistic interactions when present

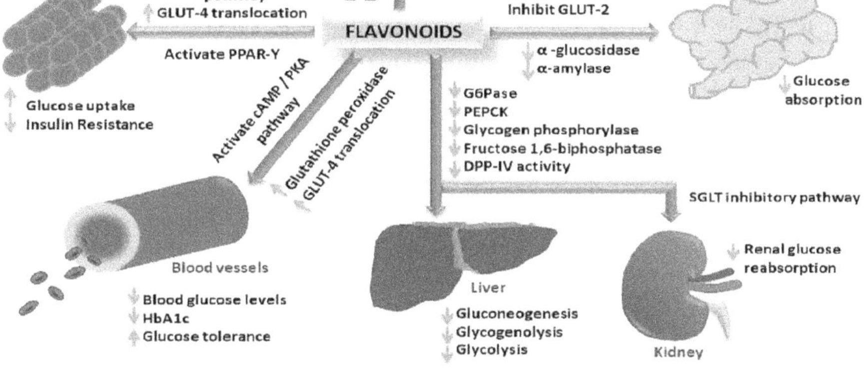

FIGURE 4.1 Mechanism of antibacterial action of plant-derivative compounds. (From Khameneh et al. 2021.)

in combination, as occurs naturally in plant extracts, making them more effective than isolated compounds alone (Ergüden 2021). Additionally, their structural diversity allows them to act against both Gram-positive and Gram-negative bacteria, as well as antibiotic-resistant strains, offering promising alternatives or adjuncts to conventional antibiotics. Phytochemicals, influenced by their chemical structures and characteristics, may have antibacterial activity by one or more of these pathways. They are classified into primary categories of alkaloids, tannins, sugars, glycosides, terpenoids, flavonoids, steroids, and coumarins based on their structures. These chemicals have clinical significance as their bioactivity typically does not induce resistance. Here, Figure 4.1 illustrates some significant plant-derived chemicals exhibiting antibacterial properties and their mechanisms of action. Polyphenols exhibit antibacterial activity against a wide range of microorganisms. Flavanols, flavonols, and phenolic acids demonstrate the most significant activities by (I) inhibiting bacterial virulence factors such as enzymes and toxins, (II) interacting with the cytoplasmic membrane or lowering pH levels, (III) suppressing biofilm formation, (IV) exhibiting synergistic effects with conventional antibiotics, and (V) diminishing extracellular polysaccharide (EP) activity while functioning as EP inhibitors (EPIs). Phenolic chemicals generated in substantial amounts exhibit promise EPI action against pathogenic microorganisms. They may impede cell wall biosynthesis and essential enzymes, including urease, sortase A, and dihydrofolate reductase (Khameneh et al. 2021).This natural defence system, evolved over millions of years to protect plants from pathogenic threats, is increasingly recognised for its potential in developing novel, eco-friendly antimicrobial agents for pharmaceutical, food preservation, and agricultural applications (Muteeb et al. 2023; Singha, Singh, and Soni 2024).

4.3.2 Anticancerous Activity

Cancer is a formidable disease that constitutes one of the most significant healthcare challenges for humanity, necessitating a proactive approach for treatment. Plants serve as reservoirs for novel chemical compounds and present a viable avenue for cancer research. Phytocompounds exhibit significant anticancer efficacy by targeting many pathways implicated in cancer start, development, and metastasis (Choudhari et al. 2020). Phytocompounds such as polyphenols, terpenoids, alkaloids, carotenoids, and organosulfur compounds exhibit anticancer properties through various mechanisms, including the induction of apoptosis, inhibition of aberrant cell proliferation, suppression of angiogenesis, and disruption of metastasis-related processes (Vikas et al. 2024). They have capacity to intervene critical signalling pathways, including NF-κB, MAPK, PI3K/Akt, and p53, therefore controlling cell cycle checkpoints and improving DNA repair mechanisms (Choudhary et al. 2023; Fakhri et al. 2022; Belwal et al. 2025). Curcumin from turmeric suppresses tumour cell proliferation by downregulating pro-inflammatory cytokines and growth factors, whereas resveratrol from grapes promotes apoptosis via mitochondrial malfunction and control of oxidative stress (Bisht, Wagner, and Bulmer 2010). Some phytochemicals serve as antioxidants, mitigating oxidative DNA damage that may initiate carcinogenesis, while others work as pro-oxidants in cancer cells, producing reactive oxygen species (ROS) in excess, resulting in the demise of cancer cells (Kocyigit, Guler, and Dikilitas 2018). Phytocompounds demonstrate selective toxicity, successfully attacking cancer cells while sparing normal cells, hence reducing adverse effects relative to traditional chemotherapy. Their synergistic effects, when utilised in conjunction with traditional anticancer agents, can augment treatment results, surmount drug resistance, and boost patient prognosis. Due to their multi-targeted mechanisms, comparatively low toxicity, and capacity to influence the tumour microenvironment, phytocompounds are being extensively investigated as viable options for cancer prevention and treatment in integrative and precision oncology (Choudhari et al. 2020).

4.3.3 Antidiabetic Activity

Phytocompounds exhibit efficient antidiabetic activity by targeting various biochemical and physiological pathways involved in glucose homeostasis and insulin regulation. These bioactive constituents such as alkaloids, flavonoids, polyphenols, terpenoids, saponins, and glycosides help in lowering elevated blood glucose levels and improving insulin sensitivity through multiple mechanisms (Ansari et al. 2022; Prabhakar and Banerjee 2020a; Situmorang et al. 2025). Many phytocompounds enhance insulin secretion from pancreatic β-cells, stimulate the regeneration of these cells, and protect them from oxidative stress-induced damage, while others increase glucose uptake in peripheral tissues by upregulating glucose transporter proteins like GLUT4 (Alam et al. 2022). Some other phytocompounds inhibit carbohydrate-digesting enzymes like α-amylase and α-glucosidase, thereby slowing down glucose absorption from the intestine and preventing postprandial hyperglycaemia (Prabhakar et al. 2023a). Phenolic compounds, a diverse group of plant-derived secondary metabolites that include flavonoids, phenolic acids, tannins, and lignans, play a significant role in the prevention and management of diabetes through their multifaceted biochemical actions. These compounds possess potent antioxidant effects, neutralising ROS and diminishing oxidative stress, a principal factor in pancreatic β-cell dysfunction and insulin resistance. Phenolic substances can augment insulin production, boost insulin sensitivity, and regulate glucose absorption by upregulating glucose transporter proteins, such as GLUT4, in muscular and adipose tissues. They also impede essential digestive enzymes such as α-amylase and α-glucosidase, consequently decelerating carbohydrate degradation and absorption in the gastrointestinal tract, which subsequently diminishes postprandial hyperglycaemia. Furthermore, phenolics can influence lipid metabolism, reduce inflammation by inhibiting pro-inflammatory cytokines, and alter signalling pathways like AMPK, PI3K/Akt, and PPARγ, which are essential for glucose and lipid homeostasis. Certain phenolic acids, such as chlorogenic acid, decrease hepatic gluconeogenesis, hence diminishing excessive glucose synthesis by the liver. By concurrently targeting multiple pathways, phenolic compounds assist in sustaining glycaemic control and safeguarding against diabetes-related complications, including neuropathy, nephropathy, and cardiovascular damage, thereby positioning them as promising agents in functional foods and antidiabetic therapeutics. Polyphenolic compounds such as quercetin and catechins exert antioxidant and anti-inflammatory effects, reducing chronic low-grade inflammation linked to insulin resistance (Clemente-Suárez et al. 2025; Nakadate et al. 2025; Rudrapal et al. 2022). Flavonoids demonstrate antidiabetic effects through many molecular pathways. Flavonoids enhance insulin secretion and optimise β-cell functionality via the PI3K/AKT signalling cascade; promote GLUT-4 translocation through AMPK activation to augment glucose uptake in adipose tissue and skeletal muscle; stimulate PPAR-γ expression to mitigate insulin resistance; activate the cAMP/PKA pathway to lower blood glucose levels and enhance glucose tolerance; elevate glutathione peroxidase activity to diminish HbA1c levels; reduce G-6-Pase, PEPCK, glycogen phosphorylase, fructose 1,6-bisphosphatase, and DPP-IV activity in the liver to curtail gluconeogenesis, glycogenolysis, and glycolysis; inhibit the SGLT pathway in the kidneys to decrease renal glucose reabsorption; and suppress GLUT-2, α-amylase, and α-glucosidase activity to lower glucose absorption in the small intestine as shown in Figure 4.2 (Al-Ishaq et al. 2019; Ansari et al. 2022).

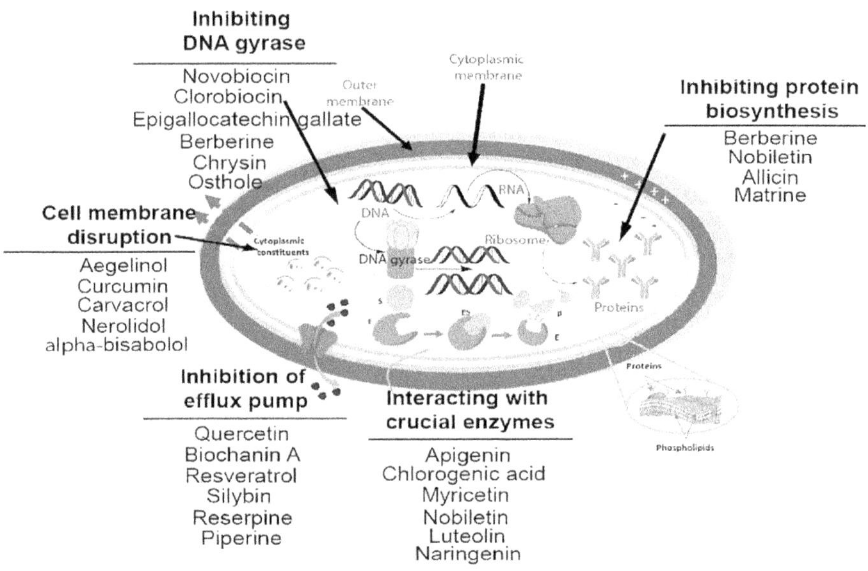

FIGURE 4.2 Mechanism of flavonoid in the management of diabetes by intervening various molecular pathways to control the high blood glucose level. Flavonoids exerting antidiabetic activity via different mechanistic pathways: Flavonoids increase insulin secretion and improve β-cell function via the PI3K/AKT signaling pathway; increase GLUT-4 translocation through AMPK activation to increase glucose uptake in adipose tissues and skeletal muscles; activate PPAR-γ expression to decrease insulin resistance; activate cAMP/PKA pathway to reduce blood glucose levels and improve glucose tolerance; increase glutathione peroxidase activity to reduce HbA1c levels; decrease G-6-Pase, PEPCK, glycogen phosphorylase, fructose 1,6-biphosphatase and DPP-IV activity in liver to decrease gluconeogenesis, glycogenolysis, and glycoslysis; inhibit SGLT pathway in kidney to decrease renal glucose reabsorption; inhibit GLUT-2, α-amylase and α-glucosidase activity to decrease glucose absorption in the small intestine. (From Ansari et al. 2022.)

Some of the triterpenoids and saponins have ability to modulate key signalling pathways like AMPK and PPARγ, promoting lipid metabolism and reducing obesity-associated metabolic dysfunction (Che et al. 2018). Therefore, phytocompounds also help in improving lipid profiles, enhance hepatic glycogen storage, and suppress hepatic gluconeogenesis, contributing to better overall glycaemic control. The synergistic interactions of these compounds, when consumed in whole plant extracts or functional foods, often yield more potent antidiabetic effects than isolated molecules. Due to their multi-targeted approach, relative safety, and minimal side effects, phytocompounds are increasingly being recognised as valuable therapeutic agents for the prevention and management of type 2 diabetes, either alone or in combination with conventional antidiabetic drugs (Clemente-Suárez et al. 2025).

4.3.4 Cardioprotective Activities

Phytocompounds are essential for maintaining cardiovascular health by addressing many pathways associated with heart disease, such as oxidative stress, inflammation, lipid dysregulation, endothelial dysfunction, and platelet aggregation. Polyphenols, including resveratrol (derived from grapes) and quercetin (found in onions and apples), augment endothelial nitric oxide synthesis, facilitating vascular relaxation and decreasing blood pressure, while simultaneously providing potent antioxidant properties that mitigate oxidative damage to blood vessels (Alum 2025; Islam et al. 2021). Flavonoids from green tea (EGCG) and cocoa enhance lipid metabolism by decreasing LDL cholesterol and triglycerides, while elevating HDL cholesterol, therefore mitigating atherosclerotic risk (Chang et al. 2015). Alkaloids such as berberine activate AMPK, enhancing insulin sensitivity, controlling cholesterol levels, and safeguarding against cardiovascular problems linked with metabolic syndrome (Feng et al. 2019). Curcumin derived from turmeric and boswellic acids from frankincense mitigate inflammatory signalling pathways, including NF-κB, which are associated with vascular inflammation and plaque development (Hu et al. 2023; Sethi et al. 2022; Peng et al. 2019). Saponins from ginseng and legumes diminish cholesterol absorption in the intestine and enhance myocardial function, whereas omega-3-rich phytochemicals from flaxseed and chia seeds regulate lipid profiles and avert arrhythmias (Marrelli et al. 2016). Some of the phytocompounds have demonstrated antithrombotic and anti-platelet characteristics, hence diminishing the likelihood of thrombus formation and ischaemic occurrences (Adili, Hawley, and Holinstat 2018). The cardioprotective effects of phytocompounds stem from their capacity to influence multiple interconnected molecular pathways, rendering them highly effective in the prevention and management of cardiovascular diseases, while providing a natural and relatively safe alternative or adjunct to conventional therapies.

4.3.5 Hepatoprotective Activities

Phytocompounds demonstrate significant hepatoprotective properties by protecting the liver from chemical, metabolic, and oxidative damage, while facilitating detoxification and regenerative processes. The liver, integral to metabolism and xenobiotic elimination, is particularly susceptible to oxidative stress, inflammation, and damage from drugs or toxins; phytochemicals mitigate these effects through multi-faceted mechanisms (Prajapati, Bhatt, and Rawal 2024). Silymarin, derived from *Silybum marianum* or milk thistle, is one of the most extensively researched hepatoprotective drugs, fortifying hepatocyte membranes, augmenting antioxidant enzyme activity, and facilitating liver cell regeneration (Vargas-Mendoza et al. 2014). Curcumin, derived from turmeric, diminishes lipid peroxidation, inhibits NF-κB-mediated inflammation, and averts fibrosis, rendering it efficacious in illnesses such as non-alcoholic fatty liver disease (NAFLD) and cirrhosis (Buonomo et al. 2019). Kaempferol, a flavonoid found in grapes, can promote mitochondrial activity and reduce fat accumulation in the liver by activating the SIRT1 and AMPK pathways. Studies suggest that kaempferol can improve liver health by enhancing mitochondrial function, reducing inflammation, and potentially alleviating steatosis (Alkandahri et al. 2023). Berberine modulates lipid and glucose metabolism, so mitigating metabolic stress on the liver, whilst catechins (derived from green tea) have potent antioxidant properties that safeguard against hepatotoxicity generated by alcohol and toxins (Yang et al. 2022). Moreover, glycyrrhizin (derived from liquorice) demonstrates antiviral and anti-inflammatory properties, rendering it beneficial in the treatment of viral hepatitis. Many phytocompounds also enhance phase I and II detoxification pathways, improve glutathione levels, and prevent collagen deposition in chronic liver injury. Phytocompounds provide extensive hepatoprotection through their antioxidant, anti-inflammatory, antifibrotic, and metabolic regulatory activities, underscoring their potential as adjuncts or alternatives in the treatment of liver illnesses (Lam et al. 2016; Li et al. 2014).

4.3.6 Antioxidants Activity

Phytocompounds exhibit potent antioxidant activity, playing a crucial role in protecting biological systems from oxidative stress-induced damage caused by an excess of ROS and reactive nitrogen species (RNS) (Kumar et al. 2023). A broad classes of phenolics, flavonoids, tannins, terpenoids, alkaloids, carotenoids, organosulfur compounds, and saponins are acknowledged for their antioxidant properties; however, a thorough assessment indicates that their effectiveness is contingent upon various structural, biochemical, and physiological factors (Kumar et al. 2022). Phenolic compounds, consist of simple phenols, phenolic acids, and flavonoids, demonstrate significant radical-scavenging activity via hydrogen or electron donation and metal-chelating properties; nevertheless, their bioavailability is frequently constrained by extensive metabolism in the gastrointestinal tract and liver, and certain phenolics may exhibit pro-oxidant effects at elevated concentrations by facilitating the redox cycling of metals (Tungmunnithum et al. 2018).

Flavonoids, a subclass of phenolics, are distinguished by their capacity to stabilise ROS and regulate redox-sensitive signalling pathways like Nrf2; yet, their *in vivo* efficacy is significantly influenced by glycosylation patterns, solubility, and cellular absorption (Zahra et al. 2024). Tannins, possessing several hydroxyl groups, exhibit significant in vitro antioxidant capacity through protein binding and radical scavenging; nevertheless, their elevated molecular weight may impede absorption, restricting systemic benefits despite substantial local antioxidant activity in the gastrointestinal tract (Guan et al. 2025). Terpenoids, including carotenoids and tocopherols, safeguard cellular membranes by neutralising singlet oxygen and preventing lipid peroxidation; however, their efficacy is affected by variables such as light, oxygen exposure, and the lipid milieu, with some carotenoids potentially functioning as pro-oxidants in elevated oxygen conditions (Bufka et al. 2024). Alkaloids, primarily recognised for their pharmacological properties, also play a role in antioxidant defence by neutralising free radicals and blocking oxidative enzymes; yet, their advantages may be counterbalanced by possible toxicity or limited therapeutic windows (Chaudhary et al. 2023). The antioxidant effects *in vivo* arise not only from direct radical scavenging but also from the indirect modulation of cellular defence mechanisms, enzyme regulation, and gene expression. Moreover, interactions among several classes of phytocompounds in whole plant extracts may result in synergistic, additive, or antagonistic effects, hence confounding efficacy estimates. Consequently, although all principal categories of phytocompounds provide antioxidant protection, their genuine health advantages are contingent upon structural configuration, dosage, bioavailability, matrix interactions, and individual metabolic reactions highlighting the necessity for comprehensive clinical and mechanistic investigations prior to converting laboratory results into therapeutic or dietary guidelines (Gulcin 2025; Tumilaar et al. 2024).

4.3.7 Antimalarial Activity

Phytocompounds have been widely acknowledged for their antimalarial potential from early ages, with many serving as the foundation for contemporary antimalarial medications; however, their efficacy and practical application require thorough assessment. Historically, chemicals such as quinine derived from Cinchona bark and artemisinin extracted from Artemisia annua transformed malaria treatment by specifically targeting Plasmodium parasites at distinct stages of their lifecycle (Mojab 2012). Alkaloids like quinine disrupt heme detoxification in the parasite's digestive vacuole, leading to toxic heme buildup, whereas artemisinin and its derivatives generate ROS upon interacting with intra-parasitic iron, resulting in damage to essential proteins and membranes. Likewise, certain terpenoids and flavonoids can obstruct parasite enzymes, impede merozoite invasion, or regulate host immunity (de Villiers and Egan 2021). Consequently, although phytocompounds are an essential source for antimalarial drug discovery, their conversion from traditional remedies or *in vitro* candidates into clinically approved, globally applicable therapies necessitate addressing challenges related to bioavailability, formulation stability, scalability, and resistance management (Siqueira-Neto et al. 2023).

4.3.8 Anti-inflammatory Activities

Phytocompounds, including phenolics, flavonoids, terpenoids, alkaloids, tannins, and lignans, have demonstrate anti-inflammatory effects via various molecular mechanisms, including the inhibition of pro-inflammatory enzymes such as cyclooxygenase (COX) and lipoxygenase (LOX), suppression of pro-inflammatory cytokines like tumour necrosis factor-alpha (TNF-α), interleukin-1β (IL-1β), and interleukin-6 (IL-6), as well as the downregulation of critical signalling pathways such as nuclear factor kappa B (NF-κB) and mitogen-activated protein kinases (MAPKs) (Gonfa et al. 2023). Flavonoids, including quercetin and luteolin, block NF-κB translocation, thus diminishing the transcription of inflammation-related genes, whereas terpenoids such as curcumin control COX-2 production and oxidative stress, yielding a dual anti-inflammatory and antioxidant effect (Zahra et al., 2024). Alkaloids, such as berberine, have demonstrated the ability to inhibit inflammatory mediators and regulate AMP-activated protein kinase (AMPK) signalling, which is essential in metabolic inflammation. Nonetheless, a thorough assessment of these data uncovers multiple constraints that impede the conversion of phytocompounds into standardised anti-inflammatory treatments (Li et al. 2015). Ultimately, although phytocompounds exhibit considerable potential as multi-targeted, relatively safe modulators of inflammation, their practical therapeutic efficacy will only be achieved through stringent clinical validation, optimised formulation, and meticulous standardisation to reconcile traditional applications with contemporary evidence-based medicine (Nisar et al. 2023).

4.3.9 Phytochemicals in Neural Disorders

Phytochemicals have attracted growing attention for their potential in preventing and managing neural disorders, including Alzheimer's disease (AD), Parkinson's disease (PD), Huntington's disease (HD), depression, anxiety, and other neurodegenerative or neuroinflammatory conditions. Their neuroprotective actions are multifaceted, involving antioxidant defence, anti-inflammatory activity, modulation of neurotransmitter systems, inhibition of neurotoxic protein aggregation, enhancement of mitochondrial function, and regulation of neuronal signalling pathways (Prabhakar, Verma, and Banerjee 2024c). For example, alkaloids such as galantamine, derived from Galanthus species, have been successfully developed into approved drugs for AD by inhibiting acetylcholinesterase and enhancing cholinergic transmission (Babashpour-Asl and Barez 2023). Flavonoids like quercetin and epigallocatechin gallate (EGCG) can scavenge ROS, upregulate endogenous antioxidant enzymes via Nrf2 activation, and modulate synaptic plasticity through brain-derived neurotrophic factor (BDNF) signalling (Jomova et al. 2025). Terpenoids like ginkgolides from Ginkgo biloba improve cerebral blood flow and protect neurons from ischemic damage, while curcumin from turmeric shows potential in inhibiting amyloid-beta aggregation and reducing neuroinflammation. However, a critical assessment reveals substantial challenges that limit the translation of these promising preclinical findings into clinically effective therapies (Assi et al. 2023).

4.3.10 Phytochemicals in Wound Healing

Phytochemicals play a significant role in wound healing by targeting multiple stages of the repair process haemostasis, inflammation, proliferation, and remodelling through their diverse bioactive properties. Compounds such as flavonoids, phenolic acids, tannins, terpenoids, alkaloids, saponins, and essential oils contribute to wound closure by combining antimicrobial, antioxidant, anti-inflammatory, and collagen-stimulating activities. Flavonoids (e.g., quercetin, kaempferol) and phenolic acids (e.g., gallic acid, caffeic acid) help reduce oxidative stress at the wound site by scavenging free radicals, which protects cells from oxidative damage and supports faster tissue regeneration (Cedillo-Cortezano et al. 2024). Tannins promote wound contraction and form a protective layer over the wound, while terpenoids like asiaticoside from Centella asiatica stimulate fibroblast proliferation, angiogenesis, and collagen synthesis, leading to improved tensile strength of the healed tissue (Yang et al. 2025). Many phytochemicals like alkaloid and saponins also modulate growth factor such as vascular endothelial growth factor (VEGF) and transforming growth factor-beta (TGF-β) that are essential for new blood vessel formation and extracellular matrix deposition. Additionally, their anti-inflammatory properties, often mediated through inhibition of cyclooxygenase (COX) and suppression of pro-inflammatory cytokines like TNF-α and IL-6, help shorten the inflammatory phase, allowing quicker progression to tissue repair (Mekky et al. 2025; Rodríguez-Negrete et al. 2024). However, while several *in vitro* and animal studies demonstrate the efficacy of phytochemicals in wound healing, clinical evidence in humans is more limited, and issues like standardisation of extracts, dosage optimisation, and stability remain challenges. Integrating phytochemicals into topical formulations such as ointments, hydrogels, and nanofiber dressings offers promising potential for accelerating wound closure, minimising scarring, and reducing infection risks in both acute and chronic wounds (Barani et al. 2021).

4.3.11 Antivenomous Effect of Phytochemicals

Venoms are complex mixtures of enzymes, peptides, and toxins that cause local tissue damage, systemic toxicity, and sometimes death; key components include proteases, phospholipases A_2 (PLA_2), hyaluronidases, and neurotoxins. Plant-derived bioactive compounds such as phenolics, flavonoids, tannins, alkaloids, terpenoids, and saponins can act against these toxins in several ways. Phenolic compounds and flavonoids can chelate metal ions required for venom enzyme activity, inhibit oxidative damage induced by venom toxins, and stabilise cell membranes against lysis (Morsy et al. 2023; Shivashankar, Murali, and Sangeetha 2019). Tannins precipitate proteins, which may inactivate venom enzymes, while alkaloids like berberine and reserpine can interfere with neurotoxic signalling at synapses. Some phytochemicals also enhance endogenous defence systems by boosting antioxidant enzyme activity or modulating immune responses, which helps limit secondary tissue damage from venom-induced oxidative stress and inflammation (Adamczyk et al. 2017). In traditional medicine, plants like *Mikania glomerata*, *Hemidesmus indicus*, *Andrographis paniculata*, and *Azadirachta indica* have been used to treat snakebite, with studies showing that their extracts can partially neutralise venom enzymes in vitro and in animal models (Félix-Silva et al. 2017). However, a critical limitation is that while these phytochemicals may delay or reduce venom toxicity, they often cannot fully reverse systemic envenomation once toxins are widely distributed in the body. Bioavailability, dose optimisation, and the speed of administration are crucial, as delays in treatment sharply reduce effectiveness. Therefore, phytochemicals are best considered as adjuncts to conventional antivenoms rather than complete replacements. Their real value may lie in first-aid applications specially in remote areas lacking immediate access to antivenom where plant-based formulations could slow venom spread, limit tissue damage, and buy time until definitive treatment is available. Therefore, the clinical integration, standardised extracts, precise dosing, and thorough safety evaluations are needed, alongside further studies on synergistic combinations of phytochemicals and antivenoms to enhance neutralisation efficiency (Gbolade 2021; Prabhakar et al. 2025c; Trim et al. 2020).

4.4 PHYTOCHEMICAL INTERVENTION IN DISEASE PROGRESSION

Phytochemicals such as polyphenols, terpenoids, and alkaloid molecules intervene in disease progression through a network of converging molecular programmes that dampen chronic inflammation, restore redox balance, rewire disordered metabolism, modulate cell-death decisions, and remodel microbe-host crosstalk across gut, liver, vasculature, brain, and tumour niches. Phytochemicals diverse plant-derived small molecules spanning polyphenols, isothiocyanates, alkaloids, terpenoids, and organosulfur compounds modulate the progression of complex diseases not as blunt "antioxidants" but as network-level regulators that retune redox homeostasis, inflammatory tone, metabolic signalling, proteostasis, and host–microbe crosstalk across organ systems from liver and vasculature to brain and tumour microenvironments; mechanistically, many compounds converge on stress-sensor transcriptional programmes (notably KEAP1–NRF2 and NF-κB), energy/nutrient-sensing axes (AMPK/SIRT1–mTOR–PGC-1α), context-dependent autophagy, and microbiome-derived metabolites (SCFAs, TMAO) to interrupt self-reinforcing pathological loops that otherwise drive fibrosis, neurodegeneration, atherothrombosis, insulin resistance, and malignant progression (Bessetti, LaGoy, and Joe 2025; Fakhri et al. 2024; Li et al. 2025; Pandey, Arora, and Mehta 2025; Wan et al. 2025).

In neurobiology, sulforaphane (SFN) exemplifies how a dietary isothiocyanate can cross the blood–brain barrier, activate NRF2, augment mitochondrial quality control, and temper microglial overactivation thereby addressing oxidative–inflammatory drivers of cognitive decline and parkinsonian pathology; recent reviews synthesise these NRF2-centric mechanisms and highlight translational signals for SFN in neurological disorders (Bessetti, LaGoy, and Joe 2025). In the inflammatory-oncologic axis, flavonoids and related polyphenols frequently inhibit NF-κB signalling (via IKK blockade, IκB stabilisation, or p65 interference), reducing transcription of COX-2/iNOS and cytokines (IL-1β, IL-6, TNF-α) that fuel tumour-promoting inflammation and therapy resistance (Pandey, Arora, and Mehta 2025). At the level of cellular energetics, numerous phytochemicals activate AMPK and reshape downstream growth and metabolism dampening mTOR, enhancing fatty-acid oxidation and glucose handling, and, in tumours, countering the Warburg-like phenotype by re-engaging mitochondrial programmes coordinated by PGC-1α; the 2024 synthesis by Fakhri and colleagues offers a pathway-resolved map linking phytochemical AMPK activation to cancer metabolic control and therapy sensitisation (Fakhri et al. 2024). Clinically, berberine remains a human-relevant anchor for the metabolic story: a 2025 systematic review and meta-analysis of randomised placebo-controlled trials finds that purified berberine significantly improves several components of metabolic syndrome (e.g., fasting glucose, triglycerides, waist circumference) with an acceptable safety profile—an effect consistent with AMPK activation, mTOR restraint, bile-acid/gut signalling, and anti-inflammatory cross-talk (Liu et al. 2025).

Because pharmacokinetics constrains many phytochemicals, formulation science is increasingly pivotal: nano- and targeted delivery are being engineered to elevate bioavailability, enhance tumour or tissue selectivity, and translate robust pathway engagement into clinical benefit, as emphasised in 2025 reviews on polyphenol therapeutics and delivery design (Lu et al. 2025; Wang et al. 2025). Beyond host cells, the gut microbiome is both partner and gatekeeper for phytochemical action: a 2025 systematic review and meta-analysis reports that prebiotic and phytochemical interventions reduce circulating TMAO (Trimethylamine N-oxide) a pro-atherogenic, kidney- and mortality-linked metabolite while shifting microbial diversity (enrichment of taxa such as *Akkermansia* and *Bifidobacterium*), offering a plausible mechanistic bridge between dietary phytochemicals and vascular risk modification (Wan et al. 2025). When these axes are viewed together, a consistent pattern emerges: NRF2 activation lowers ROS, indirectly suppressing NF-κB; AMPK reprogrammes metabolism and promotes mitophagy/autophagy, easing oxidative load and normalising insulin signalling; microbiome remodelling attenuates endotoxemia and TMAO-driven vascular injury; and delivery advances raise effective tissue exposure collectively yielding a systems-level "reset" that slows disease trajectories across neurodegeneration, cardiometabolic disorders, and cancer (Bessetti, LaGoy, and Joe 2025; Fakhri et al. 2024; Li et al. 2025; Liu et al. 2025; Lu et al. 2025; Pandey, Arora, and Mehta 2025; Wan et al. 2025; Wang et al. 2025). While in established tumours with constitutively active NRF2, blanket NRF2 stimulation may foster survival and chemoresistance, so tumour-type and stage dictate whether phytochemical strategies should activate or restrain specific nodes; likewise, autophagy can be cytoprotective in stressed normal tissues yet pro-survival in cancers, requiring targeted, disease-aware modulation (Fakhri et al. 2024; Li et al. 2025). The practical translation therefore leans towards biomarker-guided, combination designs stacking complementary phytochemicals (e.g., an NRF2 activator plus an AMPK modulator), pairing them with exercise and dietary patterns that independently improve mitochondrial and inflammatory tone, and embedding them in delivery systems that maximise bioavailability and target engagement while monitoring pathway-proximal readouts (NQO1/HO-1 induction, AMPK phosphorylation, cytokine panels, TMAO and SCFAs) to identify responders and titrate dose–duration; executed this way, phytochemical interventions advance from folklore to mechanism-anchored adjuncts capable of measurably altering disease progression (Bessetti, LaGoy, and Joe 2025; Fakhri et al. 2024; Li et al. 2025; Liu et al. 2025; Lu et al. 2025; Pandey, Arora, and Mehta 2025; Wan et al. 2025; Wang et al. 2025).

4.5 PHARMACOKINETICS AND BIOAVAILABILITY OF PHYTOCHEMICALS

Pharmacokinetics and bioavailability of phytochemicals play a decisive role in determining the real therapeutic benefits of these naturally occurring bioactive compounds, as they dictate the fate of these molecules within the human body from the moment of ingestion to their eventual excretion. Phytochemicals, which include diverse classes such as polyphenols, flavonoids, alkaloids, terpenoids, carotenoids, glucosinolates, and saponins, have long been studied for their potential in preventing and managing chronic diseases like cancer, diabetes, cardiovascular disorders, and neurodegenerative conditions (Aqil et al. 2013). However, the challenge that often hinders their successful clinical translation is not their pharmacological activity per se, but rather their poor pharmacokinetic profiles, which reduce the levels of active compounds that can reach the systemic circulation and target tissues in sufficient amounts to exert meaningful biological effects. Pharmacokinetics broadly covers four key processes, such as absorption, distribution, metabolism, and excretion (ADME), each of which can profoundly influence the bioavailability of phytochemicals. Absorption, the first critical step, is often limited by the chemical nature of phytochemicals. Many of them are lipophilic, poorly soluble in water, and possess large molecular structures that hinder passive diffusion across the intestinal epithelium (Baryakova et al. 2023). Compounds like curcumin, resveratrol, and quercetin, although pharmacologically potent, show very low absorption rates due to these structural constraints. Their absorption can also be influenced by food matrix effects, interactions with other dietary components, and the physiological environment of the gastrointestinal tract, such as pH, enzymatic activity, and the presence of bile salts. Once absorbed, phytochemicals face the second major barrier, first-pass metabolism in the gut wall and liver. During this process, enzymes such as cytochrome P450s, glucuronosyltransferases, and sulfotransferases rapidly modify these molecules into more hydrophilic metabolites, which are often less active than the parent compounds. For example, quercetin undergoes rapid glucuronidation and sulfation, which limits the availability of free quercetin in circulation (Inchingolo et al. 2022). However, in some cases, metabolites themselves may possess biological activity or act as precursors that regenerate active forms under specific conditions, highlighting the complexity of phytochemical pharmacokinetics (Boronat et al. 2021). Distribution of phytochemicals is influenced by their affinity for plasma proteins and their ability to cross cellular barriers. Lipophilic phytochemicals can accumulate in lipid-rich tissues such as adipose tissue, while hydrophilic ones tend to remain in plasma and interstitial fluids. Certain phytochemicals, particularly flavonoids and terpenoids, show an ability to cross the blood–brain barrier, which has attracted attention for their potential in neuroprotection and the management of neurodegenerative diseases like Alzheimer's and Parkinson's (Venkatesan, Ji, and Kim 2015; Yang et al. 2023).

Phytocompounds like polyphenols are extensively metabolised by colonic bacteria into smaller phenolic acids, which can be more readily absorbed and sometimes display stronger bioactivity than the parent compounds. This interplay between phytochemicals and gut microbiota highlights the significance of individual variability in therapeutic responses, as differences in gut microbial composition can lead to variations in bioavailability and pharmacological effects among individuals (Plamada and Vodnar 2021). Excretion is the final pharmacokinetic step, and most phytochemicals or their conjugated metabolites are eliminated through urine or bile. Rapid excretion is a double-edged sword: while it prevents accumulation and potential toxicity, it also limits the duration of exposure to the active compound, thereby reducing sustained therapeutic efficacy (Rathaur and SR 2019). To overcome these limitations, significant research efforts have been directed towards improving phytochemical pharmacokinetics through advanced drug delivery strategies.

Nanotechnology-based systems such as nanoparticles, liposomes, micelles, and solid lipid carriers have shown remarkable success in enhancing solubility, protecting phytochemicals from premature metabolism, and facilitating targeted delivery to tissues (Prabhakar et al. 2020b, 2025a).

4.6 FUTURE DIRECTIONS

4.6.1 Precision Phytotherapy

Precision phytotherapy is an advanced methodology in herbal medicine that incorporates the tenets of precision medicine, customising treatment based on individual differences in genetics, metabolism, microbiome, and disease mechanisms. This approach seeks to transcend the conventional "one-size-fits-all" paradigm of herbal therapy by creating interventions specifically designed for molecular endotypes, metabolic profiles, and clinical biomarkers. In contrast to traditional phytotherapy, which applies identical plant extracts or herbal formulations to a wide demographic, precision phytotherapy recognises that individual differences in the absorption, metabolism, distribution, and excretion of phytochemicals, along with variations in gut microbiome composition and host genetics, can significantly affect therapeutic results (Efferth et al. 2017). For example, individuals with Nrf2-deficient oxidative phenotypes, characterised by impaired antioxidant defence and higher oxidative stress, may benefit more from phytochemicals such as sulforaphane from broccoli sprouts, curcumin from turmeric, and resveratrol from grapes, all of which upregulate the Nrf2/ARE pathway and restore redox balance, reducing oxidative damage and susceptibility to inflammatory or degenerative diseases (Ma 2013).

Similarly, the concept of microbiome metabotypes provides another key dimension for precision phytotherapy, since different individuals' gut microbiota transform phytochemicals into distinct metabolites with variable bioactivities: for example, soy isoflavones (daidzein) are converted into equol, a potent estrogenic metabolite, only in those individuals harbouring specific bacterial strains, making them "equol producers," while non-producers gain less benefit unless supplemented with equol directly; likewise, ellagitannins from pomegranates and walnuts are metabolised into urolithins, particularly urolithin A, which improves mitochondrial health and reduces inflammation, but only in individuals with the right microbial metabotype, whereas non-producers may need direct supplementation with purified urolithin A to achieve comparable effects (Lv et al. 2024; Setchell, Brown, and Lydeking-Olsen 2002; Soukup et al. 2023). Beyond oxidative and microbiome-linked endotypes, precision phytotherapy also addresses inflammatory phenotypes, where pathways such as NF-κB, JAK/STAT, and NLRP3 inflammasome drive chronic inflammation; in such patients, compounds like quercetin (onions, apples), epigallocatechin gallate (green tea), and boswellic acids (frankincense) can be prioritised because of their targeted anti-inflammatory actions, with clinical biomarker monitoring (CRP, IL-6, TNF-α) guiding personalised dosing (Kelley et al. 2019). In the realm of metabolic disorders, where patients may exhibit endotypes defined by insulin resistance, AMPK inactivation, or PPAR dysregulation, phytochemicals like berberine (Berberis species), which activates AMPK and improves glucose uptake, resveratrol, which influences the SIRT1–AMPK axis and enhances fatty acid oxidation, and chlorogenic acid (coffee, artichoke), which improves glucose tolerance, can be matched to patient-specific metabolic signatures for maximum therapeutic impact (Entezari et al. 2022; Hu et al. 2018; Kemelo and Moseki 2025). Precision phytotherapy also shows promise in neurodegenerative conditions, where the ability of phytochemicals to cross the blood–brain barrier becomes crucial; for example, Panax ginseng and curcumin in nanoparticle formulations has been shown to inhibit amyloid plaque formation and neuroinflammation, resveratrol supports mitochondrial biogenesis and neuronal survival, and ginsenosides from ginseng modulate neurotransmitter activity and protect against neuroinflammation, making them particularly suitable for individuals with early AD or PD biomarkers as well as various cancers, including gliomas. as shown in (Del Prado-Audelo et al. 2019). The implementation of precision phytotherapy relies on multi-omics profiling like genomics, proteomics, metabolomics, and microbiomics to identify which phytochemicals will be most effective in each patient, supported by advances in nanotechnology (e.g., liposomes, micelles, phytosomes) to overcome poor solubility and bioavailability of many plant compounds (Lim et al. 2022), as shown in Figure 4.3. Taken together, precision phytotherapy represents the convergence of traditional plant-based healing and modern

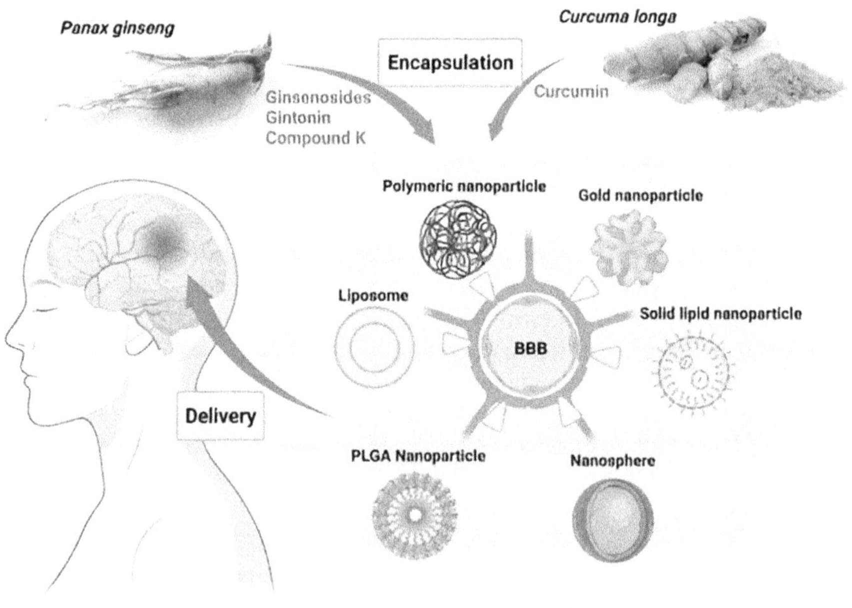

FIGURE 4.3 Phytocompounds from Curcuma longa and Panax ginseng encapsulated into different nanoparticle carriers enable targeted transport to specific brain regions from the bloodstream, promote solubility, and improve retention rates. (From Lim et al. 2022.)

systems biology, offering a scientifically rigorous, individualised approach that maximises therapeutic efficacy, minimises inter-individual variability, and paves the way for integrating phytochemicals into the framework of personalised medicine, where the right plant compound is given to the right patient at the right time and in the right form (Luo et al. 2022).

4.6.2 Next-gen Formulations: Targeted Delivery and Sustained-Release Systems

Next-generation formulations in phytotherapy represent a paradigm shift in overcoming the limitations of conventional herbal medicine, particularly poor solubility, instability, rapid metabolism, and low bioavailability, by developing targeted delivery systems and sustained-release technologies that direct phytochemicals precisely to affected tissues such as the brain, joints, or gut while maintaining effective concentrations over extended periods (Awlqadr et al. 2025). For gut-specific delivery, particularly in diseases such as inflammatory bowel disease (IBD), ulcerative colitis, or microbiome-related dysfunctions, pH-sensitive capsules, polysaccharide-coated nanoparticles, and probiotic-conjugated systems are being developed to protect phytochemicals like EPIGALLOCATECHIN GALLATE, berberine, and ellagitannins from premature degradation in the stomach, ensuring they are released in the colon where they can interact effectively with gut microbiota and inflamed tissues (Zhang and Merlin 2018). Similarly, sustained-release systems such as phytochemical-loaded nanospheres, biodegradable polymers, and phytosomes are being engineered to provide controlled release of compounds like catechins and chlorogenic acid over prolonged periods, maintaining therapeutic plasma concentrations, reducing dosing frequency, and improving patient compliance. These novel delivery platforms not only enhance pharmacokinetic properties but also enable precision phytotherapy, as formulations can be tailored to the molecular endotype of the patient (Sharafan et al. 2025). Overall, next-gen targeted and sustained-release formulations represent a critical innovation in phytomedicine, bridging the gap between promising preclinical efficacy and reliable clinical outcomes by ensuring that phytochemicals reach the right site, at the right dose, and for the right duration.

4.6.3 Network Pharmacology & Artificial Intelligence

Network pharmacology, combined with artificial intelligence (AI), is transforming phytotherapy by departing from the conventional single-compound/single-target model and investigating the multi-component, multi-target characteristics of phytochemicals, which inherently function within synergistic networks in the body (Yang et al. 2025). Network pharmacology methods enable researchers to delineate the connections among several phytochemicals and different molecular targets, thereby constructing a comprehensive understanding of how herbal formulations may re-establish biological equilibrium in intricate disorders such as cancer, diabetes, and neurodegeneration (Noor et al. 2022). Integrating AI-driven algorithms enables the analysis of extensive molecular datasets, omics profiles, and clinical evidence to forecast synergistic combinations of phytochemicals that exhibit superior efficacy in conjunction than in isolation, for instance, the combination of curcumin with piperine to enhance bioavailability, or resveratrol with quercetin for synergistic antioxidant and anti-inflammatory properties (Patel et al. 2025). Simultaneously, AI-driven *in silico* toxicity screening and drug–herb interaction modelling mitigate risk in formulations by forecasting undesirable interactions between phytochemicals and conventional medications, or among phytochemicals themselves, thus ensuring safety prior to clinical application (Prabhakar et al. 2023b; Zhang et al. 2022). These computational methodologies, like network pharmacology and machine learning, have transformed the discovery of novel pharmaceuticals that facilitate the study of extensive compound libraries and enhance the identification of the most promising candidates for drug development. For example, cellular senescence is a state in which cells permanently stop dividing but remain metabolically active, typically triggered by stressors such as DNA damage, oxidative stress, or telomere shortening. In a research work, Santiago-de-la-Cruz and colleagues implemented these methodologies in the current study to find prospective serotherapeutic agents. They developed drug-protein interaction networks associated with cellular senescence. Subsequently, employing three machine learning algorithms (*Random Forest, Support Vector Machine, and K-Nearest Neighbours*), they categorised these chemicals according to their therapeutic efficacy against senescence. These processes facilitated the identification of 714 compounds with potential therapeutic efficacy against senescence, of which 270 demonstrated favourable medicinal chemistry features, ultimately resulting in the development of an interactive web application freely available to the scientific community. It was proposed that flavonoids represent the most prevalent class of compounds, with 18 of them having never been documented as possible drugs for cellular senescence (Figure 4.4) (Santiago-de-la-Cruz et al. 2025).

Other methods like deep learning and machine learning algorithms can discover patient-specific characteristics, including genetic polymorphisms, microbiome metabotypes, and metabolic signatures, to propose personalised combinations of phytocompounds, thereby transforming phytotherapy into a precision-guided treatment. Additionally, AI-assisted molecular docking, pharmacophore modelling, and dynamic simulation enable researchers to visualise the simultaneous binding of numerous phytochemicals to multiple protein targets, facilitating the logical design of multi-component herbal formulations with enhanced efficacy (Azeem, Mustafa, and Mahrosh 2022). Network pharmacology, in combination with AI, optimises the therapeutic efficacy of phytocompounds by elucidating synergistic interactions while concurrently mitigating risks by identifying potential antagonism or toxicity, thereby establishing a secure, evidence-based, and precision-oriented framework for advanced phytotherapy (Yang et al. 2025).

4.6.4 Real-world Evidences: Pragmatic Trials Embedded in Dietary Programmes

The medicinal value of phytochemicals is increasingly being validated by real-world evidence, which goes beyond controlled laboratory settings to evaluate their efficacy in real-world situations. The intricacy of dietary practices, lifestyle variables,

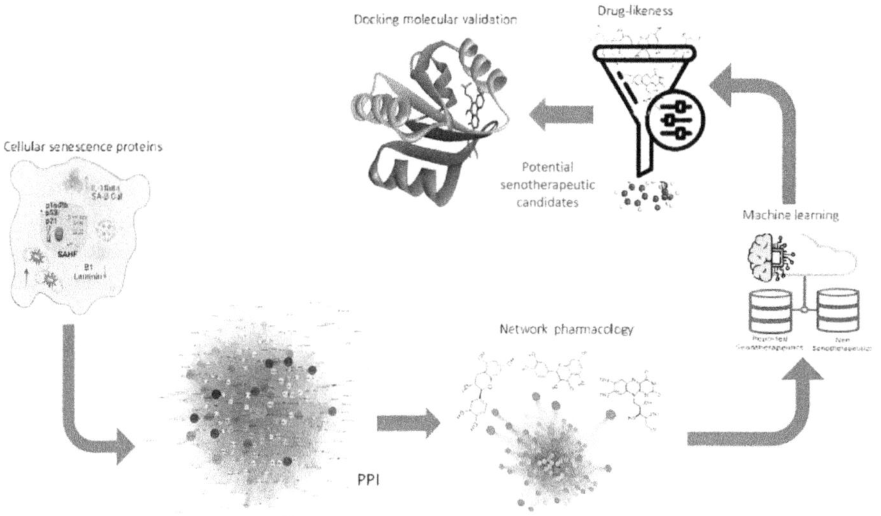

FIGURE 4.4 Implementation of network pharmacology and machine learning to screen potential flavonoid molecules best suited as a therapeutic against cell senescence. (From Santiago-de-la-Cruz et al. 2025.)

and long-term adherence patterns that affect how phytocompounds function in real populations is frequently overlooked by conventional randomised controlled trials (RCTs), despite their rigorous nature. By directly incorporating phytochemical interventions into actual dietary regimens, community-based health initiatives, or clinical lifestyle modification programmes, pragmatic trials integrated into dietary programmes fill this gap and produce more broadly applicable data on safety, efficacy, and cost-effectiveness (Al-Sahlany et al. 2025; Kaushik et al. 2021). These studies not only show how phytochemicals are really consumed by humans (via functional foods, supplements, or entire diets), but they also shed light on the real-world difficulties associated with adherence. In addition, developments in digital phenotyping, the continuous monitoring of behavioural and physiological markers using smartphones, wearable technology, and health apps, allow researchers to measure outcomes like glucose control, inflammation, or cognitive function in real time, track adherence to phytochemical-rich diets, and dynamically customise interventions (Kumar et al. 2023; Niamah et al. 2024). Wearable sensors can assess physical activity and improvements in sleep quality associated with resveratrol or quercetin supplementation, while continuous glucose monitors can record the metabolic effects of berberine-enriched diets in diabetic patients. Real-world evidence offers a more comprehensive, patient-centred understanding of how phytochemicals function across a range of populations by fusing digital phenotyping with pragmatic trial designs. This ensures that interventions are not only scalable, practical, and successful in enhancing population-level health outcomes, but also scientifically sound (Hossain et al. 2025).

4.7 CONCLUSION

Phytocompounds exhibit a wide range of therapeutic effects owing to their diverse chemical structures and ability to act on multiple molecular targets simultaneously. These natural molecules such as polyphenols, flavonoids, alkaloids, terpenoids, carotenoids, and saponins demonstrate antioxidant, anti-inflammatory, antimicrobial, anticancer, antidiabetic, cardioprotective, hepatoprotective, and neuroprotective activities, making them valuable in both preventive and therapeutic medicine. Phytocompounds represent a vast and powerful reservoir of bioactive molecules with proven therapeutic effects across a spectrum of chronic and degenerative diseases, yet their clinical translation requires a shift from traditional herbal practices to modern, science-driven frameworks. Their pharmacokinetic limitations such as poor solubility, rapid metabolism, and low systemic bioavailability have long restricted their therapeutic promise, but advances in next-generation formulations like nanoparticles, liposomes, phytosomes, and sustained-release systems are now enabling targeted delivery to the brain, joints, and gut, ensuring that active compounds reach disease sites in sufficient concentrations and for extended durations. The network pharmacology integrated with artificial intelligence is revolutionising the way we view phytocompounds: no longer as isolated chemicals but as multi-component, multi-target agents capable of synergistic effects across biological networks. These innovations, however, must be tested not only under controlled laboratory conditions but also in real-world environments, where factors such as diet, adherence, and lifestyle affect outcomes. Collectively, these approaches converge into a vision where phytocompounds are no longer viewed as crude traditional remedies but as scientifically validated, precision-guided, and technologically enhanced therapeutics, capable of addressing oxidative stress, inflammation, metabolic dysfunction, and neurodegeneration in a personalised manner. The future of phytotherapy thus lies in this integration of ancient botanical wisdom with modern systems biology, nanotechnology, artificial intelligence, and real-world analytics, creating a holistic yet evidence-based therapeutic paradigm.

CONFLICT OF INTEREST

Authors declare no conflict of research interests.

ACKNOWLEDGMENT

Authors are thankful to the Indian Institute of Technology Kharagpur, Kharagpur, India, and Woxsen University, Hyderabad, India, for providing necessary support for the research works.

REFERENCES

Abdallah, E. M., B. Y. Alhatlani, R. de Paula Menezes, and C. H. G. Martins. 2023. "Back to Nature: Medicinal Plants as Promising Sources for Antibacterial Drugs in the Post-Antibiotic Era." *Plants* 12 (17): 3077.

Adamczyk, B., J. Simon, V. Kitunen, S. Adamczyk, and A. Smolander. 2017. "Tannins and Their Complex Interaction With Different Organic Nitrogen Compounds and Enzymes: Old Paradigms Versus Recent Advances." *ChemistryOpen* 6 (5): 610–4.

Adili, R., M. Hawley, and M. Holinstat. 2018. "Regulation of Platelet Function and Thrombosis by Omega-3 and Omega-6 Polyunsaturated Fatty Acids." *Prostaglandins & Other Lipid Mediators* 139: 10–8.

Ahmed, S. N., M. Ahmad, M. Zafar, G. Yaseen, N. Iqbal, N. Rashid, and … A. Haroon. 2023. "Herbal Drugs: Safety, Cost-Effectiveness, Regulation, Current Trends, and Future Directions." In *Bioprospecting of Tropical Medicinal Plants*, 1479–93. Cham: Springer Nature Switzerland.

Akinboye, A. J., K. Kim, S. Choi, I. Yang, and J. G. Lee. 2023. "Alkaloids in Food: A Review of Toxicity, Analytical Methods, Occurrence and Risk Assessments." *Food Science and Biotechnology* 32 (9): 1133–58.

Alam, S., M. M. R. Sarker, T. N. Sultana, M. N. R. Chowdhury, M. A. Rashid, N. I. Chaity, and I. N. Mohamed. 2022. "Antidiabetic Phytochemicals from Medicinal Plants: Prospective Candidates for New Drug Discovery and Development." *Frontiers in Endocrinology* 13: 800714.

Aldossary, S. A. 2022. "Review on Pharmacology of Atropine, Clinical Use and Toxicity." *Biomedical and Pharmacology Journal* 15 (2): 1021–30.

Al-Ishaq, R. K., M. Abotaleb, P. Kubatka, K. Kajo, and D. Büsselberg. 2019. "Flavonoids and Their Anti-Diabetic Effects: Cellular Mechanisms and Effects to Improve Blood Sugar Levels." *Biomolecules* 9(9): 430.

Alkandahri, M. Y., B. T. Pamungkas, Z. Oktoba, M. Z. Shafirany, L. Sulastri, M. Arfania, and S. Z. Hamidah. 2023. "Hepatoprotective Effect of Kaempferol: A Review of the Dietary Sources, Bioavailability, Mechanisms of Action, and Safety." *Advances in Pharmacological and Pharmaceutical Sciences* 2023 (1): 1387665.

Al-Sahlany, S. T. G., A. K. Niamah, D. K. Verma, P. Prabhakar, A. R. Patel, M. Thakur, and S. Singh. 2025. "Applications of Green Synthesis of Nanoparticles Using Microorganisms in Food and Dairy Products." *Processes* 13 (5): 1560.

Alum, E. U. 2025. "Role of Phytochemicals in Cardiovascular Disease Management: Insights into Mechanisms, Efficacy, and Clinical Application." *Phytomedicine Plus* 5 (1): 100695.

Angelini, P. 2024. "Plant-Derived Antimicrobials and Their Crucial Role in Combating Antimicrobial Resistance." *Antibiotics* 13 (8): 746.

Ansari, P., S. Akther, J. M. A. Hannan, V. Seidel, N. J. Nujat, and Y. H. Abdel-Wahab. 2022. "Pharmacologically Active Phytomolecules Isolated from Traditional Antidiabetic Plants and Their Therapeutic Role for the Management of Diabetes Mellitus." *Molecules* 27 (13): 4278. https://doi.org/10.3390/nu15143266

Aqil, F., R. Munagala, J. Jeyabalan, and M. V. Vadhanam. 2013. "Bioavailability of Phytochemicals and Its Enhancement by Drug Delivery Systems." *Cancer Letters* 334 (1): 133–41.

Aryal, D., S. Joshi, N. K. Thapa, P. Chaudhary, S. Basaula, U. Joshi, and … N. Parajuli. 2024. "Dietary Phenolic Compounds as Promising Therapeutic Agents for Diabetes and Its Complications: A Comprehensive Review." *Food Science & Nutrition* 12 (5): 3025–45.

Assi, A. A., M. M. Farrag, D. M. Badary, E. A. Allam, and M. A. Nicola. 2023. "Protective Effects of Curcumin and *Ginkgo biloba* Extract Combination on a New Model of Alzheimer's Disease." *Inflammopharmacology* 31 (3): 1449–64.

Atanasov, A. G., B. Waltenberger, E. M. Pferschy-Wenzig, T. Linder, C. Wawrosch, P. Uhrin, and … H. Stuppner. 2015. "Discovery and Resupply of Pharmacologically Active Plant-Derived Natural Products: A Review." *Biotechnology Advances* 33 (8): 1582–614.

Aware, C. B., D. N. Patil, S. S. Suryawanshi, P. R. Mali, M. R. Rane, R. G. Gurav, and J. P. Jadhav. 2022. "Natural Bioactive Products as Promising Therapeutics: A Review of Natural Product-Based Drug Development." *South African Journal of Botany* 151: 512–28.

Awlqadr, F. H., K. R. Majeed, A. B. Altemimi, A. M. Hassan, S. A. Qadir, M. N. Saeed, and T. G. Abedelmaksoud. 2025. "Nanotechnology-Based Herbal Medicine: Preparation, Synthesis, and Applications in Food and Medicine." *Journal of Agriculture and Food Research* 19: 101661.

Azeem, M., G. Mustafa, and H. S. Mahrosh. 2022. "Virtual Screening of Phytochemicals by Targeting Multiple Proteins of Severe Acute Respiratory Syndrome Coronavirus 2: Molecular Docking and Molecular Dynamics Simulation Studies." *International Journal of Immunopathology and Pharmacology* 36: 03946320221142793.

Babashpour-Asl, M., and S. R. Barez. 2023. "Therapeutic and Medicinal Effects of Snowdrop (*Galanthus* Spp.) in Alzheimer's Disease: A Review." *Journal of Education and Health Promotion* 12 (1): 128.

Ballard, C. R., and M. R. M. Junior. 2019. "Health Benefits of Flavonoids." In *Bioactive Compounds*, 185–201. Duxford: Woodhead Publishing.

Barani, M., E. Sangiovanni, M. Angarano, M. A. Rajizadeh, M. Mehrabani, S. Piazza, and M. H. Nematollahi. 2021. "Phytosomes as Innovative Delivery Systems for Phytochemicals: A Comprehensive Review of Literature." *International Journal of Nanomedicine* 16: 6983–7022.

Baryakova, T. H., B. H. Pogostin, R. Langer, and K. J. McHugh. 2023. "Overcoming Barriers to Patient Adherence: The Case for Developing Innovative Drug Delivery Systems." *Nature Reviews Drug Discovery* 22 (5): 387–409.

Belwal, M. R., R. Dixit, S. K. Mewara, H. Singh, M. Megha, M. A. S. Yadav, ... & V. Chauhan. 2025. "Emerging Role of Phytochemicals in Cancer Therapy: Mechanism and Clinical Perspectives." *Journal of Carcinogenesis* 24(6S): 727–34.

Bessetti, R. N., A. LaGoy, and E. Joe. 2025. "Broccoli for the Brain: A Review of the Neuroprotective Mechanisms of Sulforaphane." *Frontiers in Cellular Neuroscience* 19: 1601366.

Bisht, K., K. H. Wagner, and A. C. Bulmer. 2010. "Curcumin, Resveratrol and Flavonoids as Anti-Inflammatory, Cyto- and DNA-Protective Dietary Compounds." *Toxicology* 278 (1): 88–100.

Bohn, T. 2019. "Carotenoids and Markers of Oxidative Stress in Human Observational Studies and Intervention Trials: Implications for Chronic Diseases." *Antioxidants* 8 (6): 179.

Boronat, A., J. Rodriguez-Morató, G. Serreli, M. Fitó, R. F. Tyndale, M. Deiana, and R. de la Torre. 2021. "Contribution of Biotransformations Carried Out by the Microbiota, Drug-Metabolizing Enzymes, and Transport Proteins to the Biological Activities of Phytochemicals Found in the Diet." *Advances in Nutrition* 12 (6): 2172–89.

Bribi, N. 2018. "Pharmacological Activity of Alkaloids: A Review." *Asian Journal of Botany* 1 (1): 1–6.

Bufka, J., L. Vaňková, J. Sýkora, and V. Křížková. 2024. "Exploring Carotenoids: Metabolism, Antioxidants, and Impacts on Human Health." *Journal of Functional Foods* 118: 106284.

Buonomo, A. R., R. Scotto, S. Nappa, M. Arcopinto, A. Salzano, A. M. Marra, and … I. Gentile. 2019. "The Role of Curcumin in Liver Diseases." *Archives of Medical Science* 15 (6): 1608–20.

Câmara, J. S., R. Perestrelo, R. Ferreira, C. V. Berenguer, J. A. Pereira, and P. C. Castilho. 2024. "Plant-Derived Terpenoids: A Plethora of Bioactive Compounds With Several Health Functions and Industrial Applications—A Comprehensive Overview." *Molecules* 29 (16): 3861.

Cedillo-Cortezano, M., L. R. Martinez-Cuevas, J. A. M. López, I. L. Barrera López, S. Escutia-Perez, and V. L. Petricevich. 2024. "Use of Medicinal Plants in the Process of Wound Healing: A Literature Review." *Pharmaceuticals* 17 (3): 303.

Chaachouay, N., and L. Zidane. 2024. "Plant-Derived Natural Products: A Source for Drug Discovery and Development." *Drugs and Drug Candidates* 3 (1): 184–207.

Chang, C. W., Y. J. Hsu, Y. M. Chen, W. C Huang, C. C. Huang, and M. C. Hsu. 2015. "Effects of Combined Extract of Cocoa, Coffee, Green Tea and Garcinia on Lipid Profiles, Glycaemic Markers and Inflammatory Responses in Hamsters." *BMC Complementary and Alternative Medicine* 15 (1): 269.

Chaudhary, P., P. Janmeda, A. O. Docea, B. Yeskaliyeva, A. F. Abdull Razis, B. Modu, and … J. Sharifi-Rad. 2023. "Oxidative Stress, Free Radicals and Antioxidants: Potential Crosstalk in the Pathophysiology of Human Diseases." *Frontiers in Chemistry* 11: 1158198.

Che, Y., Q. Wang, R. Xiao, J. Zhang, Y. Zhang, W. Gu, and … H. Kuang. 2018. "Kudinoside-D, a Triterpenoid Saponin Derived from *Ilex kudingcha*, Suppresses Adipogenesis through Modulation of the AMPK Pathway in 3T3-L1 Adipocytes." *Fitoterapia* 125: 208–16.

Choudhari, A. S., P. C. Mandave, M. Deshpande, P. Ranjekar, and O. Prakash. 2020. "Phytochemicals in Cancer Treatment: From Preclinical Studies to Clinical Practice." *Frontiers in Pharmacology* 10: 1614.

Choudhary, N., S. Bawari, J. T. Burcher, D. Sinha, D. Tewari, and A. Bishayee. 2023. "Targeting Cell Signaling Pathways in Lung Cancer by Bioactive Phytocompounds." *Cancers* 15 (15): 3980.

Ciupei, D., A. Colişar, L. Leopold, A. Stănilă, and Z. M. Diaconeasa. 2024. "Polyphenols: From Classification to Therapeutic Potential and Bioavailability." *Foods* 13 (24): 4131.

Clemente-Suárez, V. J., A. Martín-Rodríguez, A. I. Beltrán-Velasco, A. Rubio-Zarapuz, I. Martínez-Guardado, R. Valcarcel-Martin, and J. F. Tornero-Aguilera. 2025. "Functional and Therapeutic Roles of Plant-Derived Antioxidants in Type 2 Diabetes Mellitus: Mechanisms, Challenges, and Considerations for Special Populations." *Antioxidants* 14 (6): 725.

Del Prado-Audelo, M. L., I. H. Caballero-Florán, J. A. Meza-Toledo, N. Mendoza-Muñoz, M. González-Torres, B. Florán, and … G. Leyva-Gómez. 2019. "Formulations of Curcumin Nanoparticles for Brain Diseases." *Biomolecules* 9 (2): 56.

de Villiers, K. A., and T. J. Egan. 2021. "Heme Detoxification in the Malaria Parasite: A Target for Antimalarial Drug Development." *Accounts of Chemical Research* 54 (11): 2649–59.

Efferth, T., M. E. Saeed, E. Mirghani, A. Alim, Z. Yassin, E. Saeed, and … S. Daak. 2017. "Integration of Phytochemicals and Phytotherapy into Cancer Precision Medicine." *Oncotarget* 8 (30): 50284.

Entezari, M., D. Hashemi, A. Taheriazam, A. Zabolian, S. Mohammadi, F. Fakhri, and … S. Samarghandian. 2022. "AMPK Signaling in Diabetes Mellitus, Insulin Resistance and Diabetic Complications: A Pre-Clinical and Clinical Investigation." *Biomedicine & Pharmacotherapy* 146: 112563.

Ergüden, B. 2021. "Phenol Group of Terpenoids Is Crucial for Antibacterial Activity upon Ion Leakage." *Letters in Applied Microbiology* 73 (4): 438–45.

Fakhri, S., S. Z. Moradi, A. Yarmohammadi, F. Narimani, C. E. Wallace, and A. Bishayee. 2022. "Modulation of TLR/NF-κB/NLRP Signaling by Bioactive Phytocompounds: A Promising Strategy to Augment Cancer Chemotherapy and Immunotherapy." *Frontiers in Oncology* 12: 834072.

Fakhri, S., E. Sobarzo-Sánchez, K. Krukiewicz, H. Khan, and G. Sethi. 2024. "Phytochemicals Regulate Cancer Metabolism through Modulation of the AMPK/PGC-1α Signaling Pathway." *BMC Cancer* 24: 12715.

Félix-Silva, J., A. A. Silva-Junior, S. M. Zucolotto, and M. D. F. Fernandes-Pedrosa. 2017. "Medicinal Plants for the Treatment of Local Tissue Damage Induced by Snake Venoms: An Overview from Traditional Use to Pharmacological Evidence." *Evidence-Based Complementary and Alternative Medicine* 2017 (1): 5748256.

Feng, X., A. Sureda, S. Jafari, Z. Memariani, D. Tewari, G. Annunziata, and A. Z. Shen. 2019. "Berberine in Cardiovascular and Metabolic Diseases: From Mechanisms to Therapeutics." *Theranostics* 9 (7): 1923.

Gamil, N. M., H. A. Elsayed, R. M. Hamed, E. T. Salah, A. M. Ahmed, H. A. Mostafa, and … D. Abou-Hussein. 2025. "Insights from Herb Interactions Studies: A Foundational Report for Integrative Medicine." *Future Journal of Pharmaceutical Sciences* 11 (1): 46.

Gbolade, A. A. 2021. "Nigerian Medicinal Plants with Anti-Snake Venom Activity: A Review." *Journal of Malaria Research and Phytomedicine* 1 (1): 29–44.

Gonfa, Y. H., F. B. Tessema, A. Bachheti, N. Rai, M. G. Tadesse, A. N. Singab, and R. K. Bachheti. 2023. "Anti-Inflammatory Activity of Phytochemicals from Medicinal Plants and Their Nanoparticles: A Review." *Current Research in Biotechnology* 6: 100152.

Guan, Q., Y. Cao, Y. Zhang, B. Zhou, J. Wang, D. Baranenko, and … W. Lu. 2025. "Dual Roles of Tannins as Polyphenols and Metal-Complexing Agents in the Prevention and Treatment of Alzheimer's Disease: Exploring the Potential of Dietary Interventions." *Trends in Food Science & Technology* 163: 105177.

Gulcin, İ. 2025. "Antioxidants: A Comprehensive Review." *Archives of Toxicology* 99: 1–105.

Gupta, V., G. Jamwal, G. K. Rai, S. K. Gupta, R. M. Shukla, B. K. Dadrwal, and C. N. Aguilar. 2024. "Biosynthesis of Biomolecules from Saffron as an Industrial Crop and Their Regulation, with Emphasis on the Chemistry, Extraction Methods, Identification Techniques, and Potential Applications in Human Health and Food: A Critical Comprehensive Review." *Biocatalysis and Agricultural Biotechnology* 59: 103260.

Hashempur, M. H., F. Ghorat, F. Karami, A. Jahanbin, H. Nouraei, M. Abbasi, and … Z. Zareshahrabadi. 2025. "Topical Delivery Systems for Plant-Derived Antimicrobial Agents: A Review of Current Advances." *International Journal of Biomaterials* 2025 (1): 4251091.

Heinrich, M., J. Mah, and V. Amirkia. 2021. "Alkaloids Used as Medicines: Structural Phytochemistry Meets Biodiversity—An Update and Forward Look." *Molecules* 26 (7): 1836.

Heinrich, M., F. Scotti, A. Andrade-Cetto, M. Berger-Gonzalez, J. Echeverría, F. Friso, and … R. Spadafora. 2020. "Access and Benefit Sharing under the Nagoya Protocol—Quo Vadis? Six Latin American Case Studies Assessing Opportunities and Risk." *Frontiers in Pharmacology* 11: 765.

Hossain, M. S., M. A. Wazed, S. Asha, M. R. Amin, and I. M. Shimul. 2025. "Dietary Phytochemicals in Health and Disease: Mechanisms, Clinical Evidence, and Applications—A Comprehensive Review." *Food Science & Nutrition* 13 (3): e70101.

Hu, P., K. Li, X. X. Peng, Y. Kan, T. J. Yao, Z. Y. Wang, and … D. Cai. 2023. "Curcumin Derived from Medicinal Homologous Foods: Its Main Signals in Immunoregulation of Oxidative Stress, Inflammation, and Apoptosis." *Frontiers in Immunology* 14: 1233652.

Inchingolo, A. D., A. M. Inchingolo, G. Malcangi, P. Avantario, D. Azzollini, S. Buongiorno, and … G. Dipalma. 2022. "Effects of Resveratrol, Curcumin and Quercetin Supplementation on Bone Metabolism—A Systematic Review." *Nutrients* 14 (17): 3519.

Islam, S. U., M. B. Ahmed, H. Ahsan, and Y. S Lee. 2021. "Recent Molecular Mechanisms and Beneficial Effects of Phytochemicals and Plant-Based Whole Foods in Human Health." *Antioxidants* 10(7): 1016.

Jomova, K., S. Y. Alomar, R. Valko, J. Liska, E. Nepovimova, K. Kuca, and M. Valko. 2025. "Flavonoids and Their Role in Oxidative Stress, Inflammation, and Human Diseases." *Chemico-Biological Interactions* 413: 111489.

Kaushik, B., J. Sharma, P. Kumar, and A. Shourie. 2021. "Phytochemical Properties and Pharmacological Role of Plants: Secondary Metabolites." *Biosciences Biotechnology Research Asia* 18 (1): 23.

Kemelo, M., and P. Moseki. 2025. "The AMPK-SIRT1 Axis: Redefining Type 2 Diabetes Mellitus Management." *Results in Chemistry* 14: 102139.

Kelley, N., D. Jeltema, Y. Duan, and Y. He. 2019. "The NLRP3 Inflammasome: An Overview of Mechanisms of Activation and Regulation." *International Journal of Molecular Sciences* 20 (13): 3328.

Khameneh, B., N. M. Eskin, M. Iranshahy, and B. S. Fazly Bazzaz. 2021. "Phytochemicals: A Promising Weapon in the Arsenal against Antibiotic-Resistant Bacteria." *Antibiotics* 10 (9): 1044. https://doi.org/10.3390/antibiotics10091044

Kocyigit, A., E. M. Guler, and M. Dikilitas. 2018. "Role of Antioxidant Phytochemicals in Prevention, Formation and Treatment of Cancer." In *InTech*. https://doi.org/10.5772/intechopen.72217

Kukula-Koch, W. A., and J. Widelski. 2017. "Alkaloids." In *Pharmacognosy*, 163–98. Cambridge: Academic Press.

Kumar, A., N. P, M. Kumar, A. Jose, V. Tomer, E. Oz, and … F. Oz. 2023. "Major Phytochemicals: Recent Advances in Health Benefits and Extraction Method." *Molecules* 28 (2): 887.

Kumar, A., R. K. Rout, P. S. Rao, and P. Prabhakar. 2022. "Optimization of Pulsed Mode Sonication and *In Silico* Molecular Docking Study for Antioxidant Properties of Mandarin (*Citrus reticulata* Blanco) Peels." *Journal of Food Process Engineering* 46 (6): e14111.

Lam, P., F. Cheung, H. Y. Tan, N. Wang, M. F. Yuen, and Y. Feng. 2016. "Hepatoprotective Effects of Chinese Medicinal Herbs: A Focus on Anti-Inflammatory and Anti-Oxidative Activities." *International Journal of Molecular Sciences* 17 (4): 465.

Li, J. Y., H. Y. Cao, P. Liu, G. H. Cheng, and M. Y. Sun. 2014. "Glycyrrhizic Acid in the Treatment of Liver Diseases: Literature Review." *BioMed Research International* 2014 (1): 872139.

Li, M., X. Chen, Y. Zhang, and Y. Wang. 2025. "Natural Products Targeting AMPK Signaling Pathway: Advances in Antidiabetic Drug Discovery." *Frontiers in Pharmacology* 16: 1534634.

Li, W., B. Hua, M. Saud, H. Lin, W. Hou, M. S. Matter, and M. R. Young. 2015. "Berberine Regulates AMP-Activated Protein Kinase Signaling Pathways and Inhibits Colon Tumorigenesis in Mice." *Molecular Carcinogenesis* 54 (10): 1096–109.

Lim, C. L., C. S. Raju, T. Mahboob, S. Kayesth, K. K. Gupta, G. K. Jain, and V. Nissapatorn. 2022. "Precision and Advanced Nano-Phytopharmaceuticals for Therapeutic Applications." *Nanomaterials* 12 (2): 238. https://doi.org/10.3390/nano12020238

Liu, D., H. Zhao, Y. Zhang, J. Hu, and H. Xu. 2025. "Efficacy and Safety of Berberine on the Components of Metabolic Syndrome: A Systematic Review and Meta-Analysis of Randomized Placebo-Controlled Trials." *Frontiers in Pharmacology* 16: 1572197.

Lu, Y., X. Zhao, L. Chen, and J. Wang. 2025. "Advancements in Delivery Systems for Dietary Polyphenols for Radioprotection and Anti-Inflammatory Therapy." *NPJ Science of Food* 9: 19.

Luo, F., Z. Yu, Q. Zhou, and A. Huang. 2022. "Multi-Omics-Based Discovery of Plant Signaling Molecules." *Metabolites* 12 (1): 76.

Lv, J., S. Jin, Y. Zhang, Y. Zhou, M. Li, and N. Feng. 2024. "Equol: A Metabolite of Gut Microbiota with Potential Antitumor Effects." *Gut Pathogens* 16 (1): 35.

Ma, Q. 2013. "Role of Nrf2 in Oxidative Stress and Toxicity." *Annual Review of Pharmacology and Toxicology* 53 (1): 401–26.

Maniglia, B. C., E. A. Rebelatto, K. S. Andrade, A. Zielinski, and C. J. de Andrade. 2021. "Polyphenols." In *Food Bioactives and Health*, 1–39. Cham: Springer International Publishing.

Manisha, D. R. B., A. M. Begam, K. S. Chahal, and M. A. Ashok. 2025. "Medicinal Plants and Traditional Uses and Modern Applications." *Journal of Neonatal Surgery* 14 (3):162–175.

Marrelli, M., F. Conforti, F. Araniti, and G. A. Statti. 2016. "Effects of Saponins on Lipid Metabolism: A Review of Potential Health Benefits in the Treatment of Obesity." *Molecules* 21 (10): 1404.

Mekky, R. H., M. E. Abo-El Fetoh, S. A. Faheem, A. F. Radwan, M. H. Fawzy, A. M. Mustafa, and W. C. Cho. 2025. "Transforming Growth Factor-β Pathway: Biological Functions and Therapeutic Targets." *MedComm* 6 (7): e70278.

Mojab, F. 2012. "Antimalarial Natural Products: A Review." *Avicenna Journal of Phytomedicine* 2 (2): 52.

Morsy, M. A., S. Gupta, C. P. Dora, V. Jhawat, M. Dhanawat, D. Mehta, and M. El-Daly. 2023. "Venoms Classification and Therapeutic Uses: A Narrative Review." *European Review for Medical and Pharmacological Sciences* 27 (4): 1633–1653.

Msomi, N. Z., and M. B. Simelane. 2019. "Herbal Medicine." In *InTech*, 215–27. Rijeka: IntechOpen.

Muteeb, G., M. T. Rehman, M. Shahwan, and M. Aatif. 2023. "Origin of Antibiotics and Antibiotic Resistance, and Their Impacts on Drug Development: A Narrative Review." *Pharmaceuticals* 16 (11): 1615.

Nabavi, S. M., I. Suntar, D. Barreca, and H. Khan, eds. 2019. *Phytonutrients in Food: From Traditional to Rational Usage*. Duxford: Woodhead Publishing.

Nakadate, K., N. Ito, K. Kawakami, and N. Yamazaki. 2025. "Anti-Inflammatory Actions of Plant-Derived Compounds and Prevention of Chronic Diseases: From Molecular Mechanisms to Applications." *International Journal of Molecular Sciences* 26 (11): 5206.

Niamah, A. K., S. T. G. Al-Sahlany, H. K. Abdul-Sada, P. Prabhakar, S. Tripathy, B. K. Dadrwal, and C. N. Aguilar. 2024. "Phytophagous Probiotic Foods: Exploring the Intersection of Characteristics, Quality Implications, Health Benefits, and Market Dynamics." *Trends in Food Science & Technology* 154: 104795.

Nisar, A., S. Jagtap, S. Vyavahare, M. Deshpande, A. Harsulkar, P. Ranjekar, and O. Prakash. 2023. "Phytochemicals in the Treatment of Inflammation-Associated Diseases: The Journey from Preclinical Trials to Clinical Practice." *Frontiers in Pharmacology* 14: 1177050.

Noor, F., M. Tahir ul Qamar, U. A. Ashfaq, A. Albutti, A. S. Alwashmi, and M. A. Aljasir. 2022. "Network Pharmacology Approach for Medicinal Plants: Review and Assessment." *Pharmaceuticals* 15 (5): 572.

Pandey, P., N. Arora, and S. Mehta. 2025. "An Updated Review Summarizing the Anticancer Potential of Flavonoids via NF-κB Inhibition." *Frontiers in Pharmacology* 16: 1513422.

Patel, J., H. Roy, P. K. Chintamaneni, R. Patel, and R. Bohara. 2025. "Advanced Strategies in Enhancing the Hepatoprotective Efficacy of Natural Products: Integrating Nanotechnology, Genomics, and Mechanistic Insights." *ACS Biomaterials Science & Engineering* 11 (5): 2528–49.

Pebam, M., M. V. Sushma, S. A. Sankaranarayanan, A. M. Thanekar, N. Koyande, and A. K. Rengan. 2022. "Antiviral Perspectives of Economically Important Indian Medicinal Plants and Spices." *Proceedings of the Indian National Science Academy* 88 (3): 392–416.

Peng, Y., M. Ao, B. Dong, Y. Jiang, L. Yu, Z. Chen, ... & R. Xu. 2021. "Anti-Inflammatory Effects of Curcumin in the Inflammatory Diseases: Status, Limitations and Countermeasures." *Drug Design, Development and Therapy*, 4503–4525.

Plamada, D., and D. C. Vodnar. 2021. "Polyphenols—Gut Microbiota Interrelationship: A Transition to a New Generation of Prebiotics." *Nutrients* 14 (1): 137.

Prabhakar, P., and M. Banerjee. 2020a. "Antidiabetic Phytochemicals: A Comprehensive Review on Opportunities and Challenges in Targeted Therapy for Herbal Drug Development." *International Journal of Pharmaceutical Research* 12 (2): 1673–96.

Prabhakar, P., and M. Banerjee. 2020b. "Nanotechnology in Drug Delivery System: Challenges and Opportunities." *Journal of Pharmaceutical Sciences and Research* 12 (4): 492–8.

Prabhakar, P., A. Kumar, and M. Banerjee. 2024b. "Plant Phytochemicals as Potential Therapeutics in Neurological Disorders." In *Plant-Based Foods and Their Implications in Brain Health*, 239–61. Boca Raton: CRC Press.

Prabhakar, P., and B. Mamoni. 2021. "Technical Problems, Regulatory and Market Challenges in Bringing Herbal Drug into Mainstream of Modern Medicinal Practices." *Research Journal of Biotechnology* 16 (3): 212–22.

Prabhakar, P., S. Mukherjee, A. Kumar, S. Kumar, D. K. Verma, S. Dhara, and … M. Banerjee. 2023a. "Optimization of MAE for *Carica papaya* Phytochemicals, and Its *In Silico*, *In Vitro*, and *Ex Vivo* Evaluation: For Functional Food and Drug Applications." *Food Bioscience* 54: 102861.

Prabhakar, P., S. Mukherjee, A. Kumar, R. K. Rout, S. Kumar, D. K. Verma, and … M. Banerjee. 2024a. "*In Silico*, *In Vitro* and *Ex Vivo* Evaluation of the Antihyperglycaemic, Antioxidant and Cytotoxic Properties of *Coccinia grandis* L. Leaf Extract." *Food Technology and Biotechnology* 62 (2): 188–204.

Prabhakar, P., A. K. Pasayat, S. Kumar, and M. Banerjee. 2023b. "An *In Silico* Evaluation of Dipeptidyl Peptidase-4 (DPP-4) Inhibiting Potential of *Cajanus cajan, Coccinia grandis* and *Enhydra fluctuans*." In *2023 International Conference on Communication, Circuits, and Systems (IC3S)*, 1–6. Piscataway: IEEE.

Prabhakar, P., S. Tripathy, D. K. Verma, M. Banerjee, and P. P. Srivastav. 2025a. "Secondary Metabolites of Pigeonpea [*Cajanus cajan* (L.) Millsp.] with Their Biological and Pharmacological Activities." In *Integrated Improvement of Food Legumes*, 33–54. Cham: Springer Nature Switzerland.

Prabhakar, P., S. Tripathy, D. K. Verma, M. Banerjee, and P. P. Srivastav. 2025b. "Trends and Advances in Liposome Formulation Technology with an Emphasis on Ensuring Safety and Quality in Food and Drug Applications." *Food Bioscience* 69(7): 106913.

Prabhakar, P., D. K. Verma, and M. Banerjee. 2024c. "Advances in Enzyme-Assisted Extraction Technique for Antidiabetic Phytochemicals: An Emerging Approach for Green and Economical Technologies." In *Enzymatic Processes for Food Valorization*, 191–216. Cambridge: Academic Press.

Prabhakar, P., D. K. Verma, and M. Banerjee. 2025c. "Safety and Toxicity of Nutraceuticals: Concerns, Challenges and Global Regulatory Aspects." *In Food Bioactives and Nutraceuticals: Dietary Medicine for Human Health and Well-Beings*, 437–62. Singapore: Springer Nature Singapore.

Prabhakar, P., D. K. Verma, and M. Banerjee. 2025d. "Beta (β)-Glucan as Microbial Polymer with Nutraceutical Potential: Chemistry, Biosynthesis, Extraction, Identification, and Industrial Production of Bioactive Compound for Human Health." In *Microbial Nutraceuticals: Products and Processes*, 231–256. USA: John Wiley & Sons.

Prajapati, J., N. Bhatt, and R. Rawal. 2024. "Hepatoprotective Effects of Phytochemicals and Plant Extracts against Chemotherapy-Induced Liver Damage in Animal Models: A Systematic Review." *Archives of Toxicology* 99(3): 887–914.

Rathaur, P., and Kaid Johar, S. R. 2019. "Metabolism and Pharmacokinetics of Phytochemicals in the Human Body." *Current Drug Metabolism* 20 (14): 1085–102.

Riaz, M., R. Khalid, M. Afzal, F. Anjum, H. Fatima, S. Zia, and M. A. Aslam. 2023. "Phytobioactive Compounds as Therapeutic Agents for Human Diseases: A Review." *Food Science & Nutrition* 11 (6): 2500–29.

Rodríguez-Negrete, E. V., Á. Morales-González, E. O. Madrigal-Santillán, K. Sánchez-Reyes, I. Álvarez-González, E. Madrigal-Bujaidar, and J. A. Morales-González. 2024. "Phytochemicals and Their Usefulness in the Maintenance of Health." *Plants* 13 (4): 523.

Rudrapal, M., S. J. Khairnar, J. Khan, A. B. Dukhyil, M. A. Ansari, M. N. Alomary, and … R. Devi. 2022. "Dietary Polyphenols and Their Role in Oxidative Stress-Induced Human Diseases: Insights into Protective Effects, Antioxidant Potentials and Mechanisms of Action." *Frontiers in Pharmacology* 13: 806470.

Saggar, S., P. A. Mir, N. Kumar, A. Chawla, J. Uppal, and A. Kaur. 2022. "Traditional and Herbal Medicines: Opportunities and Challenges." *Pharmacognosy Research* 14 (2): 107–114.

Santiago-de-la-Cruz, J. A., N. A. Rivero-Segura, M. E. Alvarez-Sánchez, and J. C. Gomez-Verjan. 2025. "Network Pharmacology and Machine Learning Identify Flavonoids as Potential Senotherapeutics." *Pharmaceuticals* 18 (8): 1176. https://doi.org/10.3390/ph18081176

Setchell, K. D., N. M. Brown, and E. Lydeking-Olsen. 2002. "The Clinical Importance of the Metabolite Equol—A Clue to the Effectiveness of Soy and Its Isoflavones." *Journal of Nutrition* 132 (12): 3577–84.

Sethi, V., M. Garg, M. Herve, and A. Mobasheri. 2022. "Potential Complementary and/or Synergistic Effects of Curcumin and Boswellic Acids for Management of Osteoarthritis." *Therapeutic Advances in Musculoskeletal Disease* 14: 1759720X221124545.

Sharafan, M., A. Dziki, M. A. Malinowska, E. Sikora, and A. Szopa. 2025. "Targeted Delivery Strategies for Hydrophilic Phytochemicals." *Applied Sciences* 15 (13): 7101.

Sharma, P. K., A. Singh, and N. K Sharma. 2022. "A Socio-Ecological Critique on India's Local Health Traditions Amidst Rising Incidence of Global Pandemics." *Journal of Herbal Medicine* 34: 100578.

Shivashankar, S., A. Murali, and M. K. Sangeetha. 2019. "Molecular Interaction of Phytochemicals with Snake Venom: Phytochemicals of *Andrographis paniculata* Inhibit Phospholipase A2 of Russell's Viper (*Daboia russelli*)." *Biocatalysis and Agricultural Biotechnology* 18: 101058.

Singha, B., V. Singh, and V. Soni. 2024. "Alternative Therapeutics to Control Antimicrobial Resistance: A General Perspective." *Frontiers in Drug Discovery* 4: 1385460.

Siqueira-Neto, J. L., K. J. Wicht, K. Chibale, J. N. Burrows, D. A. Fidock, and E. A. Winzeler. 2023. "Antimalarial Drug Discovery: Progress and Approaches." *Nature Reviews Drug Discovery* 22 (10): 807–26.

Situmorang, P. C., C. F. Zuhra, A. Lutfia, K. M. Pasaribu, R. Hardiyanti, and A. P. Nugraha. 2025. "Harnessing Phytochemicals to Combat Diabetes: Insights into Molecular Pathways and Therapeutic Advances." *Journal of Functional Foods* 128: 106799.

Smith, D., M. da Silva, J. Jackson, and C. Lyal. 2017. "Explanation of the Nagoya Protocol on Access and Benefit Sharing and Its Implication for Microbiology." *Microbiology* 163 (3): 289–96.

Soukup, S. T., A. K. Engelbert, B. Watzl, A. Bub, and S. E. Kulling. 2023. "Microbial Metabolism of the Soy Isoflavones Daidzein and Genistein in Postmenopausal Women: Human Intervention Study Reveals New Metabotypes." *Nutrients* 15 (10): 2352.

Sun, T., S. Rao, X. Zhou, and L. Li. 2022. "Plant Carotenoids: Recent Advances and Future Perspectives." *Molecular Horticulture* 2 (1): 3.

Tijjani, H., M. H. Zangoma, Z. S. Mohammed, S. M. Obidola, C. Egbuna, and S. I. Abdulai. 2020. "Polyphenols: Classifications, Biosynthesis and Bioactivities." In *Functional Foods and Nutraceuticals: Bioactive Components, Formulations and Innovations*, 389–414. Cham: Springer International Publishing.

Toti, E., C. Y. O. Chen, M. Palmery, D. Villaño Valencia, and I. Peluso. 2018. "Non-Provitamin A and Provitamin A Carotenoids as Immunomodulators: Recommended Dietary Allowance, Therapeutic Index, or Personalized Nutrition?" *Oxidative Medicine and Cellular Longevity* 2018 (1): 4637861.

Trim, S. A., C. M. Trim, H. F. Williams, and S. Vaiyapuri. 2020. "The Failures of Ethnobotany and Phytomedicine in Delivering Novel Treatments for Snakebite Envenomation." *Toxins* 12 (12): 774.

Tufail, T., S. Fatima, H. Bader Ul Ain, A. Ikram, S. Noreen, M. Rebezov, and A. Shariati. 2025. "Role of Phytonutrients in the Prevention and Treatment of Chronic Diseases: A Concrete Review." *ACS Omega* 10 (13): 12724–55.

Tumilaar, S. G., A. Hardianto, H. Dohi, and D. Kurnia. 2024. "A Comprehensive Review of Free Radicals, Oxidative Stress, and Antioxidants: Overview, Clinical Applications, Global Perspectives, Future Directions, and Mechanisms of Antioxidant Activity of Flavonoid Compounds." *Journal of Chemistry* 2024 (1): 5594386.

Tungmunnithum, D., A. Thongboonyou, A. Pholboon, and A. Yangsabai. 2018. "Flavonoids and Other Phenolic Compounds from Medicinal Plants for Pharmaceutical and Medical Aspects: An Overview." *Medicines* 5 (3): 93.

Ullah, A., S. Munir, S. L. Badshah, N. Khan, L. Ghani, B. G. Poulson, and … M. Jaremko. 2020. "Important Flavonoids and Their Role as a Therapeutic Agent." *Molecules* 25 (22): 5243.

Vargas-Mendoza, N., E. Madrigal-Santillán, Á. Morales-González, J. Esquivel-Soto, C. Esquivel-Chirino, M. G. L. y González-Rubio, and J. A. Morales-González. 2014. "Hepatoprotective Effect of Silymarin." *World Journal of Hepatology* 6 (3): 144.

Venkatesan, R., E. Ji, and S. Y. Kim. 2015. "Phytochemicals That Regulate Neurodegenerative Disease by Targeting Neurotrophins: A Comprehensive Review." *BioMed Research International* 2015 (1): 814068.

Verma, D. K., P. Prabhakar, S. Tripathy, B. K. Dadrwal, P. Kumar, P. P. Srivastav, and M. Banerjee. 2024. "*In Vitro* and *In Silico* Evaluation of the Antioxidant, Anti-Microbial and Antihyperglycemic Properties of Giloy (*Tinospora cordifolia* L.) Stem Extract." *Biocatalysis and Agricultural Biotechnology* 56: 103059.

Vikas, B., V. K. Sweety, S. Anil, and S. Palaniyandi. 2024. "Natural Anticarcinogens: The Efficacy of Herbs and Spices." In *Herbs and Spices—New Perspectives in Human Health and Food Industry: New Perspectives in Human Health and Food Industry*, 245. UK: IntechOpen.

Wan, Z.-W., L.-J. Zheng, M. Zhao, and H. Chen. 2025. "Effects of Prebiotics and Phytochemicals on Serum TMAO Reduction and Gut Microbiota: A Systematic Review and Meta-Analysis." *Journal of Translational Medicine* 23: 6806.

Wang, M., Q. Li, Y. Zhao, and H. Sun. 2025. "Dietary Polyphenols for Tumor Therapy: Bioactivities, Nano-Therapeutic Systems, and Delivery Strategies." *Food & Function* 16 (10): 2052–87.

Yang, B., J. Xu, S. Liu, C. Wu, Y. Li, M. Lv, and … G. Wang. 2025. "Applications of Bioactive Herbal Extracts in Dressing Materials for Skin Wound Repair: Ingredients, Mechanisms and Innovations." *Interdisciplinary Medicine* 3(4): e20240117.

Yang, L., H. Wang, Z. Zhu, Y. Yang, Y. Xiong, X. Cui, and Y. Liu. 2025. "Network Pharmacology-Driven Sustainability: AI and Multi-Omics Synergy for Drug Discovery in Traditional Chinese Medicine." *Pharmaceuticals* 18 (7): 1074.

Yang, L., S. Yu, Y. Yang, H. Wu, X. Zhang, Y. Lei, and Z. Lei. 2022. "Berberine Improves Liver Injury-Induced Glucose and Lipid Metabolic Disorders via Alleviating ER Stress of Hepatocytes and Modulating Gut Microbiota in Mice." *Bioorganic & Medicinal Chemistry* 55: 116598.

Yang, M., N. Ahmad, M. Hussain, X. Lu, J. Xu, H. Zhong, and R. Guan. 2023. "A Review of Recent Advances on Cyanidin-3-Glucoside: The Biotransformation, Absorption, Bioactivity and Applications of Nano-Encapsulation." *Food & Function* 14 (14): 6320–6345.

Yeshi, K., D. Crayn, E. Ritmejerytė, and P. Wangchuk. 2022. Plant Secondary Metabolites Produced in Response to Abiotic Stresses Has Potential Application in Pharmaceutical Product Development. *Molecules* 27 (1): 313.

Yuan, H., Q. Ma, L. Ye, and G. Piao. 2016. "The Traditional Medicine and Modern Medicine from Natural Products." *Molecules* 21 (5): 559.

Zahra, M., H. Abrahamse, and B. P. George. 2024. "Flavonoids: Antioxidant Powerhouses and Their Role in Nanomedicine." *Antioxidants* 13 (8): 922.

Zhang, F., K. Ganesan, Y. Li, and J. Chen. 2022. "*In-Silico* Drug Toxicity and Interaction Prediction for Plant Complexes Based on Virtual Screening and Text Mining." *International Journal of Molecular Sciences* 23 (17): 10056.

Zhang, M., and D. Merlin. 2018. "Nanoparticle-Based Oral Drug Delivery Systems Targeting the Colon for Treatment of Ulcerative Colitis." *Inflammatory Bowel Diseases* 24 (7): 1401–1415.

Zia-Ul-Haq, M. 2021. "Past, Present and Future of Carotenoids Research." In *Carotenoids: Structure and Function in the Human Body*, 827–854. Cham: Springer International Publishing.

5 Biological Activity of Bioactive Compounds Derived from Animal, Plant, and Microbial Sources and Their Fate in Human Body

Soubhagya Tripathy and Prem Prakash Srivastav

5.1 INTRODUCTION

In recent decades, the role of bioactive compounds in promoting human health has gained significant scientific attention, mainly due to their promising therapeutic potential and preventive effects against chronic diseases. Bioactive compounds, often non-nutritive constituents in small quantities in food and biological materials, can exert physiological effects beyond basic nutritional functions (Nakra, Tripathy, and Srivastav 2025b; Singh, Tripathy, and Srivastav 2025a). These compounds can modulate metabolic processes, influence gene expression, and impact biochemical pathways, thus promoting health promotion and disease prevention (Bontempo et al. 2024). Their sources are diverse, spanning animal-derived products, plant-based foods, and microbial metabolites, each offering unique structural and functional characteristics (Xavier et al. 2024). The expanding understanding of these naturally occurring compounds has propelled interest in their integration into functional foods, pharmaceuticals, and nutraceuticals.

Animal sources provide a rich array of bioactive components such as omega-3 polyunsaturated fatty acids (PUFAs), conjugated linoleic acids (CLAs), bioactive peptides, taurine, creatine, carnitine, and chitosan, among others (Tufail et al. 2025). These compounds have been linked to cardiovascular health, anti-inflammatory responses, immune modulation, and neuroprotective functions (Ma, Fu, and Lee 2025). Marine organisms, in particular, have emerged as potent reservoirs of bioactive compounds, mainly due to their high content of long-chain PUFAs and antioxidant-rich substances. Additionally, chitin and chitosan extracted from crustacean shells have demonstrated significant antimicrobial and biodegradable properties with food packaging and drug delivery applications (Heredia et al. 2025).

DOI: 10.1201/9781003623748-5

Conversely, plants are prolific producers of secondary metabolites with bioactive functions, such as polyphenols, alkaloids, flavonoids, terpenoids, and glycosides. These compounds are the plant's defense mechanisms against environmental stressors, pathogens, and herbivores. They exhibit many bioactivities upon human consumption, including antioxidant, anticancer, anti-inflammatory, and antidiabetic properties (Banwo et al. 2021; Tripathy et al. 2025). Fruits, vegetables, herbs, spices, legumes, and whole grains are key dietary sources of plant-derived bioactive compounds. With the rise of phytotherapy and natural health remedies, the importance of plant-based bioactive compounds continues to grow, reinforcing the concept of "food as medicine" (Chintada and Golla 2025).

Microbial sources, including bacteria, fungi, and yeasts, have also garnered significant interest as producers of potent bioactive metabolites. These include bacteriocins, polysaccharides, enzymes, antibiotics, and vitamins, crucial in maintaining gut health, modulating immunity, and inhibiting pathogenic microorganisms (Rani et al. 2021). Probiotic strains, such as *Lactobacillus* and *Bifidobacterium*, have been extensively studied for their health-promoting effects, not only through competitive exclusion of harmful microbes but also via the production of functional metabolites that influence the host's metabolic and immune systems (Maftei et al. 2024). Moreover, fermented foods represent a unique platform for enriching bioactive compounds through microbial action, making them a vital component of various traditional and modern diets (Tripathy et al. 2023a).

Despite the vast potential of bioactive compounds, their effective utilization in the human body largely depends on their stability, absorption, and bioavailability. After ingestion, these compounds undergo complex metabolic transformations during digestion, which can affect their chemical structure and biological activity (Wojtunik-Kulesza et al. 2020). Factors such as pH, enzymatic action, interaction with dietary components, and the gut microbiota can significantly alter the fate of these compounds, ultimately influencing their physiological efficacy (Brinck et al. 2025). Therefore, understanding bioactive compounds' absorption, distribution, metabolism, and excretion (ADME) profile is essential to evaluate their true potential as health-promoting agents.

To overcome the limitations of poor stability and low bioavailability, novel strategies such as encapsulation have been developed. Encapsulation involves incorporating bioactive compounds into carrier systems like liposomes, nanoparticles, emulsions, and biopolymers, which can protect them from degradation and enhance their targeted delivery within the body (Prabhakar et al. 2025; Tripathy et al. 2023b). These advanced delivery systems improve the bioaccessibility and functional performance of bioactive compounds and offer controlled release and prolonged action, making them highly relevant for clinical and industrial applications (Clemente-Suárez et al. 2025).

This chapter aims to provide a comprehensive overview of the biological activities of bioactive compounds derived from animal, plant, and microbial sources and explore their action mechanisms in promoting human health. The chapter also highlights the significant health benefits associated with these compounds, particularly their antioxidant, anticancer, antimicrobial, and antidiabetic activities. Furthermore, it addresses the challenges related to the bioavailability and stability of these compounds in the human body. It discusses recent advancements in encapsulation technologies designed to enhance their functional efficacy. By integrating multidisciplinary insights from food science, pharmacology, and biotechnology, this chapter seeks to bridge the knowledge gap regarding bioactive compounds' natural origin, physiological functions, and metabolic fate. Understanding the intricate relationship between bioactive compounds and human health is critical for designing future functional foods, therapeutic interventions, and personalized nutrition strategies that align with the goals of preventive healthcare and sustainable development.

5.2 BIOACTIVE COMPOUNDS

5.2.1 Bioactive Compounds from Animal Sources

In recent years, a growing body of research has highlighted the health-promoting effects of bioactive compounds derived from animal sources. These compounds, found in animal-based foods or consumed as dietary supplements, contribute significantly to human well-being. Their unique chemical structures enable them to participate in essential biochemical pathways or modulate the physiological activities of various organs and tissues. Beyond their nutritional relevance, animal-origin bioactives hold immense value in food technology, biotechnology, and agriculture (Cuchillo-Hilario et al. 2024; Kussmann, Abe Cunha, and Berciano 2023). Animals serve as a rich and diverse source of bioactive molecules with crucial roles in human health (Table 5.1). These compounds may be indispensable for the animal's biological functions or produced in forms that benefit other organisms, including humans. Numerous naturally occurring animal-derived compounds have been isolated, characterized, and developed into therapeutic or nutritional supplements to prevent and manage various ailments (Mondal et al. 2021).

Marine fish, in particular, are recognized for their abundance of health-benefiting bioactive compounds, notably long-chain omega-3 polyunsaturated fatty acids (PUFAs). These fatty acids are known to integrate into cellular membranes and subsequently reduce the synthesis of pro-inflammatory eicosanoids by limiting arachidonic acid availability, thereby lowering the release of inflammatory cytokines (Kulczyński, Sidor, and Gramza-Michałowska 2019). Numerous clinical trials have affirmed the role of dietary fatty acids and oils in improving health outcomes in conditions such as diabetes, immune dysfunction, and neurodevelopmental disorders (Fornari Laurindo et al. 2025). Freshwater fish and fish liver oils, along with plant and algal sources, also provide rich supplies of these PUFAs (Yan et al. 2024). Additionally, conjugated linoleic acids (CLAs), naturally occurring trans fatty acids found in ruminant-derived foods, are produced through bacterial fermentation in the animal gut. These fatty acids are present in meats such as beef, lamb, pork, and poultry, and in dairy, eggs, and fish. CLAs have garnered attention for their potential anti-obesity, antihypertensive, antidiabetic, and anticancer effects, alongside their role in enhancing immune responses and reducing food allergies (Guillocheau, Legrand, and Rioux 2019; Xavier et al. 2024).

Meat and its derivatives are also significant sources of numerous bioactive compounds, especially peptides and specific fatty acids. Rich in essential amino acids,

TABLE 5.1
Major Bioactive Compounds Present in Animal, Plant and Microbial Source

Source	Bioactive Compounds	Health/Functional Benefits
Animal Source		
Marine and freshwater fish	Omega-3 polyunsaturated fatty acids (EPA, DHA)	Cardiovascular health, anti-inflammatory, neuroprotective
Ruminant-derived foods (meat, dairy, eggs)	Conjugated linoleic acid (CLA)	Anti-obesity, antihypertensive, antidiabetic, anticancer, immune enhancement
Meat and meat products	Taurine, L-carnitine, choline, creatine, coenzyme Q10, glutathione, α-lipoic acid, bioactive peptides	Antioxidant, anti-inflammatory, lipid-lowering, energy metabolism
Crustaceans, insects	Chitin, chitosan, chitooligosaccharides (COS)	Antimicrobial, biodegradable, prebiotic, encapsulation carrier
Plant Source		
Fruits, tea, coffee, vegetables, cereals, nuts	Polyphenols (flavonoids: catechins, isoflavones phenolic acids: caffeic, ferulic, sinapic)	Antioxidant, anticancer, antidiabetic, cardioprotective
Nuts, seeds, legumes, plant oils	Phytosterols (β-sitosterol, campesterol, stigmasterol)	Cholesterol-lowering, cardiovascular health
Citrus fruits, herbs, ginseng, Centella asiatica	Terpenoids (limonene, ginsenosides, centelloside)	Immune-modulating, anti-inflammatory, antiplasmodial
Cruciferous vegetables (broccoli, cabbage, mustard)	Glucosinolates	Anticancer, detoxification enzyme induction
Garlic, onion (Allium family)	Organosulfur compounds (allicin, diallyl sulfide/disulfide)	Antimicrobial, cardiovascular benefits
Microbial Source		
Bacteria (Streptomyces, Bacillus)	Antibiotics (chloramphenicol, tetracycline), lipopeptides, bacteriocins	Antimicrobial, anticancer, immune modulation
Probiotic bacteria (Lactobacillus, Bifidobacterium)	Exopolysaccharides, vitamins, enzymes	Gut health, immune modulation, competitive exclusion of pathogens
Marine bacteria/fungi (Aspergillus, Penicillium, Verrucosispora)	Alkaloids, polyketides, purpurides, asperphenin A	Antibacterial, anticancer, antioxidant
Edible/medicinal fungi (Ganoderma, Armillaria, Pleurotus)	Polysaccharides, triterpenoids	Antidiabetic, immunomodulatory, antioxidant
Microalgae (Dunaliella, Isochrysis, Nannochloropsis)	Carotenoids, polyphenols	Antioxidant, anticancer, metabolic regulation

Source: Banwo et al. 2021; Bhat, Kumar and Bhat 2015; Chen et al. 2022; Haruna and Yahaya 2021; Kulczyński, Sidor and Gramza-Michałowska 2019; Monciardini et al. 2014; Rani et al. 2021.

particularly sulfur-containing ones, meat proteins are biologically valuable and contribute to multiple physiological functions. Beyond its primary macronutrients, meat is known to contain biofunctional molecules such as taurine, l-carnitine, choline, alpha-lipoic acid, CLA, glutathione, creatine, coenzyme Q10, and various bioactive peptides (Martirosyan, Lampert, and Lee 2022). These substances exhibit beneficial properties, including antioxidant, anti-inflammatory, and lipid-lowering effects. In both intact and hydrolyzed forms, dietary proteins have been found to exert functional health benefits. The hydrolysis of these proteins results in bioactive peptides with proven *in vivo* and *in vitro* efficacy (Mondal et al. 2021).

Another noteworthy class of animal-derived compounds includes chitin and chitosan biopolymers composed of β-(1–4)-linked N-acetyl-D-glucosamine units. These are primarily found in the exoskeletons of marine crustaceans (e.g., shrimp, crab, lobster), insects, and some fungi. Chitin ranks as one of the most abundant polymers in nature, similar in structure to cellulose. Deacetylation converts chitin into chitosan (Kumari and Kishor 2020), which can be further broken down into chitooligosaccharides (COSs), water-soluble prebiotics with beneficial functions. Chitosan is valued for its non-toxic, biodegradable, and antimicrobial characteristics. In the food sector, it is utilized as an active packaging component, food additive, enzyme carrier, and a matrix for encapsulating sensitive bioactive compounds (da Silva, de Souza, and Lacerda 2019). The bioactive compounds derived from animal sources exhibit remarkable potential in promoting human health, preventing chronic diseases, and supporting biological functions beyond basic nutrition. These compounds, including polyunsaturated fatty acids, conjugated linoleic acids, bioactive peptides, and chitosan, offer diverse physiological benefits such as anti-inflammatory, antioxidant, antidiabetic, and immunomodulatory effects. With their unique chemical structures and multifunctional roles, they contribute to nutritional and therapeutic advancements and innovations in food processing, biotechnology, and sustainable agriculture.

5.2.2 Bioactive Compounds from Plant Sources

Phytonutrients, or phytochemicals, are naturally occurring plant bioactive compounds that contribute to human health. "Phyto" originates from the Latin word *phyta*, meaning "plant." These bioactive compounds are predominantly secondary metabolites, which, although not essential for a plant's fundamental biological processes like growth or reproduction, serve vital roles such as defense against environmental stressors and attraction of pollinators (Xavier et al. 2024). These metabolites help plants cope with UV radiation, pollutants, drought, temperature extremes, and pathogens (Table 5.1). Phytochemicals are widely distributed in plant parts such as roots, stems, leaves, flowers, fruits, and seeds, contributing to sensory characteristics like the color and flavor of fruits and vegetables (Agrawal et al. 2023). When consumed through plant-based foods such as fruits, vegetables, legumes, and whole grains, these compounds support human health by enhancing nutrient utilization and offering antioxidant and free radical-scavenging functions (Verma and Thakur, 2021).

Due to their significant antioxidant capacity, phytochemicals have attracted global attention in nutrition, pharmaceuticals, and functional food development (Nakra, Tripathy, and Srivastav 2025b). Evidence from *in vitro*, *in vivo*, and epidemiological studies highlights the positive health impacts of regular consumption of phytochemical-rich diets (Xavier et al. 2024). Typically, multiple classes of bioactive compounds coexist in a single plant source. Foods like whole grains, nuts, seeds, fruits, legumes, mushrooms, herbs, and spices are rich in various components. Specific examples include garlic, onions, bananas, berries, soy, and cruciferous vegetables, all of which contribute diverse bioactive compounds to the human diet (Mondal et al. 2021). Among these, polyphenols are the most abundant and widely studied plant bioactives. Flavonoids such as catechins, theaflavins, thearubigins, and isoflavones are prominent polyphenols with potent antioxidant properties (Maniglia et al. 2021). Structurally, polyphenols consist of aromatic rings with hydroxyl groups and are subdivided into categories like simple phenols and hydroxycinnamic acid derivatives, including coumaric, caffeic, ferulic, and sinapic acids (Santos-Buelga, González-Paramás, and González-Manzano 2023). Familiar sources include tea, coffee, fruits, vegetables, cereals, wine, and nuts.

Phytosterols, derived from plant triterpenes, are another group of bioactives that play a crucial role in reducing LDL cholesterol and supporting cardiovascular health. The key phytosterols like campesterol, β-sitosterol, and stigmasterol are abundant in nuts, seeds, legumes, and unrefined plant oils (Tripathy and Srivastav 2023, 2025). Their saturated counterparts, known as plant stanols (e.g., sitostanol), have also shown health benefits in clinical studies (Martirosyan, Lampert, and Lee 2022).

Terpenoids, representing one of the most structurally diverse phytochemical classes, are responsible for many of the aromatic characteristics of plants. Compounds such as limonene, found in citrus fruits, exemplify this group. Familiar dietary sources include tea, thyme, sage, cannabis, and citrus fruits. Terpenoids are known for various therapeutic properties, including antiplasmodial activity (Tripathy et al. 2025). Notably, ginsenosides, triterpene saponins from ginseng, exhibit immune-modulating effects and are key ingredients in traditional Chinese medicine (Wijesekara, Luo, and Xu 2024). Similarly, centelloside from *Centella asiatica* is used for skin treatments (Tripathy et al. 2022). Sulfur-containing glucosinolates are prevalent in cruciferous vegetables like cabbage, broccoli, cauliflower, and Brussels sprouts, as well as oilseeds such as mustard and rapeseed. Based on their side chain structures, glucosinolates are categorized into aliphatic, aromatic, and indolyl groups (Xavier et al. 2024). In the allium family, allicin is formed from alliin through enzymatic action. Subsequently, it breaks down into compounds such as diallyl sulfide and disulfide, contributing to garlic and onions' health-promoting effects (Mondal et al. 2021).

Phytochemicals represent a vast and diverse group of plant-derived bioactive compounds that contribute significantly to human health and wellness. These naturally occurring molecules, ranging from polyphenols and phytosterols to terpenoids and sulfur-containing glucosides, not only support plant defense and survival but also exhibit potent antioxidant, anti-inflammatory, cholesterol-lowering, and disease-preventive properties when consumed through a plant-based diet. Their presence in fruits, vegetables, grains, nuts, herbs, and spices underscores the importance of dietary diversity and whole-food consumption. As the global focus shifts toward functional foods and preventive healthcare, phytochemicals remain essential in promoting long-term health, reinforcing their value in both nutritional science and therapeutic applications.

5.2.3 Bioactive Compounds from Microbial Sources

Microorganisms and their secondary metabolites have long been recognized for their wide-ranging and potent biological activities (Table 5.1). Bacteria, in particular, are prolific producers of bioactive compounds, including antimicrobials, and both Gram-positive and Gram-negative strains contribute significantly to this chemical diversity (Ameen, AlNadhari, and Al-Homaidan 2021; Singh 2022). For example, Zhang et al. (2020) reported the isolation of a novel abyssomicin and six related compounds, including proximicin analogs, from a marine bacterium *Verrucosispora* sp. MS100137, sourced from ocean sediments. Among these, one new and two known compounds exhibited promising antiviral activity against influenza A virus (IAV). Similarly, Al-Dhabi et al. (2019) demonstrated that extracts from *Streptomyces* sp. Al-Dhabi-90, isolated from Saudi Arabian marine environments, was effective against various drug-resistant bacteria, including *Staphylococcus aureus*, *Klebsiella pneumoniae*, *Escherichia coli*, *Pseudomonas aeruginosa*, *Proteus mirabilis*, and *Enterococcus faecium*. Major constituents identified in these extracts included 3-methylpyridazine, n-hexadecanoic acid, indazol-4-one, octadecanoic acid, and 3a-methyl-6-((4-methylphenyl)sulfonyl) compounds, suggesting this actinomycete strain holds excellent potential for novel antibiotic development.

In addition to these microorganisms, probiotics and prebiotics also offer health benefits through various mechanisms. These include inhibiting the adhesion and colonization of pathogens in the gastrointestinal tract, producing beneficial metabolites, and enhancing immune function through immunoglobulin (Ig) production (Singh 2022). Most probiotic strains in use today are derived from the gut microbiota of healthy individuals, indicating a close connection between probiotics and host gut ecology. Their positive effects are primarily linked to the modulation of intestinal microbial populations or their metabolic activities. Consuming functional foods enriched with bioactive ingredients such as dietary fiber, vitamins, minerals, oligosaccharides, and probiotics can help maintain a healthy balance in gut microflora (Bauer-Estrada et al. 2023).

Marine ecosystems have proven to be rich sources of bioactive compounds capable of modulating key physiological processes, including enzyme regulation, transcription factor activation, and transporter inhibition. Marine fungi, in particular, are renowned for their ability to synthesize significant quantities of bioactive metabolites. Sugumaran et al. (2022) reviewed diverse secondary metabolites derived from marine fungi, sponges, algae, and mangrove plants, outlining their modes of action. These include flavonoids, phenolics, steroids, alkaloids, xanthones, and other phytochemicals known for their antioxidant and therapeutic properties, widely used in medical and pharmaceutical applications (Venegas-Ortega et al. 2019). Most research in this field has focused on Ascomycota and Basidiomycota, especially endophytic fungi known for their strong antioxidant potential.

In a noteworthy study, Xu et al. (2020) identified two novel cyclocitribol derivatives such as aspergillsteroid A and neocyclocitrinol B, from a sponge-associated

Aspergillus species. Aspergillsteroid A displayed significant antibacterial activity against *Vibrio harveyi* (MIC 16 μg/mL). Similarly, Ma et al. (2020) isolated several novel purpuride and isocoumarin compounds, along with 15 known metabolites, from *Penicillium minioluteum* ZZ1657. N-acetyl-L-valine-conjugated drimane sesquiterpenoids showed effective antimicrobial activity against MRSA, *E. coli*, and *Candida albicans*. At the same time, another purpuride demonstrated vigorous antiproliferative activity against human glioma cells. Wang et al. (2019) reported the discovery of two novel citrinin dimers and a monomer from *Penicillium citrinum* HDN-152–088, one of which exhibited antioxidant potential. Additionally, *Aspergillus* sp. was found to produce asperphenin A, a compound with promising anticancer effects against human colon cancer cells (Bae et al. 2020). Pathogenic fungi like *Candida albicans* and *Aspergillus* species are also known to synthesize various phytochemicals, such as phenolics, flavonoids, and triterpenoids, which possess diverse therapeutic properties. These natural bioactive molecules have shown activities including anticancer, antimicrobial, antiviral, antioxidant, and anti-inflammatory effects, underscoring their potential as alternative or adjunct therapies (Chandra, Sharma, and Arora 2020; Yadav and Negi 2021).

Bacteria and fungi from terrestrial and marine environments are prolific producers of diverse secondary metabolites with a broad spectrum of biological activities, including antimicrobial, antiviral, antioxidant, and anticancer properties. The extensive research on strains such as *Streptomyces*, *Aspergillus*, and *Penicillium* has highlighted their potential in combating multidrug-resistant pathogens and supporting human health through bioactive metabolites. Moreover, integrating probiotics and functional foods enriched with microbial bioactive compounds has opened promising avenues for enhancing gut health and immune function. Continued exploration and characterization of these microbial resources, especially marine-derived strains, hold immense potential for discovering and developing novel therapeutic agents in the pharmaceutical and nutraceutical industries.

5.3 MAJOR HEALTH BENEFITS OF BIOACTIVE COMPOUNDS

5.3.1 Antioxidant Activity

Oxygen is essential for sustaining life on Earth, yet it can paradoxically threaten cellular health by inducing oxidative stress by generating reactive oxygen species (ROS). These ROS are produced via multiple biological and environmental pathways. Antioxidants are crucial in mitigating oxidative damage by scavenging ROS, protecting cellular components, and supporting the repair of oxidative injuries. Based on their functional mechanisms, antioxidants are generally classified as primary or secondary. Primary antioxidants neutralize free radicals through hydrogen atom transfer (HAT) or single electron transfer (SET) pathways. Remarkably, even small quantities of these compounds can neutralize many free radicals (Hou et al. 2024; Nakra, Tripathy, and Srivastav 2025a; Tumilaar et al. 2024; Vignesh et al. 2024).

Plants are rich sources of natural antioxidants, particularly phenolic compounds, carotenoids, and coumarins, which act through diverse ROS-neutralizing pathways.

Polyphenols are considered the most potent due to their diverse chemical structures and biological functions. Compounds such as proanthocyanidins, commonly found in berries and grape seeds, are known to inhibit lipid peroxidation by donating hydrogen atoms and chelating metal ions (Abeyrathne et al. 2022; Chen et al. 2022; Dixit et al. 2023). Another important phenolic group is stilbenoids, such as resveratrol, abundant in grapes, berries, and peanuts. These compounds, including piceid, piceatannol, and pterostilbene, show variable antioxidant capacities—with resveratrol being particularly effective in protecting proteins and lipids from oxidative stress. Moreover, endogenous antioxidant systems like superoxide dismutase (SOD), catalase, and glutathione contribute to cellular defense, and compounds like resveratrol can further stimulate these enzymatic pathways (Nagapan et al. 2018). Carotenoids like β-carotene, lycopene, and lutein—fat-soluble pigments—quench singlet oxygen and peroxyl radicals through energy transfer and radical adduct mechanisms, thereby safeguarding cellular membranes. Similarly, coumarins, benzopyrone derivatives widely found in plants and microbes, exert antioxidant effects by scavenging ROS and activating pathways such as Nrf2 (Gulcin 2025).

Animal-based antioxidants are primarily peptides released during the enzymatic breakdown of proteins from sources like milk, eggs, blood, and connective tissues. Specific sequences such as QGAR, WNIP, and IRW, derived from egg white or collagen, have demonstrated significant radical-scavenging and metal-chelating activities due to their capacity to donate electron-proton pairs (Abeyrathne et al. 2022; Gulcin 2025). These antioxidant peptides are obtained through enzymatic hydrolysis or other protein-processing techniques, and their effectiveness largely depends on their sequence, amino acid composition, and molecular size (Benedé and Molina 2020). Various food-derived peptides, including those from fish, meat, and slaughterhouse by-products, have been shown to suppress oxidative stress by reducing lipid peroxidation and enhancing endogenous antioxidant enzyme activities such as SOD and GPx (Lassiter et al. 2024). During fermentation, proteins break down into smaller bioactive peptides through microbial enzymatic actions, forming dipeptides, tripeptides, or free amino acids with antioxidative potential (Toldrá et al. 2020). For example, peptides obtained from camel milk hydrolyzed with enzymes like pepsin and alcalase (e.g., RLDGQGRPRVWLGR, TPDNIDIWLGGIAEPQVKR) exhibited strong antioxidant behavior (Wali et al. 2020). Similarly, milk-derived peptides with sequences such as ARHPHPHLSFM and AVPYPQR were found to inhibit lipid peroxidation in human intestinal cells by activating the Keap1-Nrf2 antioxidant pathway (Tonolo et al. 2020).

Microorganisms, including various Gram-positive and Gram-negative bacteria, also serve as a source of natural antioxidants. For instance, a short hydrophobic peptide (Ser–Ser–Gln) isolated from the marine bacterium *Kocuria marina* CDMP 10, collected from the Gulf of Mannar, exhibited vigorous ROS-scavenging activity in hepatocellular carcinoma cell models, highlighting its potential as a pharmaceutical antioxidant agent (Rani et al. 2021; Tripathi et al. 2020). Furthermore, microalgal species like *Dunaliella salina*, *Tetraselmis chuii*, and *Isochrysis galbana* (clone Tahiti) have shown significant antioxidant properties based on DPPH assays. Among them, *I. galbana* demonstrated the highest free radical inhibition (61.64% at 50 ppm), attributed to its high phenolic content (17.798 mg GAE/g), followed by *D. salina* and

T. chuii (Widowati et al. 2017). Another study revealed that *Nannochloropsis oculata*, a polyphenol-rich microalga, possesses potent *in vitro* ROS-scavenging activity, with an IC_{50} value of 52.10 ± 0.85 µg/L (Wali et al. 2020).

Antioxidants originating from plants, animals, and microorganisms serve as the body's natural defense, mitigating the harmful effects of reactive oxygen species through diverse mechanisms such as radical scavenging, metal chelation, and enzyme activation. Plant-based antioxidants like polyphenols, carotenoids, and coumarins exhibit potent ROS-neutralizing capabilities. At the same time, animal-derived peptides and microbe-sourced compounds add to the growing arsenal of bioactive molecules with antioxidative potential. These natural antioxidants collectively represent a promising avenue for developing functional foods and therapeutic agents to combat oxidative stress-related diseases.

5.3.2 Anticancer Activity

Recent studies underscore the diverse anticancer potential of bioactive compounds across kingdoms. Animal-derived antimicrobial peptides (AMPs), including both terrestrial and marine sources, have demonstrated selective cytotoxicity against tumor cells via membrane disruption, apoptosis induction, angiogenesis inhibition, and immune activation, with several candidates now in clinical trials (Qu et al. 2024; Shahidi and Saeid 2025). Plant-derived phytochemicals such as capsaicin, epigallocatechin gallate, berberine, lycopene, and curcumin exert effects through modulation of key signaling pathways (STAT-3, PI3K/Akt, MAPK), induction of cell cycle arrest, apoptosis, and suppression of angiogenesis (Kangra et al. 2025). Notably, a docking analysis on *Moringa oleifera* phytochemicals (e.g., apigenin, hesperetin) revealed strong binding to BCL-2, surpassing Venetoclax *in silico* and highlighted synergistic multi-ligand interactions (Saha, Desai, and Biswas 2024). From microbial origins, marine actinomycetes (e.g., *Streptomyces* spp.) continue to yield potent cytotoxic metabolites such as guanahanolide A and verrucosamide active against breast and colon cancer lines (Carroll et al. 2022; Genilloud 2017). Soil bacteria like *Bacillus subtilis* produce secondary metabolites exhibiting anticancer effects against MCF-7 breast cancer cells, confirmed via both *in vitro* assays and *in silico* docking. Endophytic fungi from *Anethum graveolens* inhibit HepG2 hepatocellular carcinoma cell growth by up to 92–95% at 50 µg/mL (El-Zehery et al. 2024). Cyanobacterial metabolites such as dolastatin 10, curazine A, and calothrixins have demonstrated nanomolar-level cytotoxicity and apoptosis induction in HeLa and nasopharyngeal carcinoma cells (Barzkar, Sukhikh, and Babich 2024; Karthik et al. 2023). Collectively, these findings illustrate a rich pipeline of bioactive molecules with diverse mechanisms, advancing toward novel anticancer therapeutics.

5.3.3 Antimicrobial Activity

Bioactive compounds derived from plants, animals, and microbes have emerged as potent antimicrobial agents in recent years. From plant sources, diverse secondary metabolites such as flavonoids, tannins, phenols, terpenoids, alkaloids, and steroids have shown significant activity against WHO-priority Gram-positive and Gram-negative pathogens (e.g., *Staphylococcus aureus*, *Escherichia coli*, *Pseudomonas aeruginosa*) with MIC values often below 0.2 μg/mL and MBCs under 64 μg/mL (SeyedAlinaghi et al. 2025). Essential oils like thyme and oregano, as well as by-products and waste plant extracts, disrupt microbial cell membranes, inhibit biofilm formation, and modulate quorum sensing (Rasheed et al. 2024; Rout et al. 2022). For instance, phenolic compounds from plant waste have inhibited oral biofilms, including *P. aeruginosa* and *Bacillus* species at MICs ranging from 0.0625 to 0.5 mg/mL; compounds like catechin, cinnamic, and vanillic acids achieved up to ~80% biofilm inhibition (Mashal et al. 2024). Furthermore, synergistic effects have been observed when plant-derived compounds are combined with conventional antibiotics, notably flavonoids enhancing β-lactam efficacy against resistant strains such as MRSA (Duda-Madej 2024; Panda et al. 2025; Rasheed et al. 2024).

Animal- and microbe-derived bioactives also continue to offer rich antimicrobial potential. Endophytic fungi originating from plant hosts, such as *Trichoderma harzianum* from rosemary and *Aspergillus terreus* nanocomposites, have demonstrated bactericidal action against *S. aureus*, *K. pneumoniae*, *E. coli*, and *P. aeruginosa* (Abdulhadi, Hasan, and Gergees 2023; El-Gazzar et al. 2025). Mushrooms like *Suillus* and *Pleurotus* yield phenolics, β-carotenes, and β-glucans that show strong antimicrobial and antioxidant activities, with inhibition zones up to 45 mm against *P. aeruginosa* (Abdulhadi, Hasan, and Gergees 2023). Among microbial sources, bacterial genera such as *Streptomyces* and *Bacillus* remain prominent. *Streptomyces* spp. Continues to furnish clinically-relevant antibiotics like chloramphenicol and tetracycline, while *B. subtilis* produces lipopeptides (fengycins, gageostatins) and macrolactins with broad-spectrum antifungal and antibacterial effects (SeyedAlinaghi et al. 2025). Recent systematic work on actinomycetes from extreme environments highlights their ongoing promise, advocating genome mining and omics strategies to unveil novel biosynthetic gene clusters (BGCs) for next-generation antimicrobials (Helmi 2025).

5.3.4 Antidiabetic Activity

Numerous studies have highlighted the potential of phytochemicals such as flavonoids, alkaloids, terpenes, and phenolic acids in regulating blood glucose. These secondary plant metabolites, such as flavonoids and triterpenoids, show significant α-glucosidase and α-amylase inhibition, alongside antioxidant and insulin-sensitizing effects (Adelusi et al. 2023; Naveen et al. 2023; Sukhikh et al. 2023). For example, a recent review by Ansari et al. (2023) emphasized Mexican agrobiodiverse foods like chickpeas, rich in flavonoids and anthocyanins that inhibit α-amylase and α-glucosidase, enhance insulin receptor activity, and reduce postprandial glycemia. Similarly, anti-α-glucosidase and anti-inflammatory compounds from *Coriaria intermedia* and *Dracontomelon dao*, highlighting their enzyme inhibitory activities using LC-MS-guided fractionation (Fabian et al. 2024). Further, plant-based nanocarriers reported that ZnO and Ag nanoparticles synthesized using extracts from species like *Phyllanthus emblica* achieved >60 % inhibition of carbohydrate-hydrolyzing enzymes *in vitro* and significantly lowered glucose levels in diabetic mice (Zolkepli et al. 2022). A study on *Phaleria macrocarpa* fruit extracts (Metabolites, 2022) identified sixteen bioactive compounds, including lupenone, m-coumaric acid, and swertianolin, that inhibited α-glucosidase by up to ~82%, with strong molecular docking and dynamic simulation support (Mia et al. 2022). Similarly, Curcuma longa aqueous extracts demonstrated α-amylase and α-glucosidase inhibition (IC_{50} ~93 μg/mL and 45 μg/mL, respectively), with molecular docking highlighting stigmasterol and other constituents as key enzyme binders. Animal-sourced peptides and iridoid glycosides have emerged as new antidiabetic candidates. Mijiti et al. (2025) described cumin seed–derived peptides (PD-30 collection) that inhibited protein-tyrosine phosphatase-1B (PTP1B) with an IC_{50} of ~18.4 μg/mL, suggesting potential to enhance insulin signaling. From microbial sources, polysaccharides isolated from the fungus *Armillaria mellea* demonstrated noteworthy antidiabetic activity; improved glucose uptake and insulin signaling following oral administration in rodent models (Li et al. 2023).

5.4 FATE OF BIOACTIVE COMPOUNDS IN THE HUMAN BODY

Phytochemicals are naturally occurring bioactive compounds in plant-derived foods like fruits, vegetables, cereals, seeds, and nuts, renowned for their health-promoting roles, including antioxidant, antimicrobial, and anticancer effects (Xavier et al. 2024). However, their path inside the human body, from ingestion to excretion, is highly intricate. The ultimate bioefficacy of these compounds hinges on a sequence of physiological events: bioaccessibility, intestinal absorption, metabolic conversion, systemic distribution, and eventual elimination (Adeyanju 2024). Interestingly, the bioactive metabolites generated during metabolism can exhibit equal or superior bioactivity compared to their parent compounds. Upon liberation from the food matrix, phytochemicals traverse the gastrointestinal tract (GIT). They may be absorbed through the intestinal epithelium via passive diffusion or active transport mechanisms (Figure 5.1) (Guo, Liu, and Luo 2024). The polarity of these compounds determines their absorption route: hydrophilic phytochemicals typically enter the bloodstream via the portal vein, while lipophilic compounds are absorbed through the lymphatic system. Once in circulation, they are distributed to various body tissues and organs (Muscolo et al. 2024). Those phytochemicals and their metabolites that are not retained or utilized by the body are eventually excreted through urine or feces.

A primary factor influencing their initial bioavailability is the food matrix itself (Figure 5.2). Many phytochemicals demonstrate poor solubility or are chemically unstable, which can hinder their liberation and absorption in the digestive tract (Liu et al. 2024). For instance, hydrophobic carotenoids exhibit greater bioaccessibility when consumed with dietary fats. Conversely, some fibers can impede the release of these compounds; yet, specific polysaccharides like pectin may enhance their liberation (Frosi et al. 2024). Processing techniques, particularly thermal ones such as boiling or steaming, can degrade heat-sensitive phytochemicals or enhance their accessibility by disrupting cellular structures (Zhang et al. 2024).

The gastrointestinal tract is anatomically and functionally segmented into upper (mouth, stomach, and small intestine) and lower (colon) sections. In the upper tract, ingested foods containing phytochemicals undergo mechanical and enzymatic processing. Following mastication in the oral cavity, the food bolus is subjected to gastric enzymes like pepsin and lipases in the stomach (Hu et al. 2023). Absorption predominantly occurs in the small intestine. However, the inherent physicochemical

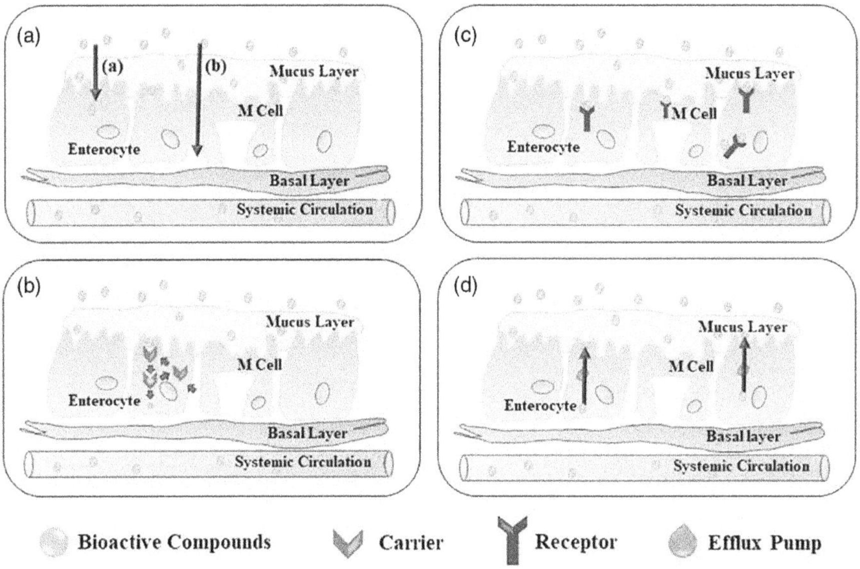

FIGURE 5.1 Transport mechanisms for bioactive components through intestinal absorption cells. (a) passive diffusion in transport mechanisms provides an energy-independent route due to osmotic pressure, either by paracellular or transcellular pathway (b) transportation mediated by carrier permits the introduction or expulsion of BACs from cells due to osmotic pressure gradients used by cellular protein transporters. These carriers include energy-dependent active transportation and easier diffusion; (c) transportation mediated by the receptor, in which BACs are directly bound and internalized by cells to cell-surface receptors; and (d) in the mechanisms of efflux pumps, the most important ATP-based efflux pump of which is a *p*-glycoprotein (*p*-GlyP) in the intestinal epithelium cells. The use of a *p*-GlyP pathway to pump out BACs from enterocytes, leading to limited biological supply. This process has been reported to specifically affect the poor bioavailability of polyphenols. (Adopted from Tripathy et al. 2021; Copyright© 2021 Tripathy, Verma, Thakur, Patel, Srivastav, Singh, Chávez-González and Aguilar.)

nature of many phytochemicals, such as their low solubility and permeability, limits their intestinal uptake. Certain compounds, like anthocyanins, are stable under the stomach's acidic conditions, promoting their progression to the small intestine for absorption. Others, such as resveratrol, degrade in acidic environments, reducing their bioaccessibility and systemic presence (Hu et al. 2023; Jo et al. 2021).

Innovative delivery systems such as nano- and microencapsulation, liposomes, and self-emulsifying drug delivery systems (SEDDS) have been employed to enhance their stability and absorption. These technologies help protect sensitive phytochemicals and increase their permeability and retention in target tissues (Prabhakar et al. 2025; Tripathy et al. 2021, 2023b). Citrus flavonoids like hesperidin and naringin are classic examples of low-bioavailability compounds that mostly reach the colon unaltered when consumed in free form (Zhang et al. 2021). Once absorbed, these bioactive compounds may associate with plasma proteins and disseminate

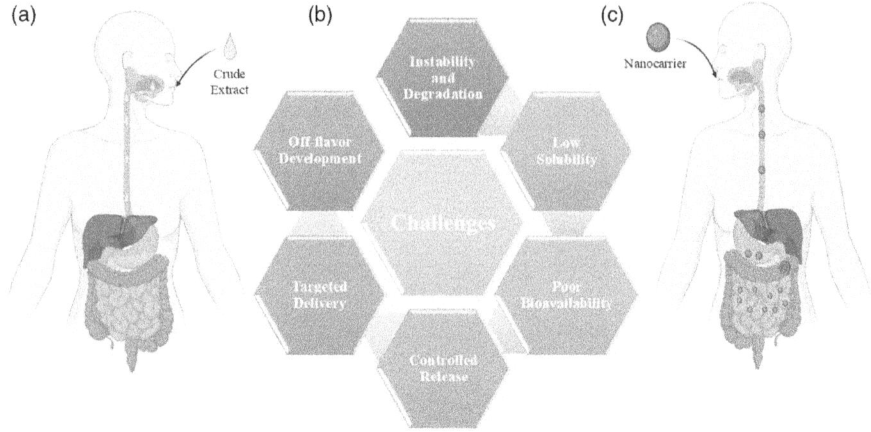

FIGURE 5.2 Different challenges of bioactive compounds in the gastrointestinal tract (GIT), (a) crude extract containing bioactive compounds if ingested directly, the acids of stomach May degrade them and very less amount will be available for intestinal part for its absorption into the human body; (b) major challenges associated with the bioactive compounds in the GIT; (c) the encapsulated crude extract containing bioactive compounds if ingested, the encapsulated matrix will protect the bioactive compounds from the acidic nature of stomach.

throughout the body. Their distribution depends on molecular characteristics such as size, lipophilicity, and affinity for specific transport proteins (Rathaur and SR 2019).

A substantial portion of ingested phytochemicals may escape absorption in the upper gastrointestinal tract and proceed to the colon, where they undergo extensive biotransformation by gut microbiota (Figure 5.3) (Domínguez-Fernández et al. 2022). This microbial fermentation leads to the generation of smaller phenolic compounds and organic acids with distinct bioactivities. Moreover, the interaction between certain phytochemicals—like anthocyanins—and gut microbiota has been shown to modulate intestinal health, influencing microbial composition and increasing the production of beneficial short-chain fatty acids (SCFAs), thereby exerting prebiotic-like effects (Wang and Chen 2013; Salehi et al. 2022). Unabsorbed phytochemicals and their derivatives are finally expelled through fecal matter at the end of the colon.

Once in systemic circulation, phytochemicals are dispersed to tissues via the bloodstream, and their distribution is often mediated by plasma proteins such as albumin (Rabbani and Ahn 2019). The extent of this binding is critical: moderate affinity allows sufficient circulation and tissue uptake, while very high binding may facilitate rapid clearance of unbound molecules through renal excretion (Alsaif et al. 2020). Water-soluble metabolites, typically formed through Phase II conjugation reactions like glucuronidation, sulfation, and methylation, are eliminated mainly through urine and bile. Bile-excreted metabolites may undergo enterohepatic recirculation or be expelled in feces (Singh and Jaglan 2024). Lipophilic metabolites, conversely, may either accumulate in lipid-rich tissues or be routed through bile for eventual elimination.

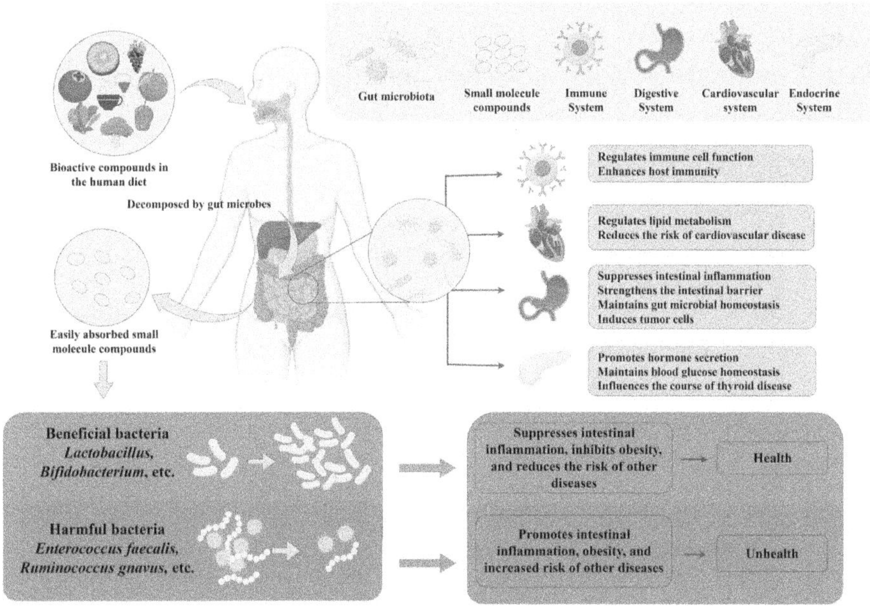

FIGURE 5.3 The relationship between gut microbes and bioactive compounds and their impact on host health. (Adopted from Wang et al. 2024; Copyright © 2024 Wang, Huang, Yang, Yang, Li, Zhong, Wen, He and Li.)

5.5 ENCAPSULATION OF BIOACTIVE COMPOUNDS TO IMPROVE BIOAVAILABILITY IN HUMAN BODY

Bioactive compounds, such as polyphenols, flavonoids, carotenoids, vitamins, and omega-3 fatty acids, are naturally occurring substances found in foods like fruits, vegetables, and herbs, known for their health-promoting properties, including antioxidant, anti-inflammatory, and anticancer effects (Banwo et al. 2021). However, their therapeutic potential is often limited by poor solubility, chemical instability, rapid metabolism, and low absorption in the gastrointestinal tract (GIT), leading to reduced bioavailability (Figure 5.2). Bioavailability refers to the fraction of an ingested compound that reaches the systemic circulation in an active form, a critical factor for its physiological efficacy (Stielow et al. 2023). Encapsulation, a technique that encases these compounds within a protective matrix or carrier, has emerged as a promising strategy to overcome these limitations by enhancing stability, solubility, and controlled release, thereby improving bioavailability and ensuring effective delivery to target tissues (Singh et al. 2025b; Tripathy, Verma, and Srivastav 2024; Tripathy and Srivastav 2023). Table 5.2 shows the advancement of encapsulation system in improving the bioavailability of bioactive compounds.

Encapsulation techniques, such as spray-drying, freeze-drying, coacervation, and nanoencapsulation, protect bioactive compounds from environmental stressors like heat, light, oxygen, and pH variations, which can degrade them during food processing,

TABLE 5.2
Recent Encapsulation Approaches for Improving Bioavailability of Bioactive Compounds

Encapsulation Methods	Encapsulation Matrix	Bioactive Compounds	Effect on Bioavailability	References
Starch Nanocarrier	Rice starch and rice protein isolates	Rosmarinic acid and β-carotene	Encapsulation increased bioaccessibility and bioavailability	Bako et al. (2025)
Nano emulsion	Tween 80, water	*Zingiber Officinale* Essential Oil	Nanoemulsion employed enhances the oral bioavailability and stability of these compounds	Kumar and Bhagwati (2025)
Solid dispersion nanoparticles	PVP VA S 630, Poloxamer 407	curcumin	The formulation shows promise for improving bioavailability	Rawat et al. (2025)
Liposomes	Soy lecithin and β-sitosterol	*Centella asiatica* polyphenols	Encapsulation enhanced the bioavailability of CALE bioactive compounds in liposomes	Tripathy and Srivastav (2025)
Nano emulsion	Miglyol 812, Tween 80, and PEG 400	*Saussurea lappa* extract	Encapsulation improved *in vitro* release and absorption.	Annisa et al. (2024)
Dispersion	Zein, caseinate, and bovine serum albumin	Quercetin (Que) and curcumin (Cur)	Zein-caseinate NPs appeared to show a higher Cur release rate but a lower Que release rate in comparison with zein-BSA NPs	Chen et al. (2024)
Nano capsule	Sodium alginate	Ganoderic acid DM	The incorporation of GA-DM into sodium alginate nanocarriers exhibited a mitigating effect on the cytotoxicity of HaCat and B16, while also demonstrating inhibitory properties against tyrosinase activity and melanin formation	Luo et al. (2024)
Metal Oxide Framework	Zeolite Imidazole Framework (ZIF-8)	2-Aminopyridine	ZIF-8 offers a pH-responsive release of 2AP loaded and 2AP@ZIF-8 caused cell death in HepG2 cells, suggesting novel therapeutic approaches against hepatocellular carcinoma.	Zochedh et al. (2024)
Double emulsion	Ethylcellulose, and dichloromethane	Olive mill pomace	The bioaccessibility (BA) of the encapsulated bioactive compounds was improved around 4 times fold compared to the non-encapsulated one.	Paulo, Tavares, and Santos (2022)

storage, or digestion (Tripathy et al. 2023b). For instance, nanoencapsulation, which involves encapsulating compounds in particles smaller than 1 μm, increases surface area, improving solubility and absorption (Rakshit, Tripathy, and Srivastav 2023). Lipid-based nanocarriers, such as nanoemulsions and nanoliposomes, are particularly effective for hydrophobic compounds like curcumin and quercetin, enhancing their solubility and protecting them from enzymatic degradation in the GIT (Tripathy and Srivastav 2023). Additionally, encapsulation can mask undesirable sensory properties, such as the bitter taste of polyphenols, improving consumer acceptance in functional foods. By controlling the release of bioactive compounds, encapsulation ensures they remain stable until they reach the small intestine, where absorption is optimal, thus maximizing their therapeutic impact (Szente, Sohajda, and Fenyvesi 2021).

Protein-based encapsulation, using materials like whey protein or sodium caseinate, is gaining attention due to its biocompatibility and ability to form stable complexes with bioactive compounds like vitamins and antioxidants. For example, whey protein's ligand-binding properties, particularly through β-lactoglobulin, allow it to form nanoparticles or emulsions that protect sensitive compounds from degradation (Gowda, Sharma, and Wairkar 2025). Similarly, polysaccharide-based carriers, such as chitosan and sodium alginate, provide robust protection and enable targeted delivery by responding to pH changes in the GIT. Recent advancements in co-encapsulation, where multiple bioactive compounds are encased together, have shown synergistic effects, further enhancing bioavailability and therapeutic outcomes (Wen et al. 2025). However, challenges remain, including high production costs, variable encapsulation efficiency, and the need for scalable methods to ensure consistent release and absorption. Ongoing research focuses on optimizing encapsulation systems to address these issues, emphasizing sustainable materials and eco-friendly techniques like supercritical micronization.

The application of encapsulated bioactive compounds extends beyond functional foods to nutraceuticals and pharmaceuticals, with significant implications for disease prevention and health promotion. For instance, encapsulated omega-3 fatty acids in dairy products prevent oxidation, maintaining their nutritional value, while encapsulated probiotics in beverages enhance gut health by surviving acidic stomach conditions (Eratte et al. 2018). Despite these advancements, gaps in understanding the long-term safety and *in vivo* behavior of nanomaterials, particularly metallic nanoparticles, necessitate further research. Regulatory frameworks are also evolving to ensure the safety of encapsulated products in food applications. As consumer demand for natural, health-promoting foods grows, encapsulation continues to be a vital tool for enhancing the bioavailability of bioactive compounds, supporting their integration into everyday diets for improved human health.

5.6 CONCLUSIONS AND FUTURE PERSPECTIVE

Bioactive compounds derived from animal, plant, and microbial sources represent a vast and diverse class of natural substances with significant potential for enhancing human health. These compounds, including polyunsaturated fatty acids, phytochemicals, microbial metabolites, peptides, polysaccharides, and polyphenols, exert a broad spectrum of biological activities such as antioxidant, anticancer, antimicrobial, and antidiabetic effects. Their ability to modulate physiological processes, regulate oxidative stress, and combat pathogenic threats makes them promising candidates for therapeutic, nutraceutical, and functional food applications. Despite their potential, the bioavailability and stability of many bioactive compounds are often limited due to degradation, poor solubility, or insufficient absorption in the human gastrointestinal tract. Understanding the fate of these compounds in the body, including digestion, absorption, metabolism, and excretion, is crucial for optimizing their health benefits. To overcome these limitations, encapsulation technologies such as nanoemulsions, liposomes, and polymeric nanoparticles have been developed, significantly improving bioactive agent delivery and controlled release.

Future research should focus on the synergistic effects of bioactive compounds from different sources, along with detailed mechanistic studies to uncover their modes of action. Integrating omics-based technologies and advanced *in vitro* and *in vivo* models can enhance our understanding of their bioefficacy and safety profiles. Sustainable and green extraction methods should also be prioritized to ensure environmental compatibility and industrial scalability. In conclusion, the strategic incorporation of bioactive compounds into the human diet, supported by advances in formulation and delivery, offers great promise for disease prevention, health promotion, and personalized nutrition. Continued interdisciplinary efforts among food scientists, nutritionists, pharmacologists, and biotechnologists will be pivotal in unlocking the full potential of these natural compounds for future healthcare and functional food innovations.

REFERENCES

Abdulhadi, S. Y., G. Q. Hasan, and R. N. Gergees. 2023. "Molecular Detection and Antimicrobial Activity of Endophytic Fungi Isolated from a Medical Plant *Rosmarinus officinalis*." *arXiv preprint* arXiv:2303.05242.

Abeyrathne, E. D. N. S., K. Nam, X. Huang, and D. U. Ahn. 2022. "Plant- and Animal-Based Antioxidants' Structure, Efficacy, Mechanisms, and Applications: A Review." *Antioxidants* 11 (5): 1025.

Adelusi, T. I., R. O. Adeyemi, M. A. Ashiru, U. C. Divine, I. D. Boyenle, A. Q. K. Oyedele, and I. M. Adewoye. 2023. "Prediction of Antidiabetic Compounds in *Curcuma longa*—In Vitro and In Silico Investigations." *Tropical Journal of Natural Product Research* 7 (10): 4937–44.

Adeyanju, A. A. 2024. "Absorption, Bioavailability, Bioaccessibility, Metabolism, and Excretion of Bioactive Food Phytochemical." In *Plant Food Phytochemicals and Bioactive Compounds in Nutrition and Health*, 99–123. Boca Raton: CRC Press.

Agrawal, R. S., R. C. Ranveer, N. B. Rathod, and N. P. Nirmal. 2023. "Phytochemicals as Bioactive Ingredients for Functional Foods." In *Recent Frontiers of Phytochemicals*, 95–108. Amsterdam: Elsevier.

Al-Dhabi, N. A., A. K. M. Ghilan, G. A. Esmail, M. V. Arasu, V. Duraipandiyan, and K. Ponmurugan. 2019. "Bioactivity Assessment of the Saudi Arabian Marine *Streptomyces* sp. Al-Dhabi-90, Metabolic Profiling and Its In Vitro Inhibitory Property Against Multidrug Resistant and Extended-Spectrum Beta-Lactamase Clinical Bacterial Pathogens." *Journal of Infection and Public Health* 12 (4): 549–56.

Alsaif, N. A., T. A. Wani, A. H. Bakheit, and S. Zargar. 2020. "Multi-Spectroscopic Investigation, Molecular Docking and Molecular Dynamic Simulation of Competitive

Interactions between Flavonoids (Quercetin and Rutin) and Sorafenib for Binding to Human Serum Albumin." *International Journal of Biological Macromolecules* 165: 2451–61.

Ameen, F., S. AlNadhari, and A. A. Al-Homaidan. 2021. "Marine Microorganisms as an Untapped Source of Bioactive Compounds." *Saudi Journal of Biological Sciences* 28 (1): 224–31.

Annisa, R., R. Muti'ah, A. A. Fitrianingsih, B. Fauziyah, N. Rahmadani, and W. Anggraini. 2024. "Evaluation of Self Nanoemulsifying Drug Delivery System from Qusthul Hindi (*Saussurea lappa*) Extract: In Vitro Release and Absorption Assessment." *Biomedical and Pharmacology Journal* 17 (4): 2329–39.

Ansari, P., J. F. Samia, J. T. Khan, M. R. Rafi, M. S. Rahman, A. B. Rahman, and V. Seidel. 2023. "Protective Effects of Medicinal Plant-Based Foods Against Diabetes: A Review on Pharmacology, Phytochemistry, and Molecular Mechanisms." *Nutrients* 15 (14): 3266.

Aslam, J., M. I. Shahzad, H. Y. Aati, K. U. R. Khan, M. Aslam, A. Jamal, and B. A. Ghalloo. 2025. "Exploring the Antidiabetic Potential of *Octhochloa compressa*: A Comprehensive Study on Chemical Profiling, In Vitro, In Vivo and In Silico Pharmacological Properties." *Frontiers in Pharmacology* 16: 1541482.

Bae, S. Y., L. Liao, S. H. Park, W. K. Kim, J. Shin, and S. K. Lee. 2020. "Antitumor Activity of Asperphenin A, a Lipopeptidyl Benzophenone from Marine-Derived *Aspergillus* sp. Fungus, by Inhibiting Tubulin Polymerization in Colon Cancer Cells." *Marine Drugs* 18 (2): 110.

Bako, H. K., I. H. Ibeogu, M. S. Yar, and C. Li. 2025. "Bioavailability of Encapsulated Rosmarinic Acid and β-Carotene in Beef Sausages: A Caco-2 Cell Study." *Food Research International* 200: 115430.

Banwo, K., A. O. Olojede, A. T. Adesulu-Dahunsi, D. K. Verma, M. Thakur, S. Tripathy, and G. L. Utama. 2021. "Functional Importance of Bioactive Compounds of Foods with Potential Health Benefits: A Review on Recent Trends." *Food Bioscience* 43: 101320.

Barzkar, N., S. Sukhikh, and O. Babich. 2024. "Study of Marine Microorganism Metabolites: New Resources for Bioactive Natural Products." *Frontiers in Microbiology* 14: 1285902.

Bauer-Estrada, K., C. Sandoval-Cuellar, Y. Rojas-Muñoz, and M. X. Quintanilla-Carvajal. 2023. "The Modulatory Effect of Encapsulated Bioactives and Probiotics on Gut Microbiota: Improving Health Status through Functional Food." *Food & Function* 14 (1): 32–55.

Benedé, S., and E. Molina. 2020. "Chicken Egg Proteins and Derived Peptides with Antioxidant Properties." *Foods* 9 (6): 735.

Bhat, Z. F., S. Kumar, and H. F. Bhat. 2015. "Bioactive Peptides of Animal Origin: A Review." *Journal of Food Science and Technology* 52 (9): 5377–92.

Bontempo, P., L. Capasso, L. De Masi, A. Nebbioso, and D. Rigano. 2024. "Therapeutic Potential of Natural Compounds Acting through Epigenetic Mechanisms in Cardiovascular Diseases: Current Findings and Future Directions." *Nutrients* 16 (15): 2399.

Brinck, J. E., A. K. Sinha, M. F. Laursen, L. O. Dragsted, J. Raes, R. V. Uribe, and T. R. Licht. 2025. "Intestinal pH: A Major Driver of Human Gut Microbiota Composition and Metabolism." *Nature Reviews Gastroenterology & Hepatology*, *22*(9), 639–656.

Carroll, A. R., B. R. Copp, R. A. Davis, R. A. Keyzers, and M. R. Prinsep. 2022. "Marine Natural Products." *Natural Product Reports* 39 (6): 1122–71.

Chandra, P., R. K. Sharma, and D. S. Arora. 2020. "Antioxidant Compounds from Microbial Sources: A Review." *Food Research International* 129: 108849.

Chen, X., S. Pan, F. Li, X. Xu, and H. Xing. 2022. "Plant-Derived Bioactive Compounds and Potential Health Benefits: Involvement of the Gut Microbiota and Its Metabolic Activity." *Biomolecules* 12 (12): 1871.

Chen, X., Z. X. Wang, Y. Zhang, W. Liu, Z. Hao-Song, Y. C. Wu, and H. J. Li. 2024. "Caseinate vs Bovine Serum Albumin on Stabilizing Zein Nanoparticles for Co-Entrapping Quercetin and Curcumin." *Colloids and Surfaces A: Physicochemical and Engineering Aspects* 683: 133041.

Chintada, V., and N. Golla. 2025. "Exploring the Therapeutic Potential of Bioactive Compounds from Plant Sources." In *Biotechnological Intervention in Production of Bioactive Compounds: Biosynthesis, Characterization and Applications*, 229–47. Cham: Springer Nature Switzerland.

Clemente-Suárez, V. J., A. Bustamante-Sanchez, A. Rubio-Zarapuz, A. Martín-Rodríguez, J. F. Tornero-Aguilera, and A. I. Beltrán-Velasco. 2025. "Biomimetic Strategies for Nutraceutical Delivery: Advances in Bionanomedicine for Enhanced Nutritional Health." *Biomimetics* 10 (7): 426.

Cuchillo-Hilario, M., M. I. Fournier-Ramírez, M. Díaz Martínez, S. Montaño Benavides, M. C. Calvo-Carrillo, S. Carrillo Domínguez, and … C. Delgadillo-Puga. 2024. "Animal Food Products to Support Human Nutrition and to Boost Human Health: The Potential of Feedstuffs Resources and Their Metabolites as Health-Promoters." *Metabolites* 14 (9): 496.

da Silva, S. B., D. de Souza, and L. D. Lacerda. 2019. "Food Applications of Chitosan and Its Derivatives." In *Chitin and Chitosan: Properties and Applications*, 315–47. Amsterdam: Elsevier.

Dixit, V., S. W. Joseph Kamal, P. Bajrang Chole, D. Dayal, K. K. Chaubey, A. K. Pal, and R. K. Bachheti. 2023. "Functional Foods: Exploring the Health Benefits of Bioactive Compounds from Plant and Animal Sources." *Journal of Food Quality* 2023 (1): 5546753.

Domínguez-Fernández, M., P. Y. T. Yang, I. A. Ludwig, M. N. Clifford, C. Cid, and A. Rodriguez-Mateos. 2022. "In Vivo Study of the Bioavailability and Metabolic Profile of (Poly)phenols after Sous-Vide Artichoke Consumption." *Food Chemistry* 367: 130620.

Duda-Madej, A. 2024. *Antimicrobial Activity of Different Plant Extracts, Plant-Derived Compounds and Synthetic Derivatives of Natural Compounds on Pathogenic Microorganisms*. Basel: MDPI.

El-Gazzar, N., R. Farouk, N. S. Diab, G. Rabie, and B. Sitohy. 2025. "Antimicrobial and Antiproliferative Activity of Biosynthesized Manganese Nanocomposite with Amide Derivative Originated by Endophytic Aspergillus Terreus." *Microbial Cell Factories* 24: 37.

El-Zehery, H. R., N. M. Ashry, A. A. Faiesal, M. S. Attia, M. A. Abdel-Maksoud, M. A. El-Tayeb, and N. K. El-Dougdoug. 2024. "Antibacterial and Anticancer Potential of Bioactive Compounds and Secondary Metabolites of Endophytic Fungi Isolated from *Anethum graveolens*." *Frontiers in Microbiology* 15: 1448191.

Eratte, D., K. Dowling, C. J. Barrow, and B. Adhikari. 2018. "Recent Advances in the Microencapsulation of Omega-3 Oil and Probiotic Bacteria through Complex Coacervation: A Review." *Trends in Food Science & Technology* 71: 121–31.

Fabian, M. C. P., R. M. N. Astorga, A. A. G. Atis, L. A. E. Pilapil, and C. C. Hernandez. 2024. "Anti-Diabetic and Anti-Inflammatory Bioactive Hits from *Coriaria Intermedia* Matsum. Stem and *Dracontomelon Dao* (Blanco) Merr. & Rolfe Bark through Bioassay-Guided Fractionation and Liquid Chromatography-Tandem Mass Spectrometry." *Frontiers in Pharmacology* 15: 1349725.

Fornari Laurindo, L., L. Fornari Laurindo, V. Dogani Rodrigues, J. da Silva Camarinha Oliveira, B. Leme Boaro, A. Cressoni Araújo, and S. M. Barbalho. 2025. "Evaluating the Effects of Seed Oils on Lipid Profile, Inflammatory and Oxidative Markers, and Glycemic Control of Diabetic and Dyslipidemic Patients: A Systematic Review of Clinical Studies." *Frontiers in Nutrition* 12: 1502815.

Frosi, I., L. Ferron, R. Colombo, and A. Papetti. 2024. "Natural Carriers: Recent Advances in Their Use to Improve the Stability and Bioaccessibility of Food Active Compounds." *Critical Reviews in Food Science and Nutrition* 64 (17): 5700–18.

Genilloud, O. 2017. "Actinomycetes: Still a Source of Novel Antibiotics." *Natural Product Reports* 34 (10): 1203–32.

Gowda, C. M., S. Sharma, and S. Wairkar. 2025. "Advanced Drug Delivery Systems Utilizing β-Lactoglobulin: An Efficient Protein-Based Drug Carrier." *Biopolymers* 116 (2): e70005.

Guillocheau, E., P. P. Legrand, and V. V. Rioux. 2019. "Benefits of Natural Dietary Trans Fatty Acids towards Inflammation, Obesity and Type 2 Diabetes: Defining the n-7 Trans Fatty Acid Family." *OCL Oilseeds and Fats Crops and Lipids* 26: 46.

Gulcin, İ. 2025. "Antioxidants: A Comprehensive Review." *Archives of Toxicology*, *99*(5), 1893–1997.

Guo, Z., Y. Liu, and Y. Luo. 2024. "Mechanisms of Carotenoid Intestinal Absorption and the Regulation of Dietary Lipids: Lipid Transporter-Mediated Transintestinal Epithelial Pathways." *Critical Reviews in Food Science and Nutrition* 64 (7): 1791–816.

Haruna, A., and S. M. Yahaya. 2021. "Recent Advances in the Chemistry of Bioactive Compounds from Plants and Soil Microbes: A Review." *Chemistry Africa* 4 (2): 231–48.

Helmi, N. R. 2025. "Exploring the Diversity and Antimicrobial Potential of Actinomycetes Isolated from Different Environments in Saudi Arabia: A Systematic Review." *Frontiers in Microbiology* 16: 1568899.

Heredia, J. B., E. P. Gutiérrez-Grijalva, and L. A. Cabanillas-Bojórquez, eds. 2025. *Bioactive Compounds Extraction from Marine Resources and Wastes*. Cham: Springer.

Hou, Y., H. Wang, J. Wu, H. Guo, and X. Chen. 2024. "Dissecting the Pleiotropic Roles of Reactive Oxygen Species (ROS) in Lung Cancer: From Carcinogenesis toward Therapy." *Medicinal Research Reviews* 44 (4): 1566–95.

Hu, Y., Q. Lin, H. Zhao, X. Li, S. Sang, D. J. McClements, and … C. Qiu. 2023. "Bioaccessibility and Bioavailability of Phytochemicals: Influencing Factors, Improvements, and Evaluations." *Food Hydrocolloids* 135: 108165.

Jo, M., C. Ban, K. K. Goh, and Y. J. Choi. 2021. "Enhancement of the Gut-Retention Time of Resveratrol Using Waxy Maize Starch Nanocrystal-Stabilized and Chitosan-Coated Pickering Emulsions." *Food Hydrocolloids* 112: 106291.

Kangra, K., S. Kakkar, V. Mittal, V. Kumar, N. Aggarwal, H. Chopra, and … V. Garg. 2025. "Incredible Use of Plant-Derived Bioactives as Anticancer Agents." *RSC Advances* 15 (3): 1721–46.

Karthik, Y., M. I. Kalyani, S. Krishnappa, K. Ramakrishna, S. M. Sayed, O. M. Aharthy, and … M. Mushtaq. 2023. "Promising Bioactive Metabolites of Mangrove Inhabitant *Streptomyces tauricus* and Prostate Cancer PC3 Cell Inhibition by Antimicrobial Peptides." *Frontiers in Microbiology* 14: 1152985.

Kulczyński, B., A. Sidor, and A. Gramza-Michałowska. 2019. "Characteristics of Selected Antioxidative and Bioactive Compounds in Meat and Animal Origin Products." *Antioxidants* 8 (9): 335.

Kumar, V., and A. Bhagwati Devi. 2025. "Formulation of *Zingiber Officinale* Essential Oil Based Nanoemulsion and Evaluation of Its Neuroprotective Effect against Hydrogen Peroxide Induced Cytotoxicity in SH-SY5Y Cell Line." *Cuestiones de Fisioterapia* 54 (2): 2867–80.

Kumari, S., and R. Kishor. 2020. "Chitin and Chitosan: Origin, Properties, and Applications." In *Handbook of Chitin and Chitosan*, 1–33. Amsterdam: Elsevier.

Kussmann, M., D. H. Abe Cunha, and S. Berciano. 2023. "Bioactive Compounds for Human and Planetary Health." *Frontiers in Nutrition* 10: 1193848.

Lassiter, K., G. Mullenix, S. Rochell, G. Tellez, M. T. Kidd, S. Dridi, and W. Bottje. 2024. "Effects of a Low Protein Soybean Meal Diet with and without *Spirulina platensis* Freshwater Microalgae on Antioxidant Systems in Broiler Liver and Muscle Tissue." *Medical Research Archives* 12 (10): 1–15.

Li, X., J. Zhu, T. Wang, J. Sun, T. Guo, L. Zhang, and … X. Xia. 2023. "Antidiabetic Activity of *Armillaria mellea* Polysaccharides: Joint Ultrasonic and Enzyme Assisted Extraction." *Ultrasonics Sonochemistry* 95: 106370.

Liu, L., D. J. McClements, X. Liu, and F. Liu. 2024. "Overcoming Biopotency Barriers: Advanced Oral Delivery Strategies for Enhancing the Efficacy of Bioactive Food Ingredients." *Advanced Science* 11 (44): 2401172.

Luo, S., Y. Song, Z. Zhou, X. Y. Xu, N. Jiang, Y. J. Gao, and X. Luo. 2024. "Optimization, Characterization and Evaluation of Sodium Alginate Nanoparticles for Ganoderic Acid DM Encapsulation: Inhibitory Activity on Tyrosinase Activity and Melanin Formation." *International Journal of Biological Macromolecules* 271: 132717.

Ma, M., H. Ge, W. Yi, B. Wu, and Z. Zhang. 2020. "Bioactive Drimane Sesquiterpenoids and Isocoumarins from the Marine-Derived Fungus *Penicillium minioluteum* ZZ1657." *Tetrahedron Letters* 61 (7): 151504.

Ma, Z. F., C. Fu, and Y. Y. Lee. 2025. "The Modulatory Role of Bioactive Compounds in Functional Foods on Inflammation and Metabolic Pathways in Chronic Diseases." *Foods* 14 (5): 821.

Maftei, N. M., C. R. Raileanu, A. A. Balta, L. Ambrose, M. Boev, D. B. Marin, and E. L. Lisa. 2024. "The Potential Impact of Probiotics on Human Health: An Update on Their Health-Promoting Properties." *Microorganisms* 12 (2): 234.

Maniglia, B. C., E. A. Rebelatto, K. S. Andrade, A. Zielinski, and C. J. de Andrade. 2021. "Polyphenols." In *Food Bioactives and Health*, 1–39. Cham: Springer.

Martirosyan, D., T. Lampert, and M. Lee. 2022. "A Comprehensive Review on the Role of Food Bioactive Compounds in Functional Food Science." *Functional Food Science* 2 (3): 64–78.

Mashal, S., A. Siddiqua, N. Ullah, R. Baloch, M. Khan, S. Z. U. Hasnain, and … A. Amin. 2024. "Bioactive Plant Waste Components Targeting Oral Bacterial Pathogens as a Promising Strategy for Biofilm Eradication." *Frontiers in Chemistry* 12: 1406869.

Mia, M. A. R., Q. U. Ahmed, S. Ferdosh, A. B. M. Helaluddin, M. S. Awal, M. N. Sarian, and Z. A. Zakaria. 2022. "In Vitro, in Silico and Network Pharmacology Mechanistic Approach to Investigate the α-Glucosidase Inhibitors Identified by Q-ToF-LCMS from *Phaleria macrocarpa* Fruit Subcritical CO_2 Extract." *Metabolites* 12 (12): 1267.

Mijiti, Y., S. Subuer, N. Ullah, E. Ismoil, H. Taoerdahong, and P. Rouzi. 2025. "Research on Cumin Peptides Using PBS Extraction and Their Multifunctional Bioactivities." *arXiv* preprint arXiv:2502.08264.

Monciardini, P., M. Iorio, S. Maffioli, M. Sosio, and S. Donadio. 2014. "Discovering New Bioactive Molecules from Microbial Sources." *Microbial Biotechnology* 7 (3): 209–220.

Mondal, S., N. P. P. Soumya, S. Mini, and S. K. Sivan. 2021. "Bioactive Compounds in Functional Food and Their Role as Therapeutics." *Bioactive Compounds in Health and Disease* 4 (3): 24–39.

Muscolo, A., O. Mariateresa, T. Giulio, and R. Mariateresa. 2024. "Oxidative Stress: The Role of Antioxidant Phytochemicals in the Prevention and Treatment of Diseases." *International Journal of Molecular Sciences* 25 (6): 3264.

Nagapan, T. S., A. R. Ghazali, D. F. Basri, and W. N. Lim. 2018. "Photoprotective Effect of Stilbenes and Its Derivatives against Ultraviolet Radiation-Induced Skin Disorders." *Biomedical and Pharmacology Journal* 11 (3): 1199–208.

Nakra, S., S. Tripathy, and P. P. Srivastav. 2025a. "Drying as a Preservation Strategy for Medicinal Plants: Physicochemical and Functional Outcomes for Food and Human Health." *Phytomedicine Plus* 5(1): 100762.

Nakra, S., S. Tripathy, and P. P. Srivastav. 2025b. "Green and Sustainable Extraction of Bioactive Compounds from *Centella asiatica* Leaves Using Microwave Pretreatment and Ultrasonication: Kinetics, Process Optimization, and Biological Activity." *Food Biophysics* 20 (1): 1–18.

Naveen, K. V., K. Saravanakumar, A. Sathiyaseelan, and M. H. Wang. 2023. "Comparative Analysis of the Antioxidant, Antidiabetic, Antibacterial, Cytoprotective Potential and Metabolite Profile of Two Endophytic *Penicillium* spp." *Antioxidants* 12 (2): 248.

Panda, S. R., A. Meher, G. Prusty, S. Behera, and B. R. Prasad. 2025. "Antibacterial Properties and Therapeutic Potential of Few Medicinal Plants: Current Insights and Challenges." *Discover Plants* 2 (1): 21.

Paulo, F., L. Tavares, and L. Santos. 2022. "In Vitro Digestion, Bioaccessibility, and Release Kinetics Studies of Encapsulated Bioactive Compounds Obtained from Olive Mill Pomace." *Journal of Food Measurement and Characterization* 16 (6): 4880–95.

Prabhakar, P., S. Tripathy, D. K. Verma, S. Singh, M. Thakur, A. K. Singh, and C. N. Aguilar. 2025. "Trends and Advances in Liposome Formulation Technology with an Emphasis on Ensuring Safety and Quality in Food and Drug Applications." *Food Bioscience* 67: 106913. https://doi.org/10.1016/j.fbio.2025.106913

Qu, B., J. Yuan, X. Liu, S. Zhang, X. Ma, and L. Lu. 2024. "Anticancer Activities of Natural Antimicrobial Peptides from Animals." *Frontiers in Microbiology* 14: 1321386. https://doi.org/10.3389/fmicb.2023.1321386

Rabbani, G., and S. N. Ahn. 2019. "Structure, Enzymatic Activities, Glycation and Therapeutic Potential of Human Serum Albumin: A Natural Cargo." *International Journal of Biological Macromolecules* 123: 979–90. https://doi.org/10.1016/j.ijbiomac.2018.11.053

Rakshit, M., S. Tripathy, and P. P. Srivastav. 2023. "Encapsulation of Natural Polyphenols for Food Applications." In *Novel Processing Methods for Plant-Based Health Foods*, 123–62. Apple Academic Press.

Rani, A., K. C. Saini, F. Bast, S. Mehariya, S. K. Bhatia, R. Lavecchia, and A. Zuorro. 2021. "Microorganisms: A Potential Source of Bioactive Molecules for Antioxidant Applications." *Molecules* 26 (4): 1142. https://doi.org/10.3390/molecules26041142

Rasheed, H. A., A. Rehman, A. Karim, F. Al-Asmari, H. Cui, and L. Lin. 2024. "A Comprehensive Insight into Plant-Derived Extracts/Bioactives: Exploring Their Antimicrobial Mechanisms and Potential for High-Performance Food Applications." *Food Bioscience* 60: 104035. https://doi.org/10.1016/j.fbio.2024.104035

Rathaur, P., and S.R. Kaid J. 2019. "Metabolism and Pharmacokinetics of Phytochemicals in the Human Body." *Current Drug Metabolism* 20 (14): 1085–102. https://doi.org/10.2174/1389200221666200103090757

Rawat, A., S. Chauhan, R. P. Singh, S. Monika, V. Gupta, and Jhawat. 2025. "Development and In-vitro Optimization of Telmisartan-Curcumin Solid Dispersion Nanoparticles for the Management of Diabetic Nephropathy Using DoE Approach." *Drug Development and Industrial Pharmacy*, 1–18. https://doi.org/10.1080/03639045.2025.XXXXX

Rout, S., S. Tambe, R. K. Deshmukh, S. Mali, J. Cruz, P. P. Srivastav, and M. S. de Oliveira. 2022. "Recent Trends in the Application of Essential Oils: The Next Generation of Food Preservation and Food Packaging." *Trends in Food Science & Technology* 129: 421–39. https://doi.org/10.1016/j.tifs.2022.09.015

Saha, A., B. M. A. Desai, and P. Biswas. 2024. "Multi-Ligand Simultaneous Docking Analysis of *Moringa oleifera* Phytochemicals Reveals Enhanced BCL-2 Inhibition via Synergistic Action." In *2024 5th International Conference on Biomedical Engineering (IBIOMED)*, 51–6. IEEE. https://doi.org/10.1109/IBIOMED60756.2024.10384645

Salehi, B., Sharifi-Rad, J., Cappellini, F., Reiner, Ž., Zorzan, D., Imran, M., ... & Maroyi, A. (2020). The Therapeutic Potential of Anthocyanins: Current Approaches Based on their Molecular Mechanism of Action. *Frontiers in Pharmacology* 11, 567172.

Santos-Buelga, C., A. M. González-Paramás, and S. González-Manzano. 2023. "Phenolic Acids and Derivatives: Description, Sources, Properties, and Applications." In *Natural Secondary Metabolites: From Nature, Through Science, to Industry*, 37–72. Cham: Springer International Publishing. https://doi.org/10.1007/978-3-031-29134-8_2

SeyedAlinaghi, S., E. Mehraeen, P. Mirzapour, and … S. Jahanfar. 2025. "A Systematic Review on Natural Products with Antimicrobial Potential against WHO's Priority Pathogens." *European Journal of Medical Research* 30: 525. https://doi.org/10.1186/s40001-025-02717-x

Shahidi, F., and A. Saeid. 2025. "Bioactivity of Marine-Derived Peptides and Proteins: A Review." *Marine Drugs* 23 (4): 157. https://doi.org/10.3390/md23040157

Singh, S. M., S. Tripathy, B. M. Ghodki, and P. P. Srivastav. 2025b. "Optimization of Wall Material Composition for Encapsulating Bioactive Compounds from Giloy (*Tinospora cordifolia*) Stems." *Journal of Food Processing and Preservation* 2025 (1): e6459848. https://doi.org/10.1155/2025/6459848

Singh, S. M., S. Tripathy, and P. P. Srivastav. 2025a. "Bioactive Compound Extraction from Giloy Leaves and Steam Using Ultrasound: Bioactivity, Antimicrobial, and LC–MS/MS Study." *Food Science and Biotechnology* 34 (9): 1835–47. https://doi.org/10.1007/s10068-025-01779-2

Singh, S. P., and D. Jaglan. 2024. *Pharmacology: e-Book for D. Pharm 2nd Year | PCI | Coloured Pages*. Thakur Publication Private Limited.

Singh, V. 2022. "Microbial Bioactive Components: Sources, Applications, and Sustainability." In *Bioactive Components: A Sustainable System for Good Health and Well-Being*, 103–17. Singapore: Springer Nature Singapore. https://doi.org/10.1007/978-981-16-8381-7_6

Stielow, M., A. Witczyńska, N. Kubryń, Ł. Fijałkowski, J. Nowaczyk, and A. Nowaczyk. 2023. "The Bioavailability of Drugs—The Current State of Knowledge." *Molecules* 28 (24): 8038. https://doi.org/10.3390/molecules28248038

Sugumaran, A., R. Pandiyan, P. Kandasamy, M. G. Antoniraj, I. Navabshan, B. Sakthivel, and … C. Ngamcharussrivichai. 2022. "Marine Biome-Derived Secondary Metabolites, A Class of Promising Antineoplastic Agents: A Systematic Review on Their Classification, Mechanism of Action and Future Perspectives." *Science of the Total Environment* 836: 155445. https://doi.org/10.1016/j.scitotenv.2022.155445

Sukhikh, S., O. Babich, A. Prosekov, O. Kalashnikova, S. Noskova, A. Bakhtiyarova, and … S. Ivanova. 2023. "Antidiabetic Properties of Plant Secondary Metabolites." *Metabolites* 13 (4): 513.

Szente, L., T. Sohajda, and É. Fenyvesi. 2021. "Encapsulation for Masking Off-Flavor and Off-Tasting in Food Production." In *Functionality of Cyclodextrins in Encapsulation for Food Applications*, 223–53. Cham: Springer International Publishing.

Toldrá, F., M. Gallego, M. Reig, M. C. Aristoy, and L. Mora. 2020. "Recent Progress in Enzymatic Release of Peptides in Foods of Animal Origin and Assessment of Bioactivity." *Journal of Agricultural and Food Chemistry* 68 (46): 12842–55.

Tonolo, F., A. Folda, L. Cesaro, V. Scalcon, O. Marin, S. Ferro, and M. P. Rigobello. 2020. "Milk-Derived Bioactive Peptides Exhibit Antioxidant Activity through the Keap1-Nrf2 Signaling Pathway." *Journal of Functional Foods* 64: 103696.

Tripathi, V. C., S. Horam, A. Singh, M. Lata, T. J. Reddy, J. Arockiaraj, and M. Pasupuleti. 2020. "The Discovery of Antioxidants in Marine Microorganisms and Their Protective Effects on the Hepatic Cells from Chemical-Induced Oxidative Stress." *Free Radical Research* 54 (2–3): 150–61.

Tripathy, S., A. R. Patel, D. K. Verma, S. Singh, G. L. Utama, M. Thakur, and C. N. Aguilar. 2023a. "Strategies, Trends, and Technological Advancements in Microbial Bioreactor System for Probiotic Products." In *Microbial Bioreactors for Industrial Molecules*, 443–63. Amsterdam: Elsevier.

Tripathy, S., and P. P. Srivastav. 2023. "Encapsulation of *Centella asiatica* Leaf Extract in Liposome: Study on Structural Stability, Degradation Kinetics and Fate of Bioactive Compounds during Storage." *Food Chemistry Advances* 2: 100202.

Tripathy, S., and P. P. Srivastav. 2025. "Sustainable Liposomal Delivery of *Centella asiatica* Polyphenols: β-Sitosterol Stabilization, LC-MS/MS Profiling, and Simulated Release Study." *Sustainable Food Technology*, *3*(4), 1053–1063.

Tripathy, S., D. K. Verma, A. K. Gupta, P. P. Srivastav, A. R. Patel, M. L. C. González, and C. N. Aguilar. 2023b. "Nanoencapsulation of Biofunctional Components as a Burgeoning Nanotechnology-Based Approach for Functional Food Development: A Review." *Biocatalysis and Agricultural Biotechnology* 53: 102890.

Tripathy, S., D. K. Verma, A. K. Gupta, P. P. Srivastav, M. L. Chávez-González, and C. N. Aguilar. 2025. "Immune-Boosting Potential of Food Bioactives." In *Food Bioactives and Nutraceuticals: Dietary Medicine for Human Health and Well-Beings*, 343–82. Singapore: Springer Nature Singapore.

Tripathy, S., D. K. Verma, and P. P. Srivastav. 2024. "Encapsulation of Enzymes-Based on Biopolymers and Biochemical Systems: Progress and Perspective." In *Enzymatic Processes for Food Valorization*, 321–46. Amsterdam: Elsevier.

Tripathy, S., D. K. Verma, M. Thakur, A. R. Patel, P. P. Srivastav, S. Singh, and C. N. Aguilar. 2021. "Encapsulated Food Products as a Strategy to Strengthen Immunity Against COVID-19." *Frontiers in Nutrition* 8: 673174.

Tripathy, S., D. K. Verma, M. Thakur, N. Chakravorty, S. Singh, and P. P. Srivastav. 2022. "Recent Trends in Extraction, Identification and Quantification Methods of *Centella asiatica* Phytochemicals with Potential Applications in Food Industry and Therapeutic Relevance: A Review." *Food Bioscience* 49: 101864.

Tufail, T., H. Bader Ul Ain, J. Ashraf, S. Mahmood, S. Noreen, A. Ijaz, and M. A. Abdullahi. 2025. "Bioactive Compounds in Seafood: Implications for Health and Nutrition." *Food Science & Nutrition* 13 (4): e70181.

Tumilaar, S. G., A. Hardianto, H. Dohi, and D. Kurnia. 2024. "A Comprehensive Review of Free Radicals, Oxidative Stress, and Antioxidants: Overview, Clinical Applications, Global Perspectives, Future Directions, and Mechanisms of Antioxidant Activity of Flavonoid Compounds." *Journal of Chemistry* 2024 (1): 5594386.

Venegas-Ortega, M. G., A. C. Flores-Gallegos, J. L. Martínez-Hernández, C. N. Aguilar, and G. V. Nevárez-Moorillón. 2019. "Production of Bioactive Peptides from Lactic Acid Bacteria: A Sustainable Approach for Healthier Foods." *Comprehensive Reviews in Food Science and Food Safety* 18 (4): 1039–51.

Verma, D. K., and M. Thakur, eds. 2021. *Phytochemicals in Food and Health: Perspectives for Research and Technological Development*. Boca Raton: CRC Press.

Vignesh, A., T. C. Amal, A. Sarvalingam, and K. Vasanth. 2024. "A Review on the Influence of Nutraceuticals and Functional Foods on Health." *Food Chemistry Advances* 4: 100749.

Wali, A., G. Yanhua, U. Ishimov, A. Yili, H. A. Aisa, and S. Salikhov. 2020. "Isolation and Identification of Three Novel Antioxidant Peptides from the Bactrian Camel Milk Hydrolysates." *International Journal of Peptide Research and Therapeutics* 26: 641–50.

Wali, A. F., Y. Al Dhaheri, J. Ramakrishna Pillai, A. Mushtaq, P. G. Rao, S. A. Rabbani, and D. A. A. Farraj. 2020. "LC-MS Phytochemical Screening, *In Vitro* Antioxidant, Antimicrobial and Anticancer Activity of Microalgae *Nannochloropsis oculata* Extract." *Separations* 7 (4): 54.

Wang, L., and C. Chen. 2013. "Emerging Applications of Metabolomics in Studying Chemopreventive Phytochemicals." *The AAPS Journal* 15: 941–50.

Wang, L., C. Li, G. Yu, Z. Sun, G. Zhang, Q. Gu, and … D. Li. 2019. "Dicitrinones E and F, Citrinin Dimers from the Marine Derived Fungus *Penicillium citrinum* HDN-152-088." *Tetrahedron Letters* 60 (44): 151182.

Wang, Q., H. Huang, Y. Yang, X. Yang, X. Li, W. Zhong, and … J. Li. 2024. "Reinventing Gut Health: Leveraging Dietary Bioactive Compounds for the Prevention and Treatment of Diseases." *Frontiers in Nutrition* 11: 1491821.

Wen, C., J. Tang, L. Cao, M. Fan, X. Lin, G. Liu, and … X. Xu. 2025. "Strategic Approaches for Co-Encapsulation of Bioactive Compounds: Technological Advances and Mechanistic Insight." *Foods* 14 (12): 2024.

Widowati, I., M. Zainuri, H. P. Kusumaningrum, R. Susilowati, Y. Hardivillier, V. Leignel, and J. L. Mouget. 2017. "Antioxidant Activity of Three Microalgae *Dunaliella salina, Tetraselmis chuii* and *Isochrysis galbana* Clone Tahiti." *IOP Conference Series: Earth and Environmental Science* 55 (1): 012067.

Wijesekara, T., J. Luo, and B. Xu. 2024. "Critical Review on Anti-Inflammation Effects of Saponins and Their Molecular Mechanisms." *Phytotherapy Research* 38 (4): 2007–22.

Wojtunik-Kulesza, K., A. Oniszczuk, T. Oniszczuk, M. Combrzyński, D. Nowakowska, and A. Matwijczuk. 2020. "Influence of *In Vitro* Digestion on Composition, Bioaccessibility and Antioxidant Activity of Food Polyphenols—A Non-Systematic Review." *Nutrients* 12 (5): 1401.

Xavier, J. R., B. Sameer, D. Gupta, S. Mehta, and O. P. Chauhan. 2024. "Bioactive Compounds of Foods: Phytochemicals and Peptides." *Food and Humanity* 2: 100354.

Xu, P., L. Ding, J. Wei, Q. Li, M. Gui, X. He, and … H. Jin. 2020. "A New Aquatic Pathogen Inhibitor Produced by the Marine Fungus *Aspergillus* sp. LS116." *Aquaculture* 520: 734670.

Yadav, D., and P. S. Negi. 2021. "Bioactive Components of Mushrooms: Processing Effects and Health Benefits." *Food Research International* 148: 110599.

Yan, K., F. Guo, M. J. Kainz, F. Li, W. Gao, S. E. Bunn, and Y. Zhang. 2024. "The Importance of Omega-3 Polyunsaturated Fatty Acids as High-Quality Food in Freshwater Ecosystems with Implications of Global Change." *Biological Reviews* 99 (1): 200–18.

Zhang, J., B. Li, Y. Qin, L. Karthik, G. Zhu, C. Hou, and … X. Liu. 2020. "A New Abyssomicin Polyketide with Anti-Influenza A Virus Activity from a Marine-Derived *Verrucosispora* sp. MS100137." *Applied Microbiology and Biotechnology* 104: 1533–43.

Zhang, M., S. Zhu, W. Yang, Q. Huang, and C. T. Ho. 2021. "The Biological Fate and Bioefficacy of Citrus Flavonoids: Bioavailability, Biotransformation, and Delivery Systems." *Food & Function* 12 (8): 3307–23.

Zhang, Q. F., W. J. Wang, S. Cai, and G. D. Zheng. 2024. "Plant Secondary Metabolites in Food: Absorption, Metabolism and Effects on Glucolipid Metabolism." *Frontiers in Nutrition* 11: 1382677.

Zochedh, A., K. Chandran, K. Arumugam, M. Priya, M. Anbazhagan, A. B. Sultan, and … T. Kathiresan. 2024. "Encapsulating 2-Aminopyridine with Zeolite Imidazole Framework Nanoparticles for Induction of Cell Death via pH-Responsive Delivery in Hepatocellular Carcinoma (HepG2) Cells." *Polycyclic Aromatic Compounds* 44 (10): 7031–49.

Zolkepli, H., R. T. Widodo, S. Mahmood, N. Salim, K. Awang, N. Ahmad, and R. Othman. 2022. "A Review on the Delivery of Plant-Based Antidiabetic Agents Using Nanocarriers: Current Status and Their Role in Combatting Hyperglycaemia." *Polymers* 14 (15): 2991.

6 Herbal Drug Development

Ankit Kumar, Rahul Kumar Rout, Pawan Prabhakar, and Pavuluri Srinivasa Rao

6.1 HERBAL DRUGS: AN INTRODUCTION

Herbal medicine, also called phytomedicine or botanical medicine, involves the therapeutic use of plants and plant-derived substances. It is among the oldest healthcare practices, forming an integral part of systems such as Ayurveda, Traditional Chinese Medicine (TCM), and Unani (Sofowora, Ogunbodede, and Onayade 2013). Herbal treatments make use of various plant components, including roots, leaves, bark, seeds, and flowers, and can be prepared as teas, extracts, powders, ointments, or capsules. These remedies are frequently employed to manage chronic conditions, strengthen immunity, enhance digestion, and alleviate stress. Herbal drugs are a part of the traditional medicine system (Salm et al. 2023). Traditional medicine refers to a system of health care that involves knowledge, skills, and practices rooted in the cultural heritage, beliefs, and experiences of various communities. It is used to promote health, prevent diseases, and manage or treat physical and mental health conditions. Various traditional medicine systems exist worldwide, each shaped by the cultural, environmental, and geographical factors where they originated. Despite their differences, they generally share a common holistic philosophy, i.e., focusing on maintaining balance between the mind, body, and surroundings, and prioritizing overall wellness rather than just treating disease. These systems typically address the individual's general health status instead of targeting specific illnesses, with plant-based treatments forming a fundamental part of traditional practices (Ajazuddin and Saraf 2010).

An interesting aspect of herbal medicine is its rich diversity of bioactive constituents that drive its therapeutic potential. Natural compounds such as alkaloids, flavonoids, and terpenes interact with human biological pathways, triggering physiological responses that can help manage different health disorders (Fernández-Ochoa et al. 2021). A deep understanding of how these compounds function at the molecular level is essential for refining herbal formulations and creating more targeted therapeutic options. Progress in analytical methodologies, along with advancements in genomics and metabolomics, has greatly facilitated the discovery and characterization of these bioactive substances (Balkrishna et al. 2024). Plant phytochemicals have many beneficial bioactive properties like anticancer, antimicrobial properties, antioxidative properties, anti-analgesic, and antitumor. Clinical experimentations of these properties help in a better understanding of pharmacokinetic parameters, bioaccessibility, bioavailability, and efficacy of developed formulations. As per the

DOI: 10.1201/9781003623748-6

World Health Organization (WHO) report, around 20,000 plants with medicinal properties are found in nearly 91 countries, out of which 12 are mega biodiversity countries (Sasidharan et al. 2011).

Although herbal medicines have a long-standing history of use, modern scientific research emphasizes the necessity of comprehensive assessments to ensure their safety and effectiveness. Unlike conventional pharmaceuticals, herbal formulations often lack uniform dosage guidelines and may interact with prescribed medications. Reports of adverse reactions associated with specific herbal products highlight the urgent need for detailed clinical studies and toxicological evaluations. Unlike modern pharmaceuticals, many herbal products lack rigorous clinical validation, and improper use may lead to side effects or interactions with conventional drugs (Salm et al. 2023). For ensuring clinical safety, both spontaneous reporting and active pharmacovigilance are valuable approaches to detect important safety concerns related to therapy. Even in the regions where herbal medicines undergo pre-market evaluation, continuous pharmacovigilance remains essential to support their safe use throughout their entire product life cycle (Zhang et al. 2015). As a result, regulatory bodies are placing greater emphasis on monitoring the quality, safety, and therapeutic efficacy of herbal remedies to protect consumer health. Over the years, scientific progress has enabled the extraction and synthesis of plant-derived bioactive compounds, which has significantly contributed to the development of modern drugs. Nevertheless, the integrative approach of herbal medicine—typically involving multiple plant components remains relevant and valued (Peltzer and Pengpid 2019). In recent decades, the phytopharmaceutical sector has experienced notable expansion, offering a wide range of products now commonly used in both complementary and alternative medicine. In spite of all these dilemmas regarding herbal medicines, around 80% of the world population are still using these medicines for their natural well-being. The unprecedented emergence of the COVID-19 pandemic has significantly intensified the worldwide interest in herbal medicines, with demand exceeding levels seen before the pandemic (Balkrishna et al. 2024). This surge is largely driven by a growing public consciousness regarding the potential side effects associated with allopathic treatments and the perceived benefits of plant-based remedies (Sahoo, Manchikanti, and Dey 2010). Additionally, the continuous rise in population and the increasing incidence of chronic health conditions are further contributing to the expansion of the herbal medicine market (Tulunay et al. 2015).

This chapter further discusses in detail the use of primary processing in herbal drug formation along with the extraction techniques for herbal drug constituents. The widely used separation technologies for phytochemical recovery are also discussed. The bioavailability and storage of herbal drugs are a major concerns, so various prevailing encapsulation techniques to achieve this are also compiled in the chapter along with the release kinetics study and finally the safety aspects for the prepared drug.

6.1.1 Steps in Herbal Drug Development

Herbal drug development starts with the selection of plant materials which is based on the ethnobotanical knowledge of the researcher, the traditional uses of the materials known so far, and scientific literature available. The selection of specific species, plant parts, and harvesting conditions are very critical in selection for drug development. After selection of the plant materials, the authentication and quality control of the available raw material are required. The botanical identification can be done by taxonomists, while the quality control can be analyzed using different phytochemical, macroscopic, or microscopic tests. Adulterations need to be checked at the very basic stage to prevent the accumulation of any admixture in the final formulation (Tulunay et al. 2015). After the selection and initial authentication of plant materials, preliminary treatment is required which facilitates the next important steps of extraction. The preliminary treatments include cleaning, cutting, drying, size reduction, etc. The selection of solvent for the extraction as well as designing the most effective extraction condition is also an important step (Picos-Salas et al. 2021). There are many extraction methods which may be broadly classified as traditional methods (maceration, solvent extraction, etc.) and modern methods/advanced methods (ultrasound-assisted extraction, microwave-assisted extraction, pulsed electric field-assisted extraction, etc.) (Kumar et al. 2023). After the extraction of phytochemicals, their standardization and quantification are needed. Identification is generally done using bioactive markers such as for alkaloids and flavonoids, while the quantification is done using equipment like HPLC, GC-MS, TLC, etc. The next step in herbal drug development is the study to determine the therapeutic effects of the extracted biomass. The therapeutic effects can be determined *in vitro*, *in vivo* or *in silico* methods. The determined effects may be anti-oxidant, antimicrobial, anti-inflammatory, anti-cancer, etc. Cell lines, animal models, and mechanism studies are a part of the research. The toxicology is an important aspect for any herbal drug; hence, a detailed toxicological analysis is required, which can establish a no observed adverse effect level for the biomaterial for their safety aspects (Smeriglio et al. 2017). After all these analyses, the clear extract with all the reports is available, so the next step focuses on the development of formulation based on the dosage, bioavailability, and stability of the herbal drug. The formulation can be in the form of tablets, capsules, syrups, or lotion, which is based on the dosage and application. The developed drug formulation then goes through clinical trials which is accomplished in different phases. The clinical trials may be randomized trials, which uses double-blinded method, where both the investigators and subjects are unknown to the formulation or single-blinded where investigator knows the formulation. In herbal drug clinical trials, single-blinded method is used as it has multidimensional treatment approaches consisting of counseling, listening, explaining, lifestyle, and dietary advice, as well as prescribing herbal medicines (Parveen et al. 2015). The regulatory approval for the herbal medicine is required for its manufacturing, marketing, and sale process. The registration process for herbal medicinal products demands comprehensive documentation covering all aspects of the formulation. This includes a complete list of ingredients and the detailed composition, especially for imported herbal products. Information about the manufacturer and assembler, along with valid licenses or certifications issued by the relevant drug regulatory authorities, must also be provided (Sahoo, Manchikanti, and Dey 2010). Additional requirements include the product's brand name, its dosage form, recommended therapeutic uses, dosage instructions, method of administration, and the intended duration of use. Furthermore, data on potential adverse effects, contraindications, warnings, necessary precautions, and possible drug interactions should be disclosed whenever available. Essential manufacturing details such as the production date, expiry date, batch or lot number, and recommended storage conditions are also critical components of the registration dossier (Parveen et al. 2015). The compliance with regulatory bodies like Ayurveda, Yoga & Naturopathy, Unani, Siddha, and Homeopathy (AYUSH), Food and Drug Administration (FDA), etc., and compliance with Good Manufacturing Practices (GMP) is required. Postmarketing surveillance is also covered under the regulations which is also known as pharmacovigilance. It monitors the adverse effects and long-term safety aspects of drugs on the general population. The details of the regulations are provided in the last section of the chapter.

6.2 PRIMARY PROCESSING IN HERBAL DRUG FORMATION

6.2.1 Drying

The raw leaves are typically dried to eliminate excess moisture, thereby enhancing handling, transportation, and storage efficiency. This also decreases the likelihood of spoilage by microorganisms through a reduction in water activity. Simultaneously, it is essential to choose the optimal drying method that effectively removes moisture from the samples while preserving other valuable components (Bahloul et al. 2009). Dried leaves of high quality have potential applications in the food, nutraceutical, and pharmaceutical sectors. Traditional drying methods, such as sun and shade drying, lack control, are time-intensive, and result in uneven drying (Yap et al. 2020). To address these limitations, innovative drying techniques such as solar drying (SD), vacuum drying (VD), microwave drying (MD), and freeze drying (FD) are employed. However, methods such as FD and VD incur high costs due to increased energy consumption (Rout, Kumar, and Rao 2024). Therefore, it is essential to achieve a balance between product quality and operational costs to obtain a high-quality product at a reasonable expense, facilitating its scalability for industrial applications.

The process of drying agricultural materials is inherently complex. Experimental data are analyzed using thin-layer drying models to predict the behavior of materials under varying temperature and time conditions. Curve fitting in mathematical modelling of drying kinetics is typically categorized into two types: empirical models and semi-theoretical models (Erbay and Icier 2010). Models of a semi-theoretical nature were developed based on Newton's law of cooling and Fick's second law of diffusion. Empirical models were developed based on the analysis of experimental findings (Dai et al. 2015). Drying rate curves are critical for the implementation of thin-layer drying equations. The subsequent parameter of interest is energy consumption during the drying process. An energy-efficient drying method is advantageous for industrial applications. In any drying process, specific energy consumption (SEC) is a critical parameter to consider when selecting a particular drying operation (Jeevarathinam et al. 2022). The direct transfer of energy to the sample, as observed in microwave drying (MD), contrasts with traditional drying methods that rely on conduction and/or convection. This approach leads to accelerated drying, an increased rate of water evaporation, referred to as specific moisture extraction rate (SMER), and consequently, reduced specific energy consumption (SEC) (Gao et al. 2022). It ensures uniform temperature distribution, resulting in consistent drying. Drying efficiency (DE) is determined by the dryer capacity and specific energy consumption (SEC). A higher DE is advantageous as it guarantees improved performance. The activation energy (Ea) is a critical parameter, indicating the minimum energy necessary for a drying system to commence the drying process. This reflects the challenge of extracting moisture from a sample (Dai et al. 2015). A lower value of Ea facilitates the initiation of the drying process. An effective drying method can be selected based on the previously discussed energy parameters.

6.2.2 Size Reduction

The efficacy of extraction also depends on the size of the samples. Reducing the sample size increases the surface area of the samples, enabling solvents for better extractability. However, this efficiency is maintained up to a particular size, and a finer size reduces extractability. The reason behind this effect is the backflow phenomenon of solvents, in which the bioactive compounds return to the cell matrix, thereby reducing the extraction yield (Kumar, Rout, and Rao 2022). To identify the best particle size for extraction, the samples are ground into fine powders and are passed through sieves with different grades. These grades are checked for their extractability, and the sample yielding the highest extraction yield is the best particle size for that sample-solvent combination. Once the solvent changes, so does the extractability. So, it must be understood that the extraction yield is a function of the solvent used for extraction study.

6.2.3 Effect of Solvents

Solvents are chosen based on the polarity of the bioactive compounds. Some compounds are non-polar in nature, which needs hydrophobic solvents for efficient extraction. Some common examples of hydrophobic solvents are hexane, benzene, toluene, etc. On the other hand, polar compounds are water-soluble and are best extracted through hydrophilic solvents. Ethanol, methanol, acetone, acetonitrile, etc. are few commonly used hydrophilic solvents. Selection of suitable solvents is crucial for efficient extraction of bioactive compounds (Kumar and Srinivasa Rao 2020). Another factor to keep in mind is the use of solvents, which are 'Generally Recommended as Safe (GRAS)' by FDA. There are different classes assigned to different solvents, which are safe for human consumption when used within permissible limits. For extraction purposes, Class-I solvents should not be used, as they are most toxic for human health. Apart from these, Class-II and III solvents can be used for extraction studies, if it can be ensured that the solvents are later removed from the miscella through distillation or rotary vacuum evaporation (Do et al. 2014). Ethanol and acetone belong to Class-III solvents, and are considered least toxic, and hence, the most widely used solvents in the extraction industry. Methanol belongs to Class-II solvents, which can also be used.

6.2.4 Effect of Feed-to-Solvent Ratio

Once the solvent is fixed for a particular sample, the next essential step is the optimum feed-to-solvent ratio. Depending upon the sample quantity, the amount of solvents is varied and the optimum condition is identified by achieving the maximum extraction yield. Lesser amount of solvents results insufficient extraction, and higher solvent amount results in unnecessary expenses and wastage. Therefore, an optimum feed-to-solvent ratio is identified before proceeding for bulk extraction. The feed-to-solvent ratio is maintained further and sample size is increased accordingly, so as to ensure industrial demands (Rout, Kumar, and Rao 2021b).

6.3 ENCAPSULATION TECHNIQUES FOR MODULATING DRUG DELIVERY

6.3.1 Drying Based Applications

6.3.1.1 Freeze Drying

The best method for dehydrating any heat-sensitive substance and for microencapsulation is freeze-drying (Desai and Park 2005). It is a stabilizing multistage operation of materials through four primary processes, including desorption, freezing, sublimation, and storage. Based on the sublimation phenomenon, freeze-drying is a drying method used for the long-term preservation of heat-sensitive foods and other biological materials. Most of the original raw material properties, including shape, dimensions, appearance, taste, color, flavor, texture, and biological activity, are preserved by a good freeze-drying procedure (Ceballos, Giraldo, and Orrego 2012). The composition and structure of the wall material primarily determine the effectiveness of protection or controlled release (Young, Sarda, and Rosenberg 1993). The most widely employed wall components include whey protein, gum arabic, maltodextrin, emulsifying starches, and others. The two main drawbacks of freeze-drying are its lengthy processing time and significant energy input. Furthermore, a barrier with an open porous structure forms between the active agent and its surroundings during processing; this high-porous wall provides poor protection when prolonged release of an active agent is needed.

6.3.1.2 Spray Drying

Spray drying is the oldest and most prevalent encapsulation process in the food industry. It is a flexible, continuous, and, more importantly, an economical operation. Encapsulation using spray drying looks to be a useful method for compounds in powder form due to its advantages. Optimal qualities of protection, stability, solubility, and controlled release of bioactive substances can be achieved (Zuidam and Shimoni 2010). Various types of encapsulating agents have been employed for spray drying, including polysaccharides (starches, maltodextrins, corn syrups, and arabic gum), lipids (stearic acid, mono- and diglycerides), and proteins (gelatin, casein, milk serum, soy, and wheat) (Saénz et al. 2009). The encapsulating material is homogenized with the carrier material, typically at a ratio of 1:4. The liquid is subsequently introduced into a spray drier and atomized using a nozzle or spinning wheel. Water is evaporated by the contact of heated air with the atomized substance. The capsules are subsequently gathered once they descend to the base of the drier (Gibbs et al. 1999). The spray drying technique requires appropriate inlet and output temperatures. Moreover, it is crucial to regulate the drying temperature to reduce the activity loss of enzymes. If the inlet temperature is too low, the water will not evaporate completely in a short time, resulting in wet spray-dried powder and a low encapsulation yield. If the temperature is very high, shattering of the microcapsules occurs. Variations in particle size and morphology during spray-drying were associated with moisture content and drying temperature. During the drying process, a film forms on the droplet's surface, resulting in an increased concentration of chemicals within the drying droplet. Ultimately, a porous, dry particle was generated. Conventional spray-dried encapsulates release their active agent instantaneously upon contact with water, contingent upon the porosity of the particles. Recent innovations of more hydrophobic and cross-linked carrier materials provide a more gradual release upon dilution in water. Examples include denatured proteins, cross-linked proteins, or cross-linked biopolymers (Zuidam and Nedović 2010).

6.3.1.3 Fluidized Bed Coating

Fluid bed coating is an encapsulating technology in which a coating is applied to powder particles through either a batch or continuous process. Fluidized bed coating is a process by which coated particles are created by applying an encapsulating agent onto a fluidized powder bed (Hemati et al. 2003). Processing variables in fluid bed coating, such as solid circulating rate and nozzle atomization pressure (Kage et al. 2003), as well as coating feed rate and temperature, are significant due to their influence on particle agglomeration and film formation, thereby affecting coating efficiency. Consequently, the optimization of coating operations with fluidized bed technology relies on assessing the impact of processing conditions. Various coating ingredients are utilized for fluidized bed coating, including gums, proteins, and starches. Fluidized-bed coating increasingly provides the food industry with a diverse array of encapsulated food components and additives. The various fluidized-bed coating processes are: (1) top-spray, (2) bottom-spray, and (3) tangential spray. Various variables are affected in fluidized bed operation, which include: (a) Process variables (inlet air temperature, air velocity, spray rate, and atomization pressure), (b) environmental variables (ambient air temperature, ambient air relative humidity), and (c) thermodynamic variables (outlet air temperature and outlet air relative humidity) (Desai and Park 2005).

6.3.2 Gel-based Applications

6.3.2.1 Ionic Gelation

Hydrocolloid gel particles or hydrogel beads are utilized for the encapsulation of bioactive chemicals or for texture modulation in culinary, pharmaceutical, probiotic, medicinal, and cosmetic goods. They are created through ionic polymerization or polyelectrolyte ionic bonding. The process, known as ionic gelation or ionotropic gelation, commences with an aqueous polymeric solution containing low molecular mass ions that interact with oppositely charged polyelectrolytes, resulting in the formation of an insoluble gel (Chongsrimsirisakhol and Pirak 2023). The notion of encapsulation involves entrapping an active material and subsequently releasing it through gel phase transitions in response to external stimuli. Various triggering methods are employed to release the encapsulated active as pH fluctuations and mechanical wear occur. Ionic gelation can be performed using atomization techniques, extrusion, coextrusion, or electrostatic deposition. A polymeric or hydrocolloid solution is often dripped or atomized into an ionic solution while maintaining steady agitation. The active ingredient intended for encapsulation is dissolved in the polymeric solution. The droplets that contact the ionic solution promptly create spherical gel formations, which encapsulate the active substance throughout the polysaccharide matrix. The technique is straightforward and uncomplicated, necessitating no specialist equipment, elevated temperatures, or organic solvents, and is regarded as cost-effective (Lee and Chang 2020). Nonetheless, a notable disadvantage is the uneven gelation of gel particles resulting from the diffusion mechanism, as surface gelation frequently precedes core gelation, leading to a soft core (Pereira Silveira et al. 2023).

Alginate, low methoxyl pectin, chitin, and chitosan are commonly employed as coating agents, with Ca^{2+} being the predominant crosslinking agent. They serve as an excellent encapsulating technology for food components and controlled release, due to their non-toxicity, high biocompatibility, and mechanical strength (Norcino et al. 2022). Capsules generated using ionic gelation are extensively utilized for hydrophobic substances. The shell, composed of alginate, gelatin, agar, low methoxyl pectin, or gellan gum, which are hydrophilic, is typically utilized for hydrophobic substances or those with minimal solubility. Hydrophilic actives present greater challenges for encapsulation compared to hydrophobic ones. Hydrocolloids significantly benefit most actives; yet their miscibility with hydrophilic cores complicates achieving effective phase separation between the core and shell. Moreover, due to the porosity of hydrogel beads, the encapsulation of hydrophilic substances leads to diminished encapsulation efficiency and inadequate controlled release characteristics (Guerra-Valle, Petzold, and Orellana-Palma 2022).

6.3.2.2 Coacervation

The principle underlying coacervation microencapsulation involves the phase separation of one or many hydrocolloids from the initial solution, followed by their subsequent deposition. It applies recently generated coacervate phase around the active ingredient suspended or emulsified inside the identical reaction medium (Al-Hamayda et al. 2023). Coacervation encapsulation can be accomplished using a singular colloidal solute like gelatin, or by a more intricate method including both gelatin and gum acacia. Complex coacervation typically lacks distinct forms and is regarded as a costly technique for encapsulating food components; nonetheless, this process should be linked to the prospective advantages it may provide, particularly for high-value, sensitive functional ingredients, such as the encapsulation of polyphenols (Desai and Park 2005).

6.3.2.3 Sol-gel Process

The sol-gel encapsulation process is a multifaceted technique for producing materials that include entrapped compounds, especially pertinent for delicate biological molecules. It depends on the transformation of a liquid solution (sol) into a solid-like gel state, wherein the substance of interest is embedded inside the gel matrix (Ashraf et al. 2015).

The sol-gel method generally comprises multiple stages:

1. *Hydrolysis*: This preliminary phase entails the reaction of a precursor (often a metal alkoxide or metal salt) with water to produce a sol, which is a colloidal suspension of solid particles within a liquid. For instance, metal alkoxide M-OR interacts with water to yield metal hydroxide M-OH and alcohol R-OH (Wang et al. 2015).
2. *Condensation (or Polycondensation)*: The hydrolyzed precursor subsequently engages in a condensation reaction, interacting with additional hydrolyzed entities or non-hydrolyzed alkoxides to establish metal-oxygen-metal (M-O-M) links, therefore forming a three-dimensional network. This phase results in the creation of a gel, a semi-rigid structure consisting of a solid network that encases the liquid solvent.
3. *Aging*: The gel undergoes aging, which enhances the network via ongoing hydrolysis and condensation, hence decreasing porosity and augmenting the thickness between colloidal particles.
4. *Drying*: The liquid solvent in the gel's pores is then eliminated using processes such as ambient pressure evaporation, supercritical drying, or freeze drying. The drying technique profoundly affects the characteristics of the finished product.
5. *Calcination and Heat Treatment*: Ultimately, the dried gel may undergo calcination at high temperatures to eliminate residual organic elements and enhance structural density, perhaps resulting in the creation of thick ceramic or glass particles. Heat treatment can also improve the mechanical properties of the material (Escobar et al. 2018).

6.4 DRUG RELEASE BEHAVIOR

In recent decades, substantial progress has occurred in drug delivery through the advancement of controlled release dosage forms. A wide range of formulations exists for oral controlled drug release, along with various physical properties that affect drug release from these formulations. Release patterns can be categorized into two types: those that release the drug at a slow zero or first-order rate, and those that deliver an initial rapid dose, followed by a slow zero or first-order release of the sustained component (Paul 2011).

Controlled release formulations exhibit significant versatility in medical applications, serving as essential tools in the modern therapeutic landscape. Their primary objectives include enhancing drug efficacy, improving patient adherence, and minimizing the frequency of administration and associated side effects.

Controlled release formulations facilitate collaboration between engineers and pharmacists, aiming to develop increasingly effective products (Singh et al. 2012). Mathematical modelling is highly beneficial for this purpose, as it facilitates the prediction of release kinetics prior to the realization of release systems. It facilitates the measurement of critical physical parameters, including the drug diffusion coefficient, by employing model fitting on experimental release data. Mathematical modelling, which necessitates an understanding of the phenomena influencing drug release kinetics, holds significant value in the optimization of formulation processes (Atif et al. 2019). The following section explains the fundamentals of drug release kinetics:

6.4.1 Noyes-Whitney Rule

The fundamental equation is as follows:

$$\frac{dM}{dt} = KS\left(C_s - C_t\right) \tag{6.1}$$

where M = mass transfer at time ‘t’; K = rate constant; S = surface area of solid particles; C_s = equilibrium concentration of solute at experimental temperature; C_t = concentration of solute at time ‘t’; dM/dt represents the amount of solute dissolved per unit area per unit time ($g/cm^2.s$). When $C_t < 15\%$ of C_s, the term ($C_s - C_t$) is expressed as C_s. This condition is expressed as ‘sink’ condition (Wimalasiri et al. 2021).

6.4.2 Nernst-Brunner Film Theory

It is a modified Fick’s law of diffusion, given as:

$$K = DS / h\gamma \tag{6.2}$$

where D = diffusion coefficient; S = surface area of diffusion layer; γ = solution volume, and h = diffusion layer thickness. A critical assumption in this equation is that surface diffusion proceeds at a much faster rate than the transport process. Moreover, the concentration gradient is linear only in the solution layer adhering to the solid surface (Okazaki, Mano, and Sugano 2008).

6.4.3 Release Kinetics Modelling

A variety of kinetic models elucidate the comprehensive release of the medication from dose forms. Qualitative and quantitative modifications in a formulation can impact drug release and *in vivo* efficacy; hence, creating methods that enhance product development by minimizing the need for bio-studies is advantageous. The utilization of *in vitro* medication dissolution data to forecast *in vivo* bio-performance is regarded as a logical advancement in the production of controlled release formulations (Hu et al. 2017). Drug release kinetics is broadly categorized into three types, which are elaborated in the following sections.

6.5 STATISTICAL METHODS

6.5.1 Exploratory Data Analysis

Despite the lack of endorsement from the FDA for exploratory data analysis approaches, these techniques are beneficial for enhancing the comprehension of dissolution data in controlled release formulations, hence their application is advocated. This method can be employed initially to compare dissolution profile data both graphically and numerically. The dissolution profile data are graphically represented by showing the mean dissolution profiles for each formulation, accompanied by error bars that extend to two standard errors at each dissolution time point. The dissolution profile data are quantitatively summarized, and 95% confidence intervals for the variations in mean dissolution profiles at each time point are assessed (Mullis et al. 2019).

6.5.2 Multivariate Analysis of Variance (MANOVA)

These methods utilized repeated measures designs in which time served as the repeated component and percent dissolved constituted the dependent variable. The computed statistics for this method included Pillai's Trace, Wilks' Lambda, Hotelling's Trace, and Roy's Largest Root. A repeated measures design was employed due to the data being obtained as repeated measurements throughout time on the same experimental device. In comparison to Student's t-test and paired t-test, the primary advantage of this design is enhanced precision (Grice and Iwasaki 2009).

In repeated measures ANOVA involving components with more than two levels, additional specific assumptions must be considered. The assumptions are compound symmetry and sphericity. The MANOVA method for repeated measurements ANOVA has gained traction in recent years due to the infrequency of these conditions being met. The compound symmetry assumption necessitates that the variances and covariances of the many repeated measures are uniform. This constitutes a necessary criterion for the validity of the univariate F test for repeated measures. The sphericity assumption is both required and sufficient condition for the validity of the F test. Violation of the compound symmetry or sphericity assumptions will yield inaccurate findings in the univariate ANOVA table (Sthle and Wold 1990). The results of Mauchly's test of sphericity are utilized for the assumption of sphericity.

6.6 MODEL DEPENDENT METHODS

Model-dependent approaches rely on various mathematical functions that characterize the dissolution profile. Upon selecting an appropriate function, the dissolution profiles are assessed based on the obtained model parameters. The model-dependent methodologies encompassed zero order, first order, Higuchi, Hixson-Crowell, Korsmeyer-Peppas, Baker-Lonsdale, Weibull, Hopfenberg, Gompertz, and regression models (Rout, Kumar, and Rao 2021a).

6.6.1 Zero-Order Model

The dissolution of drugs from dosage forms that do not disaggregate and release the medication gradually can be expressed by the equation:

$$Q_0 - Q_t = K_0 t \tag{6.3}$$

where Q_t = drug released in time 't'; Q_0 = initial amount of drug present in the solution (generally 0 at the beginning); K_0 = zero-order release constant.

To analyze the release kinetics, data from *in vitro* drug release trials were graphed as the cumulative amount of drug released against time. This relationship can characterize the drug dissolution of many modified release pharmaceutical dosage forms, including certain transdermal systems, matrix tablets with poorly soluble medicines in coated forms, and osmotic systems, among others (Marcos Luciano Bruschi 2015).

6.6.2 First-Order Model

This model has also been employed to elucidate the absorption and/or disposal of some medications, but conceptualizing this mechanism theoretically poses challenges. The release of the drug exhibiting first-order kinetics can be articulated by the equation:

$$\frac{dC}{dt} = -Kc \tag{6.4}$$

where C = concentration at any time 't'; K = first-order rate constant (s^{-1}). This equation can be further modified as:

$$log\ log\ C = logC_0 - Kt/2.303 \tag{6.5}$$

where C_0 = initial drug concentration. The acquired data are represented as the logarithmic cumulative percentage of medicine remaining vs time, resulting in a linear graph with a slope of –K/2.303. This relationship can characterize drug dissolution in pharmaceutical dosage forms, particularly those including water-soluble medicines within porous matrices (Aykin-Dinçer, Dincer, and Topuz 2024).

6.6.3 Higuchi Model

This model is predicated on the assumptions that (i) the initial drug concentration within the matrix significantly exceeds the drug's solubility; (ii) drug diffusion occurs solely in one dimension (with negligible edge effects); (iii) drug particles are considerably smaller than the thickness of the system; (iv) matrix swelling and dissolution are inconsequential; (v) drug diffusivity remains constant; and (vi) optimal sink conditions are consistently achieved in the release environment (Paul 2011). The model expression is represented by the equation:

$$f_t = Q = A\sqrt{D(2C_0 - C_s)C_s t} \tag{6.6}$$

where Q = drug released in time 't' for unit area 'A'; C_0 = initial drug concentration; C_s = equilibrium drug concentration in the solution, and D = diffusion coefficient in the matrix. This relationship always remains valid, except when the complete depletion of the drug inside the therapeutic system occurs. The Higuchi model can also be broadly simplified as follows, sometimes referred to as the simplified Higuchi model:

$$f_t = Q = K_H \times t^{1/2} \tag{6.7}$$

where K_H = Higuchi dissolution constant. The data will be plotted as cumulative drug release (%) vs. square root of time. This expression can characterize the drug dissolution from various modified release pharmaceutical dosage forms, such as certain transdermal systems and matrix tablets containing water-soluble medicines.

6.6.4 Hixson-Crowell Model

Hixson and Crowell acknowledged that the regular area of the particles is proportional to the cube root of their volume, leading to the derivation of the equation:

$$W_0^{1/3} - W_t^{1/3} = \kappa t \tag{6.8}$$

where W_0 = initial drug concentration; W_t = drug remaining to be released; t = time, and κ = constant relating surface-volume relation. The equation delineates the release from systems characterized by variations in surface area and particle or tablet diameter. To analyze the release kinetics, data from *in vitro* drug release trials were graphed as the cube root of the proportion of drug remaining in the matrix against time. This term pertains to pharmaceutical dosage forms, such as tablets, where disintegration happens along planes parallel to the drug surface, provided that the tablet dimensions decrease accordingly, hence maintaining a fixed starting geometric form throughout (Marcos Luciano Bruschi 2015).

6.6.5 Korsmeyer-Peppas Model

A simple relationship was established, which characterized drug release from a polymeric system equation. The initial 60% of drug release data were analyzed using the Korsmeyer-Peppas model to elucidate the drug release process.

$$\frac{M_t}{M_\infty} = Kt^n \tag{6.9}$$

where $\frac{M_t}{M_\infty}$ = fraction of drug released for time 't'; K = rate constant, and n = release exponent.

TABLE 6.1
Relation between Release Exponent and Drug Transport Mechanism

Release Exponent (n)	Drug Release Mechanism	Time Function
0.5	Fickian diffusion	$t^{-0.5}$
$0.45 < n = 0.89$	Non-Fickian diffusion	t^{n-1}
0.89	Case-II transport	Zero-order release
Higher than 0.89	Super case-II transport	t^{n-1}

In this model, the parameter n delineates the drug release mechanism as outlined in Table 6.1. In the context of cylindrical tablets, a value of $0.45 \leq n$ indicates a Fickian diffusion mechanism, $0.45 < n < 0.89$ signifies non-Fickian transport, $n = 0.89$ denotes Case II (relaxational) transport, and $n > 0.89$ represents super Case II transport (37, 38). To get the exponent of n, only the segment of the release curve where $Mt/M\infty < 0.6$ should be utilized. To analyze the release kinetics, data from *in vitro* drug release trials were graphed as log cumulative percentage drug release against log time (Rout, Kumar, and Rao 2021a).

6.7 EXTRACTION TECHNIQUES FOR HERBAL DRUG CONSTITUENTS

The development of herbal drugs focuses on converting traditional plant-based therapies into scientifically proven medicines that align with current pharmaceutical regulations. Unlike the development of synthetic drugs, this process presents distinct challenges, primarily due to the complex composition of plant extracts and the critical need for consistent standardization (Sasidharan et al. 2011). Extraction serves as a fundamental step in the analysis of herbal medicines, as it separates the desired chemical constituents from the raw vegetal material for subsequent separation and characterization. This process typically involves preliminary steps such as cleaning, drying to a certain moisture content using methods like air drying, oven drying, or freeze-drying, and grinding the plant material to create a uniform sample. These steps help enhance the efficiency of the extraction process by increasing the surface area that comes in contact with the chosen solvent (Bitwell et al. 2023). It is essential to conduct all the treatments carefully with a minimum thermal load to prevent the loss, alteration, or degradation of bioactive compounds (Sasidharan et al. 2011).

When plants are selected based on traditional medicinal practices, it is often recommended to follow the extraction procedures used by traditional healers to closely replicate the original herbal preparation. The choice of solvent is critical and depends on the chemical nature of the targeted compounds. Polar solvents like methanol, ethanol, or ethyl acetate are commonly used to extract hydrophilic compounds, whereas non-polar or moderately polar solvents such as dichloromethane or dichloromethane-methanol mixtures are preferred for lipophilic substances. In certain cases, hexane is employed to eliminate chlorophyll and other non-essential components (Krakowska-Sieprawska et al. 2022). Since the bioactive molecules can range from non-polar to polar and may also be sensitive to heat, selecting an appropriate extraction method is crucial. Common techniques include sonication, reflux heating, Soxhlet extraction, and others. Additionally, methods like maceration and percolation, using either fresh or dried plant materials with water or organic solvents, are frequently utilized to obtain plant extracts (Naima et al. 2015). Several advanced extraction techniques have been developed in recent years, including solid-phase micro-extraction, supercritical fluid extraction, pressurized liquid extraction, microwave-assisted extraction, solid-phase extraction, and surfactant-assisted methods (Ranjha et al. 2021). These modern approaches offer multiple benefits, such as minimizing the use of organic solvents and reducing the risk of sample degradation. They also eliminate the need for additional steps like sample clean-up or concentration before chromatographic analysis. Furthermore, these techniques enhance extraction efficiency, improve selectivity, and often accelerate the extraction process. Their compatibility with automation makes them particularly suitable for large-scale or routine extraction of plant materials (Sasidharan et al. 2011). There are different methods of extraction which varies based on their mechanism as well as the desirable extract to be obtained. Some of these methods are given in Table 6.2.

TABLE 6.2
Common Extraction Methods with Their Principles and Usage

Method	Principle	Features	Applications
Maceration	It involves soaking the biomass in a solvent at room temperature (Azwanida 2015)	It is simple, cost-effective but time-consuming	It is generally used for extraction of heat-sensitive compounds
Percolation	In this method continuous solvent flow occurs through plant powder in a percolator for extraction	It is efficient for large-scale operations	Tinctures, fluid extracts
Soxhlet Extraction	It involves repeated solvent distillation and condensation over biological material (Naima et al. 2015)	It is generally used for lipophilic material extraction	Extraction of alkaloids, essential oils, etc.
Hydro-distillation	Steam or water distillation to extract volatile components	Used for essential oils extraction (Kholaf, Gomaa, and Ziena 2017)	Extraction of aromatic plants essential oil and flavor
Ultrasound-Assisted Extraction (UAE)	It uses ultrasound waves for destruction of cell matrix for extraction of bioactive compounds (Ma et al. 2008)	It is faster and energy-efficient method	Used for extraction of bioactive compounds like phenolics, flavonoids

(Continued)

TABLE 6.2 *(Continued)*
Common Extraction Methods with Their Principles and Usage

Method	Principle	Features	Applications
Microwave-Assisted Extraction (MAE)	Works on the principle of dipole rotation and ionic polarization to generate heat (Bitwell et al. 2023)	Short extraction time, eco-friendly	Thermo-stable phytochemicals
Supercritical Fluid Extraction (SFE)	Uses supercritical CO_2 or other fluids which have better solubility and diffusion property	It is used for enhancing extractability	Essential oils, lipids, bioactive compounds
Enzyme-Assisted Extraction (EAE)	Enzymes break down cell walls to release phytochemicals	Mild conditions, improves yield	Polysaccharides, glycosides
Pressurized Liquid Extraction (PLE)	High-pressure and temperature to improve solvent penetration (Zia et al. 2022)	Efficient for polar compounds	Polyphenols, alkaloids

The extraction methods can be selected based on nature of plant materials to be extracted, e.g., for fresh material maceration is preferred while for dried powder sonication can be effective. Similarly, the method varies with the part of plant materials to be extracted like seed, leaf, roots, etc. The types of bioactive materials which we desired to extract also helps in selection of method (Ma et al. 2008). For volatile and polar extract generally hydro-distillation or Soxhlet method is preferred while for polar phenolics, ultrasound or microwave extraction is generally used (Kholaf, Gomaa, and Ziena 2017). The selection of solvent also affects method of extraction, e.g., in microwave assisted extraction, polar solvent is required. Other factors like cost and scalability of the process and environmental impact of the process are very important for the method selection as it affects overall viability of the process.

6.8 SEPARATION TECHNOLOGIES FOR BETTER AND SELECTIVE PHYTOCHEMICAL RECOVERY

Following the extraction process, the separation and purification of bioactive phytochemicals are crucial steps in isolating specific compounds from the complex composition of plant materials. The use of advanced separation techniques enhances the purity, yield, and selectivity of the target compounds, which is essential for producing herbal products that are safe, therapeutically effective, and consistently standardized (Domínguez-Rodríguez, Marina, and Plaza 2017). Since plant extracts typically contain a complex mixture of bioactive compounds with varying polarities, their separation poses a significant challenge in the process of identifying and characterizing individual phytochemicals. To isolate these compounds, it is standard practice to employ a combination of separation methods. Techniques such as thin-layer chromatography (TLC), column chromatography, flash chromatography, Sephadex chromatography, and high-performance liquid chromatography (HPLC) are commonly utilized to obtain purified substances (de Rijke et al. 2006). These isolated compounds are subsequently analyzed to determine their chemical structure and evaluate their biological activities. In addition to chromatographic methods, non-chromatographic approaches like immunoassays using monoclonal antibodies, phytochemical screening tests, and Fourier-transform infrared spectroscopy (FTIR) are also valuable tools that aid in the identification and characterization of bioactive constituents (Sasidharan et al. 2011) (Table 6.3).

TABLE 6.3
Key Separation Technologies for Herbal Drug Components

Technology	Principle	Applications	Reference
Liquid-Liquid Extraction (LLE)	It relies on the varying solubility of compounds in two immiscible solvents, allowing for their effective separation	Alkaloids, flavonoids, essential oils	(Buszewski and Szultka 2012)
Membrane Separation (UF/NF/RO)	It employs semi-permeable membranes to differentiate and separate substances according to their molecular size, electrical charge, or specific physico-chemical properties	Concentration of polyphenols, removal of impurities	(Shihab, Rashid, and Toma 2022)
Chromatographic Techniques	It works on the differences in affinity, polarity, and size. It can be classified into column chromatography, high-performance liquid chromatography (HPLC), flash chromatography, gas chromatography (GC), etc.	It is used for isolation of herbal drug components like flavonoids, phenolics, alkaloids, etc.	(Sasidharan et al. 2011)
Centrifugation	Separation technology uses density differences to fractionate biological components	It helps in removal of debris, clarification of extracts, etc.	(Li et al. 2019)
Crystallization	Purification by controlled crystallization of desired compounds	Isolation of pure compounds like berberine	(Sasidharan et al. 2011)
Solid-Phase Extraction (SPE)	Adsorption of compounds onto a solid matrix followed by selective elution	Pre-concentration of minor phytochemicals	(Buszewski and Szultka 2012)
Electrophoresis (Capillary Electrophoresis)	Separation based on charge-to-size ratio under an electric field	For small charged phytochemicals	(Markham and Mabry 1975)

Modern separation technologies present multiple advantages in the isolation and purification of herbal drug components from complex matrices. These methods exhibit superior selectivity and resolution, facilitating the precise separation of target molecules from a mixture of structurally similar constituents. Furthermore, they significantly reduce the requirement for organic solvents, promoting environmentally sustainable practices consistent with green chemistry principles. The scalability of these technologies allows seamless transition from bench-scale operations to pilot and industrial-scale production, supporting both research and commercial applications. Importantly, advanced separation methods are particularly suitable for isolating thermolabile or chemically unstable phytochemicals, as they minimize exposure to harsh conditions, thereby maintaining compound integrity. Additionally, these techniques preserve the pharmacological activity of sensitive biomolecules, ensuring the therapeutic potential of the final product remains uncompromised (Sasidharan et al. 2011).

The selection of an appropriate separation strategy is governed by multiple factors. The physicochemical characteristics of the phytoconstituents—such as polarity, molecular mass, thermal and chemical stability—are critical determinants. The required purity level also influences method selection; for instance, pharmaceutical-grade products demand higher purity standards compared to nutraceutical formulations. Cost-effectiveness and production scale must be considered, distinguishing between processes optimized for laboratory research and those suited for industrial-scale manufacturing. Moreover, adherence to regulatory guidelines—whether targeting food-grade quality or meeting pharmacopeial and drug regulatory standards—is essential to ensure product compliance and market acceptability (Li et al. 2019).

6.9 HERBAL PHYTOCHEMICAL DRUG RELEASE BEHAVIOR

Contemporary medical practice of herbal drugs depends on a limited set of essential biologically active compounds derived from natural sources, which require extensive purification and detailed characterization before clinical application. Traditional herbal therapies, however, often suffer from suboptimal delivery mechanisms, leading to reduced therapeutic efficacy (Ajazuddin and Saraf 2010). Modern drug delivery systems offer an opportunity to enhance the bioavailability and therapeutic potential of herbal actives while minimizing associated side effects. This concept forms the basis for adopting advanced delivery strategies in phytomedicine (Wu et al. 2019). Innovative herbal formulations provide multiple advantages over conventional preparations. These include improved tissue penetration, sustained and controlled release of active constituents, and enhanced stability against physical and chemical degradation. Furthermore, controlled, and site-specific delivery of herbal bioactive ensures consistent dosing and maximizes therapeutic benefits.

Herbal drug release study is very important for its suitable and targeted application for disease control. There are several factors which affect the phytochemical drug release behaviour. Some of the leading factors which influence the release behaviour are given in Table 6.4.

TABLE 6.4
Factors Affecting Phytochemical Drug Release

Factor	Effect on Release
Physicochemical Properties	The solubility, molecular weight, and physicochemical stability of phytochemicals are critical parameters that significantly influence their dissolution profile and subsequent absorption in the body (Sun et al. 2019)
Formulation Type	Various pharmaceutical formulations such as capsules, tablets, emulsions, nanoparticles, and sustained-release systems are employed to modulate the release profile of bioactive compounds (Aykin-Dinçer, Dincer, and Topuz 2024)
Matrix Composition	The incorporation of polymers, lipids, and various pharmaceutical excipients enables precise control over the release kinetics of bioactive compounds, facilitating immediate, delayed, or sustained drug release as required
pH Sensitivity	Various pH affects the release of phytochemicals from the herbal drugs (Singh et al. 2012)
Particle Size	Variation in particle size affects dissolution and absorption properties due to variation in surface to volume ratio
Encapsulation Technique	Variation in encapsulation techniques like ionic gelation, lyophilization, coacervation, etc., affects the drug release from herbal formulation (Laureanti et al. 2023)

6.9.1 *In Vivo, in Vitro,* and *in Silico* Assessment of Herbal Drug Release Characteristics

In vivo evaluation of phytochemical drug release focuses on studying how active herbal constituents are released, absorbed, distributed, metabolized, and eliminated within a living organism, such as animal models or human subjects (Singh and Sharma 2023). This assessment is essential to confirm that herbal formulations achieve the desired therapeutic effects under actual physiological conditions. The *in vivo* assessment measures bioavailability of the phytochemicals in the herbal drug formulation. Bioavailability is the amount of drug out of the total drug incorporated which reaches the systemic circulation system (Emon et al. 2025). The complete pharmacokinetics consisting of ADME (absorption, distribution, metabolism, and excretion) can also be analyzed using *in vivo* assessment of herbal drugs. Furthermore, a correlation can be established with *in vitro* release data with actual biological performance using such analysis. Based on the *in vivo* release study, finally the dosage forms and delivery systems for herbal drugs can be optimized. Certain key parameters are generally measured in *in vivo* drug release study. Among these, the maximum plasma concentration (C_{max}) indicates the peak level of the phytochemical in systemic circulation, while the time taken to reach this peak (T_{max}) reflects the absorption rate. The area under the plasma concentration-time curve (AUC) represents the total drug exposure over a specific period, providing a comprehensive view of the compound's bioavailability. Another important parameter is the half-life (t½), which describes the time required for the plasma concentration to decrease by 50%, offering insight into the compound's elimination rate (Sun et al. 2019). For preclinical and clinical evaluation, various *in vivo* models are employed. Rodent models, such as rats and mice, are commonly used during the early stages to study drug release, absorption, and initial pharmacokinetics (Sabina and Rasool 2008). Larger animals, including dogs and pigs, are utilized in more advanced studies due to their physiological similarities to humans, providing more predictive pharmacokinetic data. Ultimately, human volunteers participate in clinical trials to assess the final bioavailability, therapeutic efficacy, and safety of herbal formulations (Venkataramanan, Komoroski, and Strom 2006).

Several methodologies are implemented during *in vivo* studies. Blood and plasma sampling is a standard approach where serial samples are collected after administration of the herbal formulation. The concentration of phytochemicals in these samples is measured using sophisticated analytical techniques such as High-Performance Liquid Chromatography (HPLC), Liquid Chromatography–Mass Spectrometry (LC-MS/MS), or Gas Chromatography–Mass Spectrometry (GC-MS), enabling the determination of the compound's pharmacokinetic profile. In addition, urine and fecal analysis help evaluate excretion patterns and identify metabolic by-products, providing insight into the compound's elimination pathways. Despite these advancements, *in vivo* assessment of herbal drugs presents several challenges. Herbal formulations often contain complex mixtures of multiple phytochemicals, complicating pharmacokinetic evaluations. Many herbal compounds exhibit low bioavailability due to poor solubility and limited permeability across biological membranes, as seen with compounds.

In vitro drug release testing serves as a critical preclinical approach to investigate how herbal phytochemicals are released from various dosage forms under controlled laboratory conditions that mimic physiological environments (Venkataramanan, Komoroski, and Strom 2006). This method provides valuable information on the dissolution behavior, release kinetics, and performance of herbal formulations before proceeding to *in vivo* studies. The primary objectives of *in vitro* drug release testing include evaluating the rate and pattern of bioactive compound release, which helps in predicting the compound's bioavailability and therapeutic potential (Sabina and Rasool 2008). It also plays a vital role in guiding the design and optimization of formulations, whether for immediate, sustained, or controlled release delivery systems. Moreover, *in vitro* release studies support quality assurance by ensuring batch-to-batch uniformity and consistency in product performance.

In vitro drug release studies focus on evaluating specific characteristics that define how herbal phytochemicals are released from their formulations. One of the primary parameters is the release rate, which refers to the speed at which the active compound diffuses into the surrounding medium. The release profile describes the pattern of drug liberation over time, such as immediate burst release, sustained release, or delayed onset, depending on the formulation design (Lee and Chang 2020). Dissolution efficiency measures the total amount of phytochemical dissolved within a defined timeframe, offering insight into the potential bioavailability. In certain controlled-release systems, lag time is observed, representing the intentional delay before the onset of release. Additionally, dose uniformity is critical to ensure consistent release behaviour across different product batches, supporting both efficacy and quality assurance (Bandawane and Saudagar 2019). Various mathematical models are used to analyze *in vitro* release data and understand the underlying mechanisms (Marcos Luciano Bruschi 2015). The zero-order model describes drug release at a constant rate, making it applicable to sustained-release herbal formulations. In contrast, the first-order model assumes that the release rate is directly proportional to the amount of drug remaining in the dosage form, which is typical for conventional herbal preparations. The Higuchi model explains release through diffusion from a matrix, commonly used for semi-solid and solid herbal systems. The Korsmeyer–Peppas model helps determine whether the release mechanism is dominated by diffusion, erosion, or a combination of both, which is especially useful for complex herbal delivery systems. These models collectively aid in optimizing formulation strategies and predicting *in vivo* behavior based on *in vitro* data (Rout, Kumar, and Rao 2021a).

Another method of drug release characteristics assessment is *in silico* method. It involves the application of computational models and simulations to predict the drug release, absorption, and bioavailability of phytomedicines before initiating laboratory experiments or animal studies (Chang et al. 2022). This approach plays a crucial role in the early stages of herbal drug development, allowing for formulation design and optimization while significantly reducing both time and development costs. The primary objectives of *in silico* evaluation include forecasting drug release behaviour under simulated physiological conditions, modeling the diffusion, dissolution, and potential degradation of phytochemicals, and simulating pharmacokinetic processes such as absorption, distribution, metabolism, and excretion (ADME). Several advanced computational tools are employed in this process. Physiologically Based Pharmacokinetic (PBPK) modelling is used to estimate systemic exposure and absorption profiles of herbal compounds, while Molecular Dynamics (MD) simulations analyze the interactions between phytochemicals and delivery systems such as nanoparticles or liposomes (Jia et al. 2022). Computational Fluid Dynamics (CFD) models are utilized to simulate the diffusion and dissolution of drugs within gastrointestinal fluids, offering insights into release behaviour under various physiological conditions. The Quantitative Structure-Activity Relationship (QSAR) approach predicts key properties like solubility, permeability, and release kinetics based on the molecular structure of the phytochemical (Chang et al. 2022). Finite Element Modelling (FEM) further aids in simulating the release mechanisms of complex formulations, particularly in controlled-release systems.

6.10 SAFETY ASPECTS OF HERBAL MEDICINES

Herbal medicines are commonly regarded as natural and therefore safe; however, this perception does not guarantee they are free from risks. The safety of herbal products is influenced by multiple factors, including the quality of raw materials, dosage accuracy, preparation techniques, and the health status of the individual using them (Ekor 2014). Inappropriate use or poor-quality formulations can lead to adverse effects, toxicity, or interactions with conventional drugs. Therefore, comprehensive toxicological assessments, along with evaluations of potential side effects and herb-drug interactions, are critical to ensure the safe and effective use of herbal medicines (Bandawane and Saudagar 2019).

6.11 CONCLUSION

In conclusion, the development of herbal drugs signifies a promising but intricate path in the progression of modern treatments, providing a natural and frequently safer alternative to synthetic medications. Herbal remedies, grounded in ancient knowledge systems like Ayurveda, Traditional Chinese Medicine, and Indigenous practices, have garnered increased scientific attention due to their diverse biological activity, reduced toxicity, and synergistic therapeutic effects. Due to the proliferation of chronic diseases, drug resistance, and adverse effects associated with traditional medications, herbal remedies are increasingly seen as effective complementary or independent alternatives for disease management. Nonetheless, despite their promise, the development of herbal pharmaceuticals encounters considerable obstacles, including variability in phytochemical composition influenced by extraction, formulation, and dosing, a lack of high-quality clinical trials, and intricate regulatory frameworks that vary by country. Furthermore, *in vitro*, *in vivo,* and *in silico* analyses and pharmacokinetic assessments now facilitate the identification of lead phytocompounds and the optimization of their therapeutic profiles early in the development process. The amalgamation of ancient knowledge with contemporary science, together with rigorous quality control and safety assessment processes, is crucial for the incorporation of herbal medicines into conventional healthcare.

REFERENCES

Ajazuddin, S., and Saraf. 2010. “Applications of Novel Drug Delivery System for Herbal Formulations.” *Fitoterapia* 81 (7): 680–89. https://doi.org/10.1016/j.fitote.2010.05.001

Al-Hamayda, A., B. Abu-Jdayil, M. Ayash, and J. Tannous. 2023. “Advances in Microencapsulation Techniques Using Arabic Gum: A Comprehensive Review.” *Industrial Crops and Products* 205(12): 117556. https://doi.org/10.1016/J.INDCROP.2023.117556

Ashraf, M. A., A. M. Khan, M. Ahmad, and M. Sarfraz. 2015. “Effectiveness of Silica Based Sol-Gel Microencapsulation Method for Odorants and Flavors Leading to Sustainable Environment.” *Frontiers in Chemistry* 3 (8): 42. https://doi.org/10.3389/FCHEM.2015.00042

Atif, R., T. S. Eldeen, L. Ahmed, I. Yahya, A. Omara, and M. Eltayeb. 2019. “Study the Using of Nanoparticles as Drug Delivery System Based on Mathematical Models for Controlled Release.” *International Journal of Latest Technology in Engineering* 8 (5): 52–6. www.ijltemas.in

Aykin-Dinçer, E., C. Dincer, and O. K. Topuz. 2024. “Modeling of Release Mechanism of Sage (Salvia Fruticosa Miller) Phenolics Encapsulated in Alginate Capsule: Physicochemical Properties.” *Journal of Food Processing and Preservation* 2024 (1): 7598455. https://doi.org/10.1155/2024/7598455

Azwanida, N. N. 2015. "A Review on the Extraction Methods Use in Medicinal Plants, Principle, Strength and Limitation." *Medicinal & Aromatic Plants* 04 (03). https://doi.org/10.4172/2167-0412.1000196

Bahloul, N., N. Boudhrioua, M. Kouhila, and N. Kechaou. 2009. "Effect of Convective Solar Drying on Colour, Total Phenols and Radical Scavenging Activity of Olive Leaves (Olea Europaea L.)." *International Journal of Food Science and Technology* 44 (12): 2561–67. https://doi.org/10.1111/J.1365-2621.2009.02084.X

Balkrishna, A., N. Sharma, D. Srivastava, A. Kukreti, S. Srivastava, and V. Arya. 2024. "Exploring the Safety, Efficacy, and Bioactivity of Herbal Medicines: Bridging Traditional Wisdom and Modern Science in Healthcare." *Future Integrative Medicine* 3 (1): 35–49. https://doi.org/10.14218/FIM.2023.00086

Bandawane, A., and R. Saudagar. 2019. "A Review on Novel Drug Delivery System: A Recent Trend." *Journal of Drug Delivery and Therapeutics* 9 (3): 517–21. https://doi.org/10.22270/JDDT.V9I3.2610

Bitwell, C., S. S. Indra, C. Luke, and M. K. Kakoma. 2023. "A Review of Modern and Conventional Extraction Techniques and Their Applications for Extracting Phytochemicals from Plants." *Scientific African* 19 (3): e01585. https://doi.org/10.1016/J.SCIAF.2023.E01585

Buszewski, B., and M. Szultka. 2012. "Past, Present, and Future of Solid Phase Extraction: A Review." *Critical Reviews in Analytical Chemistry* 42 (3): 198–213. https://doi.org/10.1080/07373937.2011.645413

Ceballos, A. M., G. I. Giraldo, and C. E. Orrego. 2012. "Effect of Freezing Rate on Quality Parameters of Freeze Dried Soursop Fruit Pulp." *Journal of Food Engineering* 111 (2): 360–65. https://doi.org/10.1016/J.JFOODENG.2012.02.010

Chang, Y., B. A. Hawkins, J. J. Du, P. W. Groundwater, D. E. Hibbs, and F. Lai. 2022. "A Guide to In Silico Drug Design." *Pharmaceutics* 15 (1): 49. https://doi.org/10.3390/PHARMACEUTICS15010049

Chongsrimsirisakhol, O., and T. Pirak. 2023. "Polyphenol Release and Antioxidant Activity of the Encapsulated Antioxidant Crude Extract from Cold Brew Spent Coffee Grounds Under Simulated Food Processes and an In Vitro Static Gastrointestinal Model." *Foods* 12 (5). https://doi.org/10.3390/foods12051000

Dai, J. W., J. Q. Rao, D. Wang, L. Xie, H. W. Xiao, Y. H. Liu, and Z. J. Gao. 2015. "Process-Based Drying Temperature and Humidity Integration Control Enhances Drying Kinetics of Apricot Halves." *Drying Technology* 33 (3): 365–76. https://doi.org/10.1080/07373937.2014.954667

Desai, K. G. H., and H. J. Park. 2005. "Recent Developments in Microencapsulation of Food Ingredients." *Drying Technology* 23 (7): 1361–94. https://doi.org/10.1081/DRT-200063478

Do, Q. D., A. E. Angkawijaya, P. L. Tran-Nguyen, L. H. Huynh, F. E. Soetaredjo, S. Ismadji, and Y. H. Ju. 2014. "Effect of Extraction Solvent on Total Phenol Content, Total Flavonoid Content, and Antioxidant Activity of Limnophila Aromatica." *Journal of Food and Drug Analysis* 22 (3): 296–302. https://doi.org/10.1016/J.JFDA.2013.11.001

Domínguez-Rodríguez, G., M. L. Marina, and M. Plaza. 2017. "Strategies for the Extraction and Analysis of Non-Extractable Polyphenols from Plants." *Journal of Chromatography A* 1514 (9): 1–15. https://doi.org/10.1016/J.CHROMA.2017.07.066

Ekor, M. 2014. "The Growing Use of Herbal Medicines: Issues Relating to Adverse Reactions and Challenges in Monitoring Safety." *Frontiers in Neurology* 4 (1). https://doi.org/10.3389/FPHAR.2013.00177

Emon, D. D., M. S. Islam, M. A. R. Mazumder, M. G. Aziz, and M. S. Rahman. 2025. "Recent Applications of Microencapsulation Techniques for Delivery of Functional Ingredient in Food Products: A Comprehensive Review." *Food Chemistry Advances* 6 (3): 100923. https://doi.org/10.1016/J.FOCHA.2025.100923

Erbay, Z., and F. Icier. 2010. "A Review of Thin Layer Drying of Foods: Theory, Modeling, and Experimental Results." *Critical Reviews in Food Science and Nutrition* 50 (5): 441–64. https://doi.org/10.1080/10408390802437063

Escobar, S., C. Bernal, J. M. Bolivar, B. Nidetzky, F. López-Gallego, and M. Mesa. 2018. "Understanding the Silica-Based Sol-Gel Encapsulation Mechanism of Thermomyces Lanuginosus Lipase: The Role of Polyethylenimine." *Molecular Catalysis* 449 (4): 106–13. https://doi.org/10.1016/J.MCAT.2018.02.024

Fernández-Ochoa, Á., F. J. Leyva-jiménez, M. L. Cádiz-gurrea, S. Pimentel-moral, and A. Segura-carretero. 2021. "The Role of High-resolution Analytical Techniques in the Development of Functional Foods." *International Journal of Molecular Sciences* 22 (6): 1–20. https://doi.org/10.3390/IJMS22063220

Gao, Y., X. Yang, L. Chu, Y. Zhang, and Q. Li. 2022. "Experimental Investigation and Thin-Layer Modelling of Cassava Slice Drying." *Journal of Thermal Analysis and Calorimetry* 147 (2): 1379–87. https://doi.org/10.1007/S10973-020-10401-X/FIGURES/6

Gibbs, B. F., S. Kermasha, I. Alli, and C. N. Mulligan. 1999. "Encapsulation in the Food Industry: A Review." *International Journal of Food Sciences and Nutrition* 50 (3): 213–24. https://doi.org/10.1080/096374899101256

Grice, J. W., and M. Iwasaki. 2009. "A Truly Multivariate Approach to Manova." *Applied Multivariate Research* 12 (3): 199. https://doi.org/10.22329/AMR.V12I3.660

Guerra-Valle, M., G. Petzold, and P. Orellana-Palma. 2022. "Optimization of Encapsulation by Ionic Gelation Technique of Cryoconcentrated Solution: A Response Surface Methodology and Evaluation of Physicochemical Characteristics Study." *Polymers* 14 (5): 1031. https://doi.org/10.3390/POLYM14051031/S1

Hemati, M., R. Cherif, K. Saleh, and V. Pont. 2003. "Fluidized Bed Coating and Granulation: Influence of Process-Related Variables and Physicochemical Properties on the Growth Kinetics." *Powder Technology* 130 (1–3): 18–34. https://doi.org/10.1016/S0032-5910(02)00221-8

Hu, Y., W. Zhang, Z. Ke, Y. Li, and Z. Zhou. 2017. "In Vitro Release and Antioxidant Activity of Satsuma Mandarin (Citrus Reticulata Blanco Cv. Unshiu) Peel Flavonoids Encapsulated by Pectin Nanoparticles." *International Journal of Food Science and Technology* 52 (11): 2362–73. https://doi.org/10.1111/IJFS.13520

Jeevarathinam, G., R. Pandiselvam, T. Pandiarajan, P. Preetha, T. Krishnakumar, M. Balakrishnan, V. Thirupathi, S. Ganapathy, and D. Amirtham. 2022. "Design, Development, and Drying Kinetics of Infrared-Assisted Hot Air Dryer for Turmeric Slices." *Journal of Food Process Engineering* 45 (6): e13876. https://doi.org/10.1111/JFPE.13876

Jia, Q., Q. He, L. Yao, M. Li, J. Lin, Z. Tang, X. Zhu, and X. Xiang. 2022. "Utilization of Physiologically Based Pharmacokinetic Modeling in Pharmacokinetic Study of Natural Medicine: An Overview." *Molecules* 27 (24): 8670. https://doi.org/10.3390/MOLECULES27248670

Kage, H., R. Abe, R. Hattanda, T. Zhou, H. Ogura, and Y. Matsuno. 2003. "Effect of Solid Circulation Rate on Coating Efficiency and Agglomeration in Circulating Fluidized Bed Type Coater." *Powder Technology* 130 (1–3): 203–10. https://doi.org/10.1016/S0032-5910(02)00267-X

Kholaf, G. M., E. G. Gomaa, and H. M. Ziena. 2017. "Antimicrobial Activity of Some Egyptian Citrus Peels Extracts." *Alexandria Science Exchange Journal* 38 (6): 872–83. https://doi.org/10.21608/ASEJAIQJSAE.2017.4771

Krakowska-Sieprawska, A., A. Kiełbasa, K. Rafińska, M. Ligor, and B. Buszewski. 2022. "Modern Methods of Pre-Treatment of Plant Material for the Extraction of Bioactive Compounds." *Molecules* 27 (3): 730. https://doi.org/10.3390/MOLECULES27030730

Kumar, A., R. K. Rout, and P. S. Rao. 2022. "Effect of Drying Methods on Physico-Chemical and Bioactive Compounds of Mandarin (Citrus Reticulata) Peel." *International Journal of Food Engineering* 18 (10–11): 689–700. https://doi.org/10.1515/IJFE-2022-0121

Kumar, A., R. K. Rout, P. S. Rao, and P. Prabhakar. 2023. "Optimization of Pulsed Mode Sonication and in Silico Molecular Docking Study for Antioxidant Properties of Mandarin (Citrus Reticulata Blanco) Peels." *Journal of Food Process Engineering* 46 (6): e14111. https://doi.org/10.1111/JFPE.14111

Kumar, A., and P. Srinivasa Rao. 2020. "Optimization of Pulsed-Mode Ultrasound Assisted Extraction of Bioactive Compounds from Pomegranate Peel Using Response Surface Methodology." *Journal of Food Measurement and Characterization* 14 (6): 3493–507. https://doi.org/10.1007/S11694-020-00597-9

Laureanti, E. J. G., T. S. Paiva, L. M. de Matos Jorge, and R. M. M. Jorge. 2023. "Microencapsulation of Bioactive Compound Extracts Using Maltodextrin and Gum Arabic by Spray and Freeze-Drying Techniques." *International Journal of Biological Macromolecules* 253 (12): 126969. https://doi.org/10.1016/J.IJBIOMAC.2023.126969

Lee, T., and Y. H. Chang. 2020. "Structural, Physicochemical, and in-Vitro Release Properties of Hydrogel Beads Produced by Oligochitosan and de-Esterified Pectin from Yuzu (Citrus Junos) Peel as a Quercetin Delivery System for Colon Target." *Food Hydrocolloids* 108 (6): 106086. https://doi.org/10.1016/j.foodhyd.2020.106086

Li, Q., Z. Wang, C. Dai, Y. Wang, W. Chen, X. Ju, J. Yuan, and R. He. 2019. "Physical Stability and Microstructure of Rapeseed Protein Isolate/Gum Arabic Stabilized Emulsions at Alkaline PH." *Food Hydrocolloids* 88 (3): 50–7. https://doi.org/10.1016/J.FOODHYD.2018.09.020

Ma, Y. Q., X. Q. Ye, Z. X. Fang, J. C. Chen, G. H. Xu, and D. H. Liu. 2008. "Phenolic Compounds and Antioxidant Activity of Extracts from Ultrasonic Treatment of Satsuma Mandarin (Citrus Unshiu Marc.) Peels." *Journal of Agricultural and Food Chemistry* 56 (14): 5682–90. https://doi.org/10.1021/JF072474O

Marcos Luciano Bruschi. 2015. "Mathematical Models of Drug Release." In Strategies to Modify the Drug Release from Pharmaceutical Systems, 63–86. https://doi.org/10.1016/b978-0-08-100092-2.00005-9

Markham, K. R., and T. J. Mabry. 1975. "Ultraviolet-Visible and Proton Magnetic Resonance Spectroscopy of Flavonoids." In *The Flavonoids*, 45–77. https://doi.org/10.1007/978-1-4899-2909-9_2

Mullis, A. S., S. R. Broderick, A. M. Binnebose, N. Peroutka-Bigus, B. H. Bellaire, K. Rajan, and B. Narasimhan. 2019. "Data Analytics Approach for Rational Design of Nanomedicines with Programmable Drug Release." *Molecular Pharmaceutics* 16 (5): 1917–28. https://doi.org/10.1021/ACS.MOLPHARMACEUT.8B01272

Naima, R., M. Oumam, H. Hannache, A. Sesbou, B. Charrier, A. Pizzi, and F. Charrier - El Bouhtoury. 2015. "Comparison of the Impact of Different Extraction Methods on Polyphenols Yields and Tannins Extracted from Moroccan Acacia Mollissima Barks." *Industrial Crops and Products* 70 (8): 245–52. https://doi.org/10.1016/J.INDCROP.2015.03.016

Norcino, L. B., J. F. Mendes, J. A. Figueiredo, N. L. Oliveira, D. A. Botrel, and L. H. C. Mattoso. 2022. "Development of Alginate/Pectin Microcapsules by a Dual Process Combining Emulsification and Ultrasonic Gelation for Encapsulation and Controlled Release of Anthocyanins from Grapes (Vitis Labrusca L.)." *Food Chemistry* 391 (4). https://doi.org/10.1016/j.foodchem.2022.133256

Okazaki, A., T. Mano, and K. Sugano. 2008. "Theoretical Dissolution Model of Poly-Disperse Drug Particles in Biorelevant Media." *Journal of Pharmaceutical Sciences* 97 (5): 1843–52. https://doi.org/10.1002/JPS.21070

Parveen, A., B. Parveen, R. Parveen, and S. Ahmad. 2015. "Challenges and Guidelines for Clinical Trial of Herbal Drugs." *Journal of Pharmacy and Bioallied Sciences* 7 (4): 329–33. https://doi.org/10.4103/0975-7406.168035

Paul, D. R. 2011. "Elaborations on the Higuchi Model for Drug Delivery." *International Journal of Pharmaceutics* 418 (1): 13–17. https://doi.org/10.1016/j.ijpharm.2010.10.037.

Peltzer, K., and S. Pengpid. 2019. "The Use of Herbal Medicines among Chronic Disease Patients in Thailand: A Cross-Sectional Survey." *Journal of Multidisciplinary Healthcare* 12: 573–82. https://doi.org/10.2147/JMDH.S212953

Pereira Silveira, M., F. L. Chaves Almeida, I. Dutra Alvim, and A. Silvia Prata. 2023. "Encapsulation of Pomegranate Polyphenols by Ionic Gelation: Strategies for Improved Retention and Controlled Release." *Food Research International* 174 (12): 113590. https://doi.org/10.1016/J.FOODRES.2023.113590

Picos-Salas, M. A., J. B. Heredia, N. Leyva-López, D. L. Ambriz-Pérez, and E. P. Gutiérrez-Grijalva. 2021. "Extraction Processes Affect the Composition and Bioavailability of Flavones from Lamiaceae Plants: A Comprehensive Review." *Processes* 9 (9). https://doi.org/10.3390/PR9091675

Ranjha, M. M. A. N., S. Irfan, J. M. Lorenzo, B. Shafique, R. Kanwal, M. Pateiro, and R. N. Arshad, et al. 2021. "Sonication, a Potential Technique for Extraction of Phytoconstituents: A Systematic Review." *Processes* 9 (8). https://doi.org/10.3390/PR9081406

Rijke, E., P. Out, W. M. A. Niessen, F. Ariese, C. Gooijer, and U. A. T. Brinkman. 2006. "Analytical Separation and Detection Methods for Flavonoids." *Journal of Chromatography A* 1112 (1–2): 31–63. https://doi.org/10.1016/J.CHROMA.2006.01.019

Rout, R. K., A. Kumar, and P. S. Rao. 2021a. "Encapsulation of Oregano (Origanum Vulgare) Leaf Polyphenols: Development, Characterization and in-Vitro Release Study." *Food Hydrocolloids for Health* 1 (8): 100028. https://doi.org/10.1016/j.fhfh.2021.100028

Rout, R. K., A. Kumar, and P. S. Rao. 2021b. "A Multivariate Optimization of Bioactive Compounds Extracted from Oregano (Origanum Vulgare) Leaves Using Pulsed Mode Sonication." *Journal of Food Measurement and Characterization* 15 (4): 3111–22. https://doi.org/10.1007/S11694-021-00902-0

Rout, R. K., A. Kumar, and P. S. Rao. 2024. "A Comparative Assessment of Drying Kinetics, Energy Consumption, Mathematical Modeling, and Multivariate Analysis of Indian Borage (Plectranthus Amboinicus) Leaves." *Journal of Food Process Engineering* 47 (5). https://doi.org/10.1111/JFPE.14630

Sabina, E. P., and M. Rasool. 2008. "An in Vivo and in Vitro Potential of Indian Ayurvedic Herbal Formulation Triphala on Experimental Gouty Arthritis in Mice." *Vascular Pharmacology* 48 (1): 14–20. https://doi.org/10.1016/J.VPH.2007.11.001

Saénz, C., S. Tapia, J. Chávez, and P. Robert. 2009. "Microencapsulation by Spray Drying of Bioactive Compounds from Cactus Pear (Opuntia Ficus-Indica)." *Food Chemistry* 114 (2): 616–22. https://doi.org/10.1016/J.FOODCHEM.2008.09.095

Sahoo, N., P. Manchikanti, and S. Dey. 2010. "Herbal Drugs: Standards and Regulation." *Fitoterapia* 81 (6): 462–71. https://doi.org/10.1016/J.FITOTE.2010.02.001

Salm, S., J. Rutz, M. van den Akker, R. A. Blaheta, and B. E. Bachmeier. 2023. "Current State of Research on the Clinical Benefits of Herbal Medicines for Non-Life-Threatening Ailments." *Frontiers in Pharmacology* 14 (9): 1234701. https://doi.org/10.3389/FPHAR.2023.1234701/XML

Sasidharan, S., Y. Chen, D. Saravanan, K. M. Sundram, and L. Yoga Latha. 2011. "Extraction, Isolation and Characterization of Bioactive Compounds from Plants' Extracts." *African Journal of Traditional, Complementary and Alternative Medicines*. https://doi.org/10.4314/ajtcam.v8i1.60483

Shihab, J. M., K. T. Rashid, and M. A. Toma. 2022. "A Review on Membrane Technology Application for Vegetable Oil Purification Processes." *International Journal of Food Engineering* 18 (10–11): 655–77. https://doi.org/10.1515/IJFE-2022-0058/HTML

Singh, B., V. Sharma, A. Dhiman, and M. Devi. 2012. "Design of Aloe Vera-Alginate Gastroretentive Drug Delivery System to Improve the Pharmacotherapy." *Polymer - Plastics Technology and Engineering* 51 (13): 1303–14. https://doi.org/10.1080/03602559.2012.698684

Singh, H., and G. Sharma. 2023. "Recent Development of Novel Drug Delivery of Herbal Drugs." *RPS Pharmacy and Pharmacology Reports* 2 (4): 1–8. https://doi.org/10.1093/RPSPPR/RQAD028

Smeriglio, A., D. Barreca, E. Bellocco, and D. Trombetta. 2017. "Proanthocyanidins and Hydrolysable Tannins: Occurrence, Dietary Intake and Pharmacological Effects." *British Journal of Pharmacology* 174 (11): 1244–62. https://doi.org/10.1111/BPH.13630

Sofowora, A., E. Ogunbodede, and A. Onayade. 2013. "The Role and Place of Medicinal Plants in the Strategies for Disease Prevention." *African Journal of Traditional, Complementary and Alternative Medicines* 10 (5): 210–29. https://doi.org/10.4314/AJTCAM.V10I5.2

Sthle, L., and S. Wold. 1990. "Multivariate Analysis of Variance (MANOVA)." *Chemometrics and Intelligent Laboratory Systems* 9 (2): 127–41. https://doi.org/10.1016/0169-7439(90)80094-M

Sun, S., Y. Wang, A. Wu, Z. Ding, and X. Liu. 2019. "Influence Factors of the Pharmacokinetics of Herbal Resourced Compounds in Clinical Practice." *Evidence-Based Complementary and Alternative Medicine: ECAM* 2019: 1983780. https://doi.org/10.1155/2019/1983780

Tulunay, M., C. Aypak, H. Yikilkan, and S. Gorpelioglu. 2015. "Herbal Medicine Use among Turkish Patients with Chronic Diseases." *Journal of Intercultural Ethnopharmacology* 4 (3): 217. https://doi.org/10.5455/JICE.20150623090040

Venkataramanan, R., B. Komoroski, and S. Strom. 2006. "In Vitro and in Vivo Assessment of Herb Drug Interactions." *Life Sciences* 78 (18): 2105–15. https://doi.org/10.1016/J.LFS.2005.12.021

Wang, X., N. Ahmed, G. Alvarez, M. Tuttolomondo, C. Helary, M. Desimone, and T. Coradin. 2015. "Sol-Gel Encapsulation of Biomolecules and Cells for Medicinal Applications." *Current Topics in Medicinal Chemistry* 15 (3): 223–44. https://doi.org/10.2174/1568026614666141229112734

Wimalasiri, V. W., S. P. Dunuweera, A. N. Dunuweera, and R.M.G. Rajapakse. 2021. "Noyes-Whitney Dissolution Model-Based PH-Sensitive Slow Release of Paclitaxel (Taxol) from Human Hair-Derived Keratin Microparticle Carriers." *BioMed Research International* 2021: 6657482. https://doi.org/10.1155/2021/6657482

Wu, I. Y., S. Bala, N. Škalko-Basnet, and M. P. di Cagno. 2019. "Interpreting Non-Linear Drug Diffusion Data: Utilizing Korsmeyer-Peppas Model to Study Drug Release from Liposomes." *European Journal of Pharmaceutical Sciences* 138 (7): 105026. https://doi.org/:10.1016/j.ejps.2019.105026

Yap, J. Y., C. L. Hii, S. P. Ong, K. H. Lim, F. Abas, and K. Y. Pin. 2020. "Effects of Drying on Total Polyphenols Content and Antioxidant Properties of Carica Papaya Leaves." *Journal of the Science of Food and Agriculture* 100 (7): 2932–37. https://doi.org/10.1002/JSFA.10320

Young, S. L., X. Sarda, and M. Rosenberg. 1993. "Microencapsulating Properties of Whey Proteins. 1. Microencapsulation of Anhydrous Milk Fat." *Journal of Dairy Science* 76 (10): 2868–77. https://doi.org/10.3168/JDS.S0022-0302(93)77625-0

Zhang, J., I. J. Onakpoya, P. Posadzki, and M. Eddouks. 2015. "The Safety of Herbal Medicine: From Prejudice to Evidence." *Evidence-Based Complementary and Alternative Medicine: ECAM* 2015: 316706. https://doi.org/10.1155/2015/316706

Zia, S., M. R. Khan, M. A. Shabbir, A. A. Maan, M. K. I. Khan, M. Nadeem, A. A. Khalil, A. Din, and R. M. Aadil. 2022. "An Inclusive Overview of Advanced Thermal and Nonthermal Extraction Techniques for Bioactive Compounds in Food and Food-Related Matrices." *Food Reviews International* 38 (6): 1166–96. https://doi.org/10.1080/87559129.2020.1772283

Zuidam, N. J., and V. A. Nedović. 2010. "Encapsulation Technologies for Active Food Ingredients and Food Processing." In *Encapsulation Technologies for Active Food Ingredients and Food Processing*, 1–400. https://doi.org/10.1007/978-1-4419-1008-0/COVER

Zuidam, N. J., and E. Shimoni. 2010. "Overview of Microencapsulates for Use in Food Products or Processes and Methods to Make Them." In *Encapsulation Technologies for Active Food Ingredients and Food Processing*, 3–29. https://doi.org/10.1007/978-1-4419-1008-0_2

7 Quality Control, Laws and Regulatory Affairs in Herbal Drugs

Niharika Sahoo Bhattacharya and Sukanta Shekhar Bhattacharya

7.1 INTRODUCTION

Herbal medicines are the basis of traditional healthcare systems and are becoming more and more popular in modern system of medicine (Sahoo, Manchikanti, and Dey 2010; Welz, Emberger-Klein, and Menrad 2018). Plant-derived compounds have a long history of uses across continents. Starting from traditional healing practices to modern day phytomedicines, all showcase the use of plants, plant parts, and plant-derived chemical compounds in the treatment of various ailments. There has been a growing demand of herbal medicines around the world owing to the notion that plant-derived medicines are safer as compared to synthetic chemical pharmaceuticals (Karimi, Majlesi, and Rafieian-Kopaei 2015). It is believed that herbal medicines have lesser side effects and many times it is consumed without any supervision or prescription. However, these popular medicines are not completely risk-free. There have been reported cases of side effects and contraindications and interaction with existing pharmaceuticals (Ekor 2014). Herbal medicines are available as food supplements, functional food and health product, or medicines, which makes it difficult to manage. A streamlined regulatory system can help in ensuring proper quality of herbal medicines. In this context, this chapter examines the prevalence and regulation of herbal medicines worldwide, focusing specifically on three key regions: India, the United States (US), and the European Union (EU). A comparative analysis of their historical, socio-cultural, market, and regulatory contexts offers essential understanding of the evolving role of herbal medicines and the challenges encountered by this sector.

7.2 IMPORTANCE OF HERBAL MEDICINE IN HISTORY AND CULTURE

The use of herbs and different plant materials dates back to ancient time, and it still remains an integral aspect of healthcare for many populations worldwide. In India, herbal medicine is closely linked to the old systems of Ayurveda, Siddha, and Unani, which are all backed by the Ministry of AYUSH. Ayurveda, which is more than 5000 years old, is based on traditional writings (Table 7.1) and believes in a holistic approach for physical and mental wellbeing. It focuses on using herbs

DOI: 10.1201/9781003623748-7

TABLE 7.1
Different Texts of Ayurveda

Charaka Samhita	Describes all guiding principles of Ayurveda for diagnosis and treatment of diseases
Sushruta Samhita	This samhita mainly deals with surgery, describes layers of skin, cosmetic surgery, grafting, etc. It covers more than 1000 diseases
Vagbhata Samhita	A concise treatise used by all physicians, which covers all subjects
Madhava Nidana	A book on pathology of diseases which includes causative factors, signs, and symptoms, and pathogenesis of the diseases in general and for specific diseases have been dealt with in great extent
Sharangdhara Samhita	This book describes various single and compound drug formulations and their method of preparation
Bhavaprakasha Nighantu	The material medica of Ayurveda includes description of more than 100 herbs. It describes principles and pharmacodynamics of single drugs

Source: Directorate of Ayush, 2025 (From Ayush 2025).

to keep people healthy and treat diseases by balancing the body, mind, and spirit. Ayurveda is considered as a scientific art of healing, and it is thriving because of its faith in its principles and abundance of varieties of herbs in Indian sub-continent (AYUSH 2025). Indian herbal therapeutics are not limited to rural or traditional contexts; they constitute a substantial segment of healthcare alternatives in both urban and semi-urban areas. Herbal products like neem *(Azadirachta indica)*, ashwagandha (*Withania somnifera*), tulsi (Ocimum sanctum), and turmeric (*Curcuma longa*) are well-known in Indian homes and are becoming more popular around the world because they have been shown to have therapeutic effects (Ayati et al. 2019; Bharath et al. 2023; Dipankar et al. 2025; Mehnaz et al. 2024; Prakash and Gupta 2005; Tallon, Koturbash, and Blum 2025).

In the United States, approximately 20% of the population use herbal medicines and they are mostly available as dietary supplements (Bent 2008). Ginkgo, Garlic, St. John's Wort, Soy, and kava are few of the popular herbs, which are available as dietary supplements in US. Native American healing traditions played a big role in the early American herbal knowledge. However, immigrant communities and integrative health trends have changed how herbal medicine is used in the US today. The 1994 Dietary Supplement Health and Education Act (DSHEA) made it possible for a regulated market for herbal supplements. Today, this market is worth billions of dollars and millions of people use botanicals to stay healthy and for specific medical purposes. Even though safety and manufacturing standards are important to regulators, herbal medicines are still mostly seen as supplements rather than drugs. This is because the focus is on giving consumers more options and letting them make informed choices rather than making therapeutic claims (Barnes, Bloom, and Nahin 2008).

Countries within the EU also have long histories of using standardized herbal medicines, illustrated by phytomedicines like St. John's Wort (*Hypericum perforatum*) and Valerian (*Valeriana officinalis*) that have explicit therapeutic indications

(Calapai 2008). The EU exemplifies a region with an intricate relationship between traditional herbal knowledge and strict regulatory oversight. Europe's herbal medicinal tradition includes not only folk medicine but also homeopathy and phytotherapy. These are all supported by scientific pharmacognosy. The Traditional Herbal Medicinal Products Directive (Directive 2004/24/EC) provides a system for registering and selling herbal products, which have been used safely for a long time without needing full drug approval. This protects consumers while still giving them access to these products.

7.3 SIZE AND GROWTH OF THE HERBAL MEDICINE MARKET

Worldwide the herbal medicines market has grown faster than ever before as more people are interested in natural and holistic treatments owing to the multiple reported side effects of synthetic drugs. In the early 2020s, the global market for herbal supplements was worth several billion dollars and it is expected to keep growing. There are a few big reasons why herbal medicines are still used and growing in these areas. First, the increasing number of chronic diseases and multiple illnesses around the world has made patients and healthcare providers look for alternate ways to treat them that might help with symptoms with fewer side effects. Herbal medicines frequently exhibit multi-targeted bioactivities, which are particularly effective for intricate conditions like arthritis, diabetes, and mental health disorders. Second, growing popularity of natural and eco-friendly health products are making people choose herbal products over synthetic drugs. This trend is especially strong among young people who are health conscious and live in cities around the world. Third, advances in technology have improved the scientific knowledge and quality of herbal medicines, making people less worried about their safety and effectiveness. These include advanced phytochemical analyses, standardization techniques, and clinical research that confirm traditional uses and make it easier for regulators to approve them. Lastly, the integration of traditional medicine with modern healthcare is being emphasized by various governments and the innovation and quality control aspects have been prioritized for strengthening the sector's commercial potential.

The herbal medicines and nutraceutical form a big part of the larger AYUSH industry in India. It is worth several billion dollars and is growing by more than 15% each year (FICCI 2025). Both domestic consumption and rising export demand are driving market growth. This is especially true for countries with large South Asian diaspora populations and natural health markets. India's herbal market is becoming more competitive on the world stage. The government is doing this by supporting new ideas in traditional medicine, quality certifications, export incentives, and protections for intellectual property.

The US herbal product market is very varied, with sales of more than $10 billion a year. It includes everything from dietary supplements to herbal remedies that can be bought without a prescription. In the US, people who buy wellness and disease prevention products tend to prefer botanical products. This is especially true for products that support the immune system, help with weight management, and improve mental health. The country allows dietary supplements to be marketed with structure-function claims, but it does not allow claims about treating diseases unless they go through regular drug approval pathways. This keeps herbal supplements separate from approved botanical drugs.

For herbal medicines, there is a lot of regulatory harmony between EU member states, with a focus on safety, effectiveness, and quality control. The market for herbal products is worth several billion euros, and there are many customers who value both traditional and scientifically proven products. The EU also cares about sustainability and getting botanicals from ethical sources. This affects how the market works and how products are made in this region. Regulatory pathways like the Traditional Herbal Registration scheme help people trust the products in this market (EMA 2001).

The global herbal medicine market is worth about USD 233.08 billion in 2024 and is expected to grow at a rate of 8.23% per year to reach USD 437 billion by 2032. However, this sector faces problems with quality control and regulatory harmonization. Even though herbal medicines are one of the oldest types of medicine and are used by about 80% of the world's population for primary healthcare needs, they still have problems with quality control ensuring the safety and effectiveness of the product. Quality control problems happen all over the world because of inconsistent manufacturing, gaps in regulations, contamination, and adulteration. These problems have the potential to cause side effects and put consumers at risk.

7.4 OVERVIEW OF HERBAL MEDICINE SYSTEMS

The herbal medicines, which are once rooted in traditional healing practices and folklore, are now developing into an evidence-based segment of modern healthcare. The rising consumer interest in natural health products for their physical and mental well-being is increasing the acceptance of such products as complementary and alternative medicine. Among the major global players, India, US, and EU, each represent vast and influential markets, shaped by distinct cultural, regulatory, and economic factors.

7.4.1 Traditional Herbal Medicine Systems of India

India stands as one of the world's largest consumers and producers of herbal medicines, largely due to its rich heritage in traditional medicinal systems such as Ayurveda, Siddha, Unani, and homeopathy. The country's herbal market is intrinsically linked to the cultural acceptance and governmental support through the Ministry of AYUSH (Ayurveda, Yoga and Naturopathy, Unani, Siddha, and Homeopathy).

Ayurveda, Siddha, and Unani are the primary traditional medicinal systems widely practiced in India, each with unique theoretical foundations, diagnostic methods, and herbal therapeutics. These systems collectively serve millions across the nation and contribute significantly to global alternative medicine markets. This section examines each system's history, philosophy, herbal formulations, and contemporary relevance, highlighting their intrinsic role in India's healthcare and herbal medicine sector.

7.4.1.1 Ayurveda: The Science of Life

Ayurvedic therapy emphasizes holistic healing, integrating diet, lifestyle changes, detoxification practices (Panchakarma), therapeutic massages, yoga, and, notably, herbal medicines. Hundreds of plants and minerals are used in various preparations such as decoctions, powders, oils, and tablets. Herbal formulations often include combinations designed to restore doshic balance and promote overall well-being. Iconic herbs include ashwagandha (*Withania somnifera*), renowned for its adaptogenic properties; tulsi (*Ocimum sanctum*), valued for respiratory and immune support; and turmeric (*Curcuma longa*), widely recognized for anti-inflammatory effects.

The foundational Ayurvedic textbooks, i.e., the samhitas, accurately describe the pharmacognosy of medicinal plants, their therapeutic properties, dosages, and preparation methods. These texts remain authoritative sources and form the basis for modern Ayurvedic education and practice. The Ministry of AYUSH actively promotes Ayurveda through educational institutes, research bodies, and certification of herbal products, facilitating global recognition and export (Ministry of Ayush 2025).

7.4.1.2 Siddha: The Ancient Medicine of Tamil Nadu

Siddha medicine is also one of the oldest medical systems in the world. The Siddha system is believed to have evolved from 10000 to 4000 BC. With its Dravidian origin, it serves mostly the people living in southern India, Sri Lanka, Malaysia, Singapore, Mauritius, and some of the South-East Asian countries (Society for the Journal of Indian Medicine and Homoeopathy 2025). This system of medicine is traditionally linked to the Siddhars, the mystic saints who are said to have attained spiritual enlightenment and divine knowledge, the Siddha system integrates spiritual and material knowledge to promote health (Ministry of Ayush 2025). Siddha system provides a holistic and scientific approach of treatment, which is used in preventive, curative, and promotive healthcare (CCRS, Ministry of Ayush 2025). Siddha's pharmacopoeia extensively uses herbs but also incorporates minerals and metals, prepared through elaborate purification processes aimed at detoxification and enhancing therapeutic efficacy. Herbal formulations in Siddha tend to be complex, including powders, oils, and medicinal wines, targeting chronic diseases, respiratory ailments, skin disorders, and digestive problems. Like Ayurveda, Siddha texts such as the Siddha maruthuvam literature provide detailed descriptions of medicinal plants and compound formulations. The government and academic institutions have been contributing to research and standardization of Siddha herbal drugs, striving to align traditional preparations with modern quality controls. Notably, the Central Council for Research in Siddha (CCRS) plays a pivotal role in advancing Siddha medicine through scientific validation. This organization is established to promote, co-ordinate, and conduct research in SSM, focusing on Drug Research, Literary Research, Preclinical, Clinical studies.

7.4.1.3 Unani System of Medicine

Unani medicine traces its origins to the father of medicine, Greek physician Hippocrates. This was further developed by Arab and Persian scholars such as Ibn Sina (Avicenna). It was introduced to India by the Arabs and consolidated under Mughal patronage, evolving into a major traditional medical system practiced widely particularly in northern India. Unani therapy includes herbal medicines, dietary measures, regimental therapies like massage and cupping, and surgery in some cases. The materia medica of Unani comprises hundreds of herbs, many of which overlap with Ayurvedic and Persian medicinal materials (The Unani system of medicine,

Ministry of Ayush 2025). Notable herbs in Unani practice include henna (*Lawsonia inermis*) for skin conditions, black seed (*Nigella sativa*) for immune enhancement, and licorice (*Glycyrrhiza glabra*) for respiratory and digestive issues (AYUSH 2013). The Unani pharmacopoeia embraces herbal drugs often in compound formulations like decoctions, powders, and electuaries, tailored to restore humoral balance. Educational efforts by institutions such as the Central Council of Indian Medicine focus on preserving classical Unani knowledge while incorporating modern research methodologies to verify efficacy and safety. In the area of research and development in Unani System of Medicine, the Central Council for Research in Unani Medicine (CCRUM) is an apex organization functioning under the Ministry of AYUSH, Government of India (CCRUM 2025).

7.4.1.4 Common Themes and Contemporary Integration

Ayurveda, Siddha, and Unani share many similarities despite having differences in their philosophy and methodology. All of them are holistic in nature, which takes into account the patient's life style, diet, mental health and use of herbal therapeutics. The respective herbal pharmacopeias reflect centuries of empirical knowledge and local biodiversity utilization. The Ministry of AYUSH plays an guiding role, which aims at promoting development, regulation and promotion of ASU medicines (Ministry of Ayush 2025).

7.4.2 United States Herbal Medicine Market

The advent of integrative medicine and growing acceptance by healthcare professionals have significantly contributed to the growth of herbal market. Raw herbals, standardized extracts, combined formulations, and functional foods fortified with herbal ingredients, etc., constitute major product portfolio in US market. Leading companies invest heavily in marketing, scientific research, and compliance with current Good Manufacturing Practices (cGMP) to appeal to a health-conscious consumer base.

Despite regulatory leniency on marketing claims compared to pharmaceutical drugs, the US Food and Drug Administration (FDA) actively monitors adverse event reports and manufacturing standards, enhancing consumer safety. The demand for transparency, quality certification, and evidence-based products is increasing amidst a competitive landscape fueled by natural and organic product trends (Barnes et al. 2021). Globally, consumer skepticism over synthetic drugs and antibiotic resistance also contributes to herbal product uptake (Rossiter, Fletcher, and Wuest 2017).

7.4.3 EU Herbal Medicine Market

The EU herbal medicine market is distinguished by its highly regulated environment, which emphasizes safety, efficacy, and quality control. Legislative frameworks such as the Traditional Herbal Medicinal Products Directive (Directive 2004/24/EC) provide a simplified registration pathway for traditional herbal medicines with an established safety profile, allowing products to enter the market with recognized traditional use instead of full clinical trials (EMA 2025). This legislation has resulted in a proliferation of registered herbal medicines across member states and harmonized market access.

Sustainability and ethical sourcing are paramount in the EU herbal market, driven by consumer preferences and regulatory frameworks promoting biodiversity conservation and fair trade (Gatt, Vella Bonanno, and Zammit 2024). These factors impact supply chains and product marketing, setting the EU apart as a market committed to environmental and social responsibility. In addition to traditional herbal medicines, the EU market includes a significant sector of food supplements containing herbal extracts. The European Food Safety Authority (EFSA) regulates health claims on such products stringently, affecting product formulation and advertisement (EC 2006). The EU's focus on scientific substantiation and consumer protection ensures that herbal medicine continues to be integrated cautiously yet progressively into healthcare systems.

Despite the promising growth, each region encounters specific challenges affecting market dynamics. India faces hurdles in quality control infrastructure, standardization, and regulatory coherence among traditional and contemporary herbal products. The US grapples with balancing consumer access against safety concerns, especially given the variable quality and unsubstantiated claims in a largely self-regulated supplement market. The EU must continually reconcile the need for rigorous scientific evidence with respect for traditional knowledge and consumer demand. In this context, the present chapter discusses the challenges in herbal drug regulation and quality control aspect in ensuring safe, effective, and quality medicine. It further enumerates the possible solutions and directions for future enhancement in this area.

7.5 THE GLOBAL LANDSCAPE OF HERBAL MEDICINE QUALITY CONTROL

There has been a gradual merge in the traditional medicinal knowledge and modern evidence-based treatment system. While the conventional synthetic chemically derived drugs are reproduced with uniform characteristics, the herbal drugs or compound of botanical origin exhibits inherent variation in properties. This necessitates specialized approaches for quality control and regulatory oversight for herbal medicinal products.

The complexity of herbal medicine quality control stems from multiple factors. The multi-component nature of botanical extracts, seasonal, and geographical variations in active constituents, challenges in botanical authentication, and the diverse processing methods employed across different traditional medicine systems(Balekundri and Mannur 2020, Muyumba et al. 2021). These challenges are further compounded by the significant regulatory disparities that exist between major markets, creating a regulatory hurdle, where the same herbal product may face vastly different quality standards and approval processes depending on its intended market (Sahoo, Manchikanti, and Dey 2010). The following sections explore the complex role of quality control in herbal medicines, specifying its impact on safety assurance, efficacy validation, and market acceptance. Through rigorous quality control practices, manufacturers, regulators, and consumers all benefit from enhanced confidence in herbal therapeutics.

7.5.1 Ensuring Safety Through Quality Control

Safety is the foremost criterion in the use of any medicinal product including herbal medicines, despite their natural origin, are no exception. Natural products do not automatically equate to safe products. The assumption is proven misleading by the numerous reports on botanical products containing toxic compounds, contaminants, or adulterants, which make the drug unsafe. Effective quality control helps to identify and mitigate these risks before products reach consumers. Key safety concerns addressed by quality control include contamination with heavy metals (such as lead, mercury, and arsenic) (Luo et al. 2020; Validandi et al. 2025), microbial contamination including pathogenic bacteria and fungi (Lima et al. 2020; Turkson et al. 2020; WHO 2007), residual pesticides from cultivation, and adulteration with synthetic drugs or undeclared substances(Ahiabor, Darkwah, and Donkor 2024). For example, studies have shown instances where Ayurvedic products contained harmful levels of heavy metals due to either environmental accumulation or manufacturing lapses (Mikulski et al. 2017). Therefore, standardized testing protocols for contaminants are integral to a quality control system.

Moreover, raw material authentication is critical to safety. Substitution or adulteration with incorrect species can introduce unexpected toxic effects or reduce therapeutic efficacy. Morphological, microscopic, and molecular identification techniques (including DNA barcoding) are now widely applied in quality control laboratories to verify plant identity (Balekundri and Mannur 2020).

Stability testing during the manufacturing process ensures that herbs retain their chemical integrity throughout the product development and commercialization. Additionally, microbial limit testing ensures that the products are free from harmful microbes, which is crucial for patient safety, especially in immunocompromised individuals (Fong 2002). Overall, consistent application of quality control protocols minimizes health risks and builds trust among healthcare providers and consumers alike.

7.5.2 Validating Efficacy Through Standardization and Quality Parameters

One of the greatest challenges faced by herbal medicine producers and regulators has been the variability inherent in botanical materials. Unlike single-molecule pharmaceuticals, herbs contain complex mixtures of phytochemicals whose concentrations fluctuate due to environmental, genetic, and processing factors. Quality control functions to minimize this variability through standardization, thus ensuring consistent therapeutic outcomes.

Standardization involves defining specific quality parameters such as the amount or concentration of bioactive or marker compounds within the herbal preparation.

Chromatographic techniques, such as high-performance liquid chromatography (HPLC), high-performance thin-layer chromatography (HPTLC), and gas chromatography (GC), combined with spectroscopic methods, play crucial roles in phytochemical profiling and quantification (Dinakaran, Chelle, and Avasarala 2019; Mandal et al. 2024; Moshari-Nasirkandi et al. 2023; Naseer and Adil 2025). By controlling these parameters, manufacturers can assure that the end product contains adequate levels of active principles responsible for efficacy. This is critical not only for consumer safety but also for supporting clinical claims about the medicine's therapeutic effects. For example, quality control studies on *Withania somnifera* (Ashwagandha) focus on the concentration of withanolides, its primary bioactive constituents, to ensure product efficacy (Kumar and Patnaik 2016). Efficacy validation also ties closely with pharmacological and clinical investigations. Quality control ensures that formulations tested in preclinical or clinical studies correspond precisely to marketed products, thereby providing scientific credibility and regulatory compliance.

7.5.3 The Role of Quality Control in Building Consumer and Regulatory Confidence

A medicine's commercial success is highly dependent on consumer and healthcare practitioner's confidence. The regulatory authorities have the primary responsibility to evaluate product safety and efficacy. Quality control forms the backbone of this trust. In India, adherence to Good Manufacturing Practices (GMP) as per Schedule T of the Drugs and Cosmetics Act and certification by regulatory bodies ensures the quality and reliability of a medicinal product. In addition, compliance with standards set by pharmacopoeias such as the Indian pharmacopoeia ensures quality consistency across batches(AYUSH 2025). In international markets such as the US and EU, regulatory frameworks require documented evidence of quality control for registration and marketing approval. The FDA's enforcement of cGMP for dietary supplements and the European Medicines Agency's (EMA) Traditional Herbal Medicinal Products Directive necessitate stringent quality controls. These regulatory requirements effectively function as entry barriers for low-quality or sub-standard unsafe products, ensuring the availability of standard compliant products.

Furthermore, quality seals, certifications, and transparent labeling act as market differentiators and increase consumer confidence. For instance, an herbal product that declares its standardized content of marker compounds and complies with contaminant limits is often preferred by consumers, including healthcare practitioners recommending herbal therapies. Quality control also safeguards against product recalls, legal challenges, and reputational damage. Failure to comply with quality standards can lead to distribution bans, consumer distrust, and financial losses for the manufacturers.

7.5.4 Quality Control Methods

Quality control methods for herbs focus on the accurate identification and verification of botanical material to ensure authenticity, purity, and efficacy of herbal medicines. Common techniques include macroscopic and microscopic examination, chemical profiling, chromatographic techniques, DNA barcoding, and spectroscopic methods, each serving a unique purpose in confirming the identity and quality of herbs(Chan et al. 2018; Wang et al. 2023; Yan et al. 2025).

a. **Macroscopic Examination:** Visual inspection of part features like shape, color, texture, and size to distinguish authentic herbs from adulterants. Some of the most frequently misidentified herbs by morphology alone are those with closely resembling visual features, similar vernacular names, or use of plant parts where differences are subtle. For example, species within genera such as *Cuscuta* (dodder), *Melissa* (lemon balm), *Ocimum* (basil), and *Artemisia* (wormwood) are often substituted or adulterated due to superficial similarities. Vernacular confusion also leads to mixed uses of *Borago officinalis* and *Echium amoenum*, as both are called "Gavzaban" in different regions. Additional common misidentifications occur with *Hypericum* species (e.g., St. John's Wort) and other Hypericum varieties, which are visually very similar; bamboo tea, where carnation leaves (Dianthus chinensis) have been falsely marketed as bamboo due to similar leaf appearance; and foxglove (Digitalis) and plantain plants, which share broad leaf shapes but differ dramatically in safety. Confusion is aggravated when the material is fragmented, powdered, or marketed under local/common names, making exclusive reliance on morphological features risky for these herbs (Balu and Alagesaboopathi 1995)
b. **Microscopic Examination:** Observation of cellular structures, trichomes, and other diagnostic features under a microscope to help in species-level identification
c. **Organoleptic Evaluation:** This process uses senses (sight, smell, taste, touch) to detect abnormalities or inconsistencies
d. **Chemical Profiling:** Through chemical profiling identification of marker compounds and active substances using various chemical assays are done
e. **Chromatographic Methods (TLC, HPTLC, HPLC, GC):** It helps in separation and quantification of compounds, assessment of chemical fingerprints, and detection of adulteration
f. **DNA Barcoding:** Molecular technique to authenticate species identity by sequencing unique DNA regions
g. **Spectroscopic Methods (FTIR, NIR, Raman):** Determination of chemical constituents and functional groups for rapid analysis

Table 7.2 summarizes different methods used for quality control of herbal medicines. Each method complements the others to provide a robust, multi-level approach for quality assurance in herbal identification and verification.

7.5.5 Comprehensive Quality Control Systems

A comprehensive quality control system in herbal medicines covers the entire product lifecycle of the herbal formulation starting from raw material sourcing to final product testing and post-market surveillance. The quality checking begins with raw

TABLE 7.2
Methods Employed for Quality Control in Herbal Medicines

Method	Principle/ Technique	Key Purpose	Strengths	Limitations
Macroscopic Examination	Physical features inspection	Initial species authentication	Simple, quick	Sometimes ambiguous
Microscopic Examination	Cellular structure observation	Detailed tissue identification	High specificity	Requires expertise
Organoleptic Evaluation	Sensory assessment	Detect abnormalities/ adulteration	Rapid, cost-effective	Subjective
Chemical Profiling	Marker compound identification	Verification of active substances	Specific, scientific	Needs chemical reagents
Chromatography (TLC/HPTLC/ HPLC/GC)	Separation and quantification	Detects fingerprints, contamination/ adulteration	Accurate, quantitative	Equipment intensive
DNA Barcoding	Genetic markers sequencing	Confirm botanical identity, detects adulterants	Reliable, works on processed form	Costly, database needed
Spectroscopy (FTIR/NIR/ Raman)	Molecular vibration analysis	Rapid analysis of functional groups/ constituents	Fast, portable instruments	Interference, complexity

material sourcing controls, ensuring collection from authenticated plant species and adherence to Good Agricultural and Collection Practices (GACP). In the next step, analytical testing of raw materials and finished products for identity, purity, potency, and contaminants are carried out using established protocol. Consistent processing is achieved through manufacturing controls complying with GMP standards to prevent cross-contamination. Stability analysis guarantees shelf life and product safety throughout storage and use. The reproducibility is accomplished by performing batch-to-batch consistency testing. At the end, post-market pharmacovigilance systems help to monitor adverse events and maintain product integrity in the marketplace.

7.6 REGULATORY FRAMEWORKS GOVERNING HERBAL MEDICINES

7.6.1 Regulation of Herbal Medicines in India

India, known for its millennia-old heritage of traditional medicine systems, has developed a specialized and multifaceted regulatory framework to govern herbal medicines. These regulatory mechanisms seek to ensure the quality, safety, and efficacy of herbal products, thus protecting consumers and promoting the legitimate growth of traditional healthcare. The backbone of herbal medicine regulation in

India is enshrined in the Drugs and Cosmetics Act, 1940, and its accompanying Drugs and Cosmetics Rules, 1945 (India), supplemented by guidelines developed by ministries and regulatory agencies dedicated to Ayurveda, Siddha, Unani, and other indigenous systems collectively known as AYUSH.

The Drugs and Cosmetics Act serve as the principal legislation for regulating drugs and cosmetics in India, including herbal medicinal products. While initially designed with conventional pharmaceuticals, the Act was later expanded via amendments and specialized provisions to address the distinct nature of traditional medicines. The Chapter IVA explicitly regulates Ayurvedic, Siddha, and Unani (ASU) drugs, acknowledging their unique formulation processes, raw materials, and therapeutic paradigms.

Under this regulatory system, herbal medicines are broadly categorized into classical medicines and proprietary formulations. Classical medicines refer to products prepared using ingredients and formulations described in authoritative texts such as the Charaka Samhita and Sushruta Samhita for Ayurveda or the Siddha materia medica and Unani pharmacopoeia. Proprietary Ayurvedic, Siddha, and Unani medicines, by contrast, are newly developed formulations that utilize only ingredients listed in these classical pharmacopoeias but do not appear in the classical texts. This distinction allows flexibility for innovation while maintaining traditional ingredient integrity.

7.6.1.1 Definition and Classification of Herbal and Traditional Medicines Under the Act

The Act broadly defines a "drug" (which includes herbal medicines) under Section 3(b) as any substance intended for use in the diagnosis, treatment, mitigation, or prevention of disease in humans or animals. This includes herbal preparations used internally or externally for medicinal purposes. Herbal medicines are considered drugs when they are derived from herbs, plant materials, or combinations thereof and are intended for medicinal applications.

Proprietary Ayurvedic, Siddha, and Unani medicines refer to formulations that are prepared from ingredients mentioned in the authoritative classical texts of these traditional systems of medicine, but which themselves are not described or found in those classical texts. More specifically, these are formulations that contain only ingredients mentioned in the official pharmacopoeias and authoritative books of Ayurveda, Siddha, or Unani medicine, as specified in the First Schedule of the Drugs and Cosmetics Act, 1940 (DCA 1940). However, unlike classical medicines, proprietary medicines are new formulations or combinations developed using these traditional ingredients but not specifically described in the classical texts themselves. They exclude formulations administered by parenteral routes and also exclude classical medicines already listed in the authoritative books. Proprietary medicines may include innovations such as novel dosage forms or combinations, created for commercial purposes while still relying exclusively on traditional ingredients.

For the ASU drugs, central regulatory requirement is the licensing system, governed by the Drugs and Cosmetics Rules. Manufacturers of herbal products must obtain licenses from state or central drug authorities to legally produce or market

BOX 7.1

Rule 2(ee) of the Drugs and Cosmetics Rules:

> *"Ayurvedic, Siddha or Unani proprietary medicine means a drug which is a formulation containing only such ingredients as mentioned in the formulae described in the authoritative books specified in the First Schedule to the Act, but does not include a medicine administered by the parenteral route and also does not include a formulation which is included in the authoritative books."*

BOX 7.2

The authoritative texts listing allowed formulations for Ayurvedic, Siddha, and Unani medicines under the Drugs and Cosmetics Act, 1940, are specified in the First Schedule to the Act (DCA 1940). These texts serve as official references for ingredients and classic formulations that manufacturers must adhere to when producing traditional medicines.

medicines. Licensing mandates compliance with GMP as per Schedule T, which prescribes rigorous standards for infrastructure, personnel qualifications, hygiene, equipment, quality control laboratories, and documentation. The enforcement of Schedule T ensures that herbal medicines are manufactured under conditions promoting safety and consistency.

Equally significant are the standards prescribed in the official Ayurvedic, Siddha, and Unani pharmacopoeias. These texts provide detailed monographs describing the identity, purity, and potency criteria for raw herbs and finished products, alongside recommended analytical methods. For instance, parameters such as moisture content, ash values, extracts, and limits on microbial contamination, heavy metals, and pesticide residues are specified to prevent adulteration and ensure therapeutic quality.

Labeling and packaging requirements form another critical component of regulation. The Drugs and Cosmetics Rules stipulate that herbal medicines must display clear, accurate, and comprehensive information, including the list of ingredients with their botanical names, dosage form, manufacturing license number, batch number, manufacturing and expiry dates, instructions for use, storage conditions, and necessary warnings. These provisions guard against misleading claims and promote informed consumer choice. The licensing and manufacture of proprietary Ayurvedic, Siddha, and Unani medicines are regulated under the Drugs and Cosmetics Act with requirements for safety, efficacy, and labeling under Part XVI and XVII of the Drugs and Cosmetics Rules, 1945.

The proprietary ASU medicines allow industry innovation within the traditional medicine framework by creating novel products using known traditional ingredients, while ensuring regulatory oversight for quality and safety. This definition and regulation aim to balance the protection and promotion of India's rich traditional medical heritage with the modern industry's need for product development and consumer protection.

7.6.1.2 Licensing Procedures for Herbal and Traditional Medicines

Manufacturing units of ASU medicines must obtain licenses under Section 33G. Manufacturers of ASU drugs must apply to the Licensing Authority (state or central) with details of premises, staff qualification, and manufacturing process. Structure layout, quality control measures, list of ingredients, existing GMP compliance, sample products, and batch records are required for approval of licenses. The Licensing Authority inspects the manufacturing unit for compliance with Schedule T and upon satisfactory assessment, a license is granted.

7.6.1.3 Schedule T: GMPs Specific to Ayurveda and Related Systems

GMPs are the backbone of pharmaceutical production quality worldwide, ensuring reproducibility, cleanliness, and safety. Recognizing the specificities of Ayurvedic and traditional formulations, Schedule T was incorporated into the Drugs and Cosmetics Rules tailored to the requirements for ASU drugs. Schedule T mandates that manufacturing facilities have proper ventilation, storage areas protected from moisture and contamination, and segregated zones for various manufacturing steps to avoid cross-contamination. The raw herbs and minerals used should be authenticated botanically, free from adulterants, and stored under ideal conditions to preserve potency. This includes adherence to Good Agricultural and Collection Practices (GACP). Manufacturing SOPs must be established, documented, and followed strictly for extraction, purification, drying, and compounding. Batch records must be maintained rigorously. An in-house quality control laboratory is required to test raw materials, semi-finished, and finished products for identity (microscopy, organoleptic examination), phytochemical content (e.g., marker compounds by chromatography), microbial load, heavy metals, and pesticide residues. It mandates that every worker involved in manufacturing should be trained in hygiene and GMP principles, preventing contamination and errors. Schedule T sets requirements to ensure materials and finished products are packaged in materials that protect integrity, bearing labels with batch numbers, manufacturing and expiry dates, dosage instructions, and warnings.

Since enforcement, Schedule T has motivated modernization of many Ayurvedic manufacturers, investment in quality labs, and improved manufacturing environments. However, smaller and traditional producers sometimes struggle to meet these standards due to costs and infrastructural limitations, an issue the government addresses through capacity-building schemes (Sahoo and Manchikanti 2013).

7.6.1.4 Pharmacopoeial Standards and Quality Control

Pharmacopoeial standards for herbal medicines in India set the official benchmarks for quality, purity, identity, and strength of traditional medicines derived from plant sources. These standards are critical to ensuring safety, efficacy, and consistency across batches manufactured under the Ayurvedic, Siddha, and Unani systems. The key Pharmacopoeial Standards for Herbal Medicines in India are

- The Ayurvedic Pharmacopoeia of India (API)
- The Siddha Pharmacopoeia of India
- The Unani Pharmacopoeia of India

Compliance with pharmacopoeial standards is essential for product registration and marketing approval. Interestingly, the Indian pharmacopoeial Commission has progressively integrated advanced analytical technologies to improve authentication and monitoring of herbal drugs, reflecting global trends in herbal medicine quality assurance. Monographs now encompass widely used botanicals like Ashwagandha, Guduchi, and Tulsi, extending to complex polyherbal formulations.

Beyond statutory law, the Ministry of AYUSH has issued several guidelines to enhance quality control, safety assessment, and post-market vigilance of herbal medicines. The ministry actively promotes the National Pharmacovigilance Program for ASU drugs to monitor and document adverse drug reactions, thereby enhancing pharmacovigilance tailored to the pharmacology of semi-complex herbal formulations.

Safety assessment in India balances respect for traditional use with modern scientific rigor. While longstanding use of herbal products provides a degree of reassurance, manufacturers, and regulators also recognize the importance of toxicological evaluations, heavy metal testing, and microbial safety standards. Certain herbal drugs suspected of toxicity or with insufficient safety data are subject to stricter scrutiny, including requirements for clinical trials when traditional use cannot fully justify safety claims.

7.6.2 Regulatory Framework for Botanical Products in the United States

The regulatory landscape governing botanical products in the United States is characterized by two distinct, yet interrelated frameworks. The approval is primarily based on the product's intended use, claims, and marketing approach. These frameworks are delineated by the Dietary Supplement Health and Education Act of 1994 (DSHEA), which establishes the rules for dietary supplements, and the US Food and Drug Administration's (FDA) more stringent regulatory pathways designed for botanical drugs under the Federal Food, Drug, and Cosmetic Act (FD&C Act). Together, these regulatory provisions both enable consumer access to herbal products and ensure necessary safeguards to protect public health through appropriate oversight (Bouldin et al. 1999; Snow, Leach, and Clare 2017; Wu et al. 2010).

Prior to the enactment of DSHEAs, the herbal products and supplements were often regulated under broader food or drug laws. Since the products were neither completely food nor meeting the standards of drugs, they faced regulatory uncertainty. Given the importance of dietary supplements for overall health promotion and disease prevention, in 1994 DSHEA was enacted. It created a distinct regulatory pathway for dietary supplement and dietary ingredients including vitamins, minerals, herbs, and botanicals. This arrangement allowed manufacturers to market these products which did not fit into conventional definition of food or drugs, without prior FDA approval. However, they have to meet the key requirements related to safety and labeling. Importantly, DSHEA empowered consumers by promoting access to a wide range of supplements while allowing the FDA to intervene post-market if safety concerns arose.

Under DSHEA, dietary supplements are defined expansively as orally ingested products intended to supplement the diet, containing one or more dietary ingredients such as vitamins, minerals, herbs or other botanicals, amino acids, or concentrates, metabolites, constituents, or extracts thereof. This definition encapsulates traditional herbal products and extends to various novel forms derived from botanical sources. Despite being regulated as foods, these products are distinct from conventional foods by the virtue of their method of consumption and therapeutic intent.

Under DSHEA, the responsibility for product safety rests primarily with the manufacturer or distributor. To facilitate consumer protection, DSHEA mandates that dietary supplements meet specific labeling requirements, including the disclosure of ingredients, recommended usage, and any known risks. Under the regulation, the label may carry structure/function claims, statements regarding the role of a nutrient or ingredient intended to affect normal bodily structure or function, such as "supports immune health." However, such claims must be accompanied by FDA disclaimers stating that the claim has not been evaluated by the FDA and that the product is not intended to diagnose, treat, cure, or prevent any disease.

More explicit health claims linking a dietary supplement to the diagnosis, cure, mitigation, treatment, or prevention of disease are categorically prohibited unless approved through rigorous FDA drug approval processes. DSHEA also introduced a regulatory category for "new dietary ingredients" (NDIs), defined as dietary ingredients not marketed in the United States prior to October 15, 1994. Manufacturers intending to market NDIs must notify the FDA and provide evidence supporting their safety, thereby addressing emerging supplement ingredients without compromising consumer safety.

Complementing the focus on labeling and safety, under DSHEA-specific GMPs, guidelines for dietary supplements are available. These GMP regulations formalize the processes by which manufacturers ensure the quality, purity, and strength of their products, demanding consistent production controls, documentation, and testing.

On the other hand, botanical drugs are treated as drugs and are regulated under the FDA's conventional drug regulatory framework. These products are intended for diagnosis, treatment, cure, mitigation, or prevention of disease and require a much higher evidentiary burden before they enter the market. The FDA defines botanical drugs as "products containing vegetable materials such as plants, algae, or macroscopic fungi, or combinations thereof." Recognizing the inherent complexity and variability of herbal constituent mixtures, the FDA's Botanical Drug Development Guidance permits certain scientific flexibilities absent in synthetic drug appraisal, allowing complex mixtures without full chemical characterization provided safety and efficacy are rigorously demonstrated.

To bring a botanical drug to market, sponsors must submit an Investigational New Drug (IND) application to the FDA, detailing the product's manufacturing process, preclinical safety data, and clinical trial protocols. Clinical trials are mandatory to substantiate both safety and efficacy, mirroring requirements for conventional drugs. Upon successful completion of the trials, the sponsor must file a New Drug Application (NDA) for FDA review. Throughout this process, the FDA grants latitude in chemistry, manufacturing, and controls (CMC) documentation to accommodate the complex natural mixtures inherent in botanicals, while still maintaining rigorous standards to assure patient safety.

The important differences between dietary supplements and botanical drugs reflect the underlying principles of public health policy, risk assessment, and intended use. Dietary supplements, regulated as foods, enjoy virtually unfettered access to the market, yet are limited in their claims and subject to post-market safety oversight. Meanwhile, botanical drugs are subjected to the full rigors of drug approval to substantiate therapeutic claims and protect patients from false or unsafe products.

A striking example of the botanical drug pathway's success is Veregen®, a botanical ointment derived from green tea leaves (sinecatechins), approved by the FDA in 2006 for treating genital warts. This approval marked the first time a botanical product cleared rigorous FDA scrutiny to be marketed as a prescription drug, demonstrating the feasibility of the botanical drug development framework. On the other hand, widely used supplements like Echinacea remain categorized under DSHEA, marketed with structure/function claims related to immune support rather than disease treatment.

Thus, the US regulatory system is a perfect example where consumer access and public health safeguards are balanced appropriately. While DSHEA's flexible dietary supplement provisions allow easy access to dietary supplements, safety of public health is maintained through stringent regulatory examination when therapeutic claims are made. This dual pathway underscores the need for clear regulatory distinctions based on product claims and intended use, thereby fostering innovation, consumer choice, and safety. Table 7.3 provides the distinction between the two pathways.

7.6.3 Regulation of Traditional Herbal Medicinal Products in the EU

The EU has established a distinctive and comprehensive regulatory framework to oversee traditional herbal medicinal products (THMPs), reflecting its commitment to safeguarding public health while preserving access to herbal medicines rooted in traditional practices. The Directive 2004/24/EC, which amends the earlier directive 2001/83/EC, provides a simplified pathway for the authorization and registration of traditional herbal medicinal products across member states.

Directive 2004/24/EC recognizes the unique nature of traditional herbal medicines compared to conventional pharmaceuticals. Acknowledging that many herbal products are used extensively based on long-standing traditional use rather than on modern clinical trial data, the directive creates a legal category specifically for THMPs. This category benefits from a reduced evidentiary burden while ensuring that products meet essential safety, quality, and efficacy standards appropriate to their traditional use.

TABLE 7.3
Key Differences Between Dietary Supplements Under DSHEA and Botanical Drugs Under the FDA's Drug Regulatory Pathway

Regulatory Aspect	Dietary Supplements (DSHEA)	Botanical Drugs (FDA Drug Pathway)
Legal Classification	Food (Dietary Supplement category)	Drug
Premarket Approval Required	No premarket approval Manufacturers responsible for safety	Required New Drug Application (NDA) process mandatory
Claims Allowed	Structure/function claims with disclaimers Health claims limited and regulated	Can make disease treatment claims post-approval
Safety and Efficacy Evidence	Safety responsibility on manufacturers Efficacy not required	Safety and efficacy must be demonstrated premarket
Manufacturing Standards	Dietary supplement GMPs	Pharmaceutical cGMP (current Good Manufacturing Practices)
Clinical Trials	Not required	Required for safety and efficacy
Labeling Requirements	Specific labeling with FDA disclaimer	Drug labeling standards, strict assertion of claims
Adverse Event Reporting	Mandatory reporting of serious adverse events	Routine pharmacovigilance and post-market surveillance
Market Surveillance	FDA conducts post-market monitoring and can remove unsafe products	Continuous FDA oversight, recall authority

Under the directive, a traditional herbal medicinal product is defined as a product intended for use without the supervision of a medical practitioner for diagnosing, treating, or prescribing. It must have been used medicinally for at least 30 years, including at least 15 years within the EU, with sufficient evidence of safety and plausible efficacy from traditional use. This recognition of historical use allows applicants seeking registration to rely primarily on bibliographic or expert evidence establishing the longstanding use rather than large-scale clinical trials required for new pharmaceutical drugs.

The simplified registration process under Directive 2004/24/EC reduces costs and administrative barriers, thereby facilitating the availability of safe herbal medicines on the EU market while maintaining consumer protection. However, the directive does not compromise on essential regulatory requirements. Products must comply with stringent standards for quality, including verification of botanical raw materials, manufacturing processes that satisfy GMP, and appropriate labeling that includes indications, posology, dosage form, warnings, and contraindications.

To support the implementation of this directive and harmonize regulatory decisions, the European Medicines Agency (EMA) established the Committee on Herbal Medicinal Products (HMPC). The HMPC fulfills a pivotal role in providing scientific opinions and guidelines concerning herbal medicines, thereby fostering consistency across member states.

The HMPC systematically evaluates and publishes monographs on herbal substances and preparations, which serve as scientific reference documents detailing safe traditional use, indications, contraindications, and quality parameters. These monographs are invaluable resources for regulators and pharmaceutical companies, as they inform the assessment of product dossiers submitted for traditional use registration. By centralizing and standardizing expert scientific knowledge, the HMPC enhances transparency, reduces duplication of effort between member states, and streamlines market access.

In addition to the monographs, the EMA provides comprehensive guidelines on the quality, safety, and efficacy assessments required for traditional herbal medicinal products. These guidelines emphasize several critical aspects specific to herbal products, including the importance of Good Agricultural and Collection Practice (GACP) to ensure the identity and purity of raw plant materials. The guidelines require manufacturers to implement rigorous quality controls, encompassing botanical identification through macroscopic and microscopic techniques, phytochemical profiling, and limits for contaminants such as heavy metals, pesticides, and microbial contamination.

Safety evaluation under the EMA's guidelines recognizes the challenges inherent in herbal products. While traditional use provides some reassurance, the guidelines mandate a thorough assessment of possible toxicological effects, contraindications, interactions with other medicines, and adverse reactions documented during the product's use, where necessary, non-clinical safety studies or pharmacovigilance data may be required to substantiate safety claims.

Efficacy evidence for THMPs is primarily drawn from documented traditional use, but the EMA advises that this use must be plausible and supported by bibliographic data or expert consensus. Although formal clinical trials are not always mandatory for traditional herbal medicines, the guidelines encourage the inclusion of any available clinical data to strengthen a product's dossier.

Labeling requirements stipulated by the directive and EMA guidelines mandate that consumers receive clear and accurate information. This includes the traditional indication(s), dosage instructions, route of administration, and any necessary warnings, such as restrictions on use by children or pregnant women. Such transparency is essential for enabling informed consumer choices, especially given that traditional herbal products are often used without direct healthcare supervision.

The combined regulatory framework of Directive 2004/24/EC, the HMPC, and EMA guidance has contributed significantly to the harmonization of herbal medicine regulation within the EU. It has facilitated the legal market entry of a wide range of traditional herbal medicinal products while ensuring high standards of quality and consumer safety. Importantly, this framework respects and integrates the cultural heritage associated with herbal medicine, thereby promoting pluralism in healthcare options.

In the broader global context, the EU's regulatory approach for herbal medicine represents a comprehensive, science-based, and culturally sensitive framework. This system effectively governs the quality, safety, and efficacy of traditional herbal medicines in the European market, ensuring consumer protection while facilitating access to herbal therapies grounded in tradition and experience.

Across these jurisdictions, regulatory categories for herbal medicines pivot on traditional use, safety profiles, manufacturing standards, and claims made (Table 7.4). India's system emphasizes authenticity and traditional knowledge combined with contemporary quality standards, reflecting the country's rich indigenous medicine heritage. The US's bifurcated approach distinguishes dietary supplements from botanical drugs based on intended use and marketing claims, balancing consumer access with public protection. The EU's THMP pathway harmonizes traditional use

TABLE 7.4
Comparison of Regulatory Categories, Registration Requirements, Safety/Efficacy Evaluation, Labeling, and Marketing Claim Restrictions in Herbal Medicines

Aspect	India (Traditional Medicines)	United States (Dietary Supplements vs Botanical Drugs)	European Union (THMPs)
Primary Legal Category	Traditional medicines under AYUSH & Drugs Act	Dietary supplements (DSHEA) and botanical drugs (FDA drug pathway)	Traditional Herbal Medicinal Products (Directive 2004/24/EC)
Definition Basis	Classical texts and traditional formulations	Intended use & claims distinguish Supplements = Food, botanical drugs = Drug	Minimum 30 years traditional use, non-prescription medicines
Registration Requirement	Mandatory license for manufacture and sale	*Dietary supplements*: none premarket *Botanical drugs*: Full NDA process	Simplified registration based on traditional use evidence
Quality Control Standards	Schedule T GMP Pharmacopoeial standards	*Dietary supplements*: GMP for supplements *Botanical drugs*: pharmaceutical GMP	GMP and GACP adherence; EMA monographs for quality criteria
Safety and Efficacy Evaluation	Traditional use plus toxicology/clinical when needed	*Supplements*: Post-market surveillance *Botanical drugs*: premarket clinical trials	Traditional use documentation plus safety data, limited efficacy trials
Labeling Requirements	Ingredient detail, dose, warning per Drugs Rules	*Supplements*: label with disclaimers *Botanical drugs*: drug labeling rules	Clear indication of traditional use, dosage instructions, safety warnings
Marketing Claims Restrictions	Claims linked to traditional use only No unauthorized therapeutic claims	*Supplements*: structure/function claims only *Botanical drugs*: therapeutic claims allowed if approved	Claims must reflect traditional indications No disease treatment claims without full authorization

acceptance with scientific evaluation, easing market entry while safeguarding health through centralized scientific guidance. While India relies heavily on traditional use documentation plus modern quality and safety testing; the US dietary supplements depend on manufacturer vigilance and less prescriptive efficacy data, contrasted by full clinical trials for botanical drugs; the EU depends on bibliographic traditional evidence combined with rigorous quality and safety standards. But the labeling regimes universally emphasize transparency and consumer information but differ in detail. Marketing claims are strictly controlled across all the three-jurisdiction prohibiting disease treatment assertions for traditional or dietary supplement products. So, while the regulatory systems in India, the US, and the EU vary in structure and stringency, they share common goals of ensuring quality, safety, and truthful communication about herbal medicines.

7.7 CHALLENGES IN QUALITY CONTROL OF HERBAL MEDICINES ACROSS JURISDICTION

7.7.1 Complexity of Botanical Products

The variation in quality of herbal medicines stems from the inherent complexity of plant products. Herbal medicines often comprise whole or processed plant parts containing many phytochemicals in varying proportions. The geographical source, time of harvesting, or processing may pose challenge in toxicity assessment, leading to variation in quality and integrity of the final herbal product. In bio-diversity rich country like India, medicinal plants are collected from different agro-climatic zones, which have different nutrient characteristics. Therefore, it is very difficult to get uniformity in quality of herbal material. Similarly, in the US and EU, the variability is due to the sourcing or importation of plant materials from different parts of the world. These differences pose problems to develop consistent quality parameters. Therefore, it is essential to adapt good agricultural and collection practices (GACP) to ensure traceability and minimize adulteration or contamination risks (Betz, Brown, and Roman 2011; Pelkonen et al. 2012; Pferschy-Wenzig and Bauer 2015).

Despite efforts in regulatory harmonization and scientific methods, quality control of herbal medicines confronts multifaceted issues. The manufacturers grapple with problems of raw material variability, standardization difficulties, manufacturing inconsistencies, and enforcement challenges. A thorough understanding of these obstacles is essential for stakeholders aiming to safeguard public health, preserve traditional knowledge integrity

7.7.2 Adulteration and Contamination

Adulteration is the intentional addition of unrelated substances of lower quality to the raw materials to increase the weight or volume of the medicinal product. Sometimes, adulteration happens through environmental pollution, or improper handling, or processing of the materials. Furthermore, contaminants such as heavy metals, pesticides, microbial organisms, and mycotoxins poses serious risks (Basaran, Pasli, and Basaran 2022; Chan 2003; Ernst 2004; Osman et al. 2023). Being a non-conventional form of medicine, the herbal medicines do not strictly require safety, efficacy or dosage studies nor safety and sterility regulation. Improper handling and storage of such products results in microbial or chemical contaminants. Also, the presence of heavy metals in the environment due to industry, application of medical technologies or pesticide has resulted in the accumulation of heavy metals in plants more than the permissible limits. For example, in India, the presence of heavy metals such as lead, mercury, and arsenic beyond permissible limits has been documented, sometimes driven by traditional manufacturing processes involving mineral powders (bhasma). The subsequent accumulation of heavy metals in plants and subsequently in the food chain is one of the major reasons for health concerns arising due to use of herbal medicines. The USFDA has issued warnings about pesticide residues and microbial contamination in imported herbal supplements, affecting consumer safety and market confidence. Similarly, in EU, there are reports of contaminants in traditional herbal products, which have triggered development of stringent regulatory standards for import and marketing of herbal medicinal product. Ensuring consistent screening and enforcement across these jurisdictions is thus critical but resource-intensive.

7.7.3 Lack of Universal Chemical and Biological Markers for Standardization

The lack of universal chemical and biological markers for standardization is a persistent problem for herbal medicinal products. The active ingredient in herbal medicine is a complex mixture of phytochemical constituents. Unlike synthetic drugs with single active ingredients, herbal medicines contain complex mixtures with more than one active ingredients. So, the overall therapeutic effects may result from synergistic interactions among multiple constituents.

EMA defines chemical marker as “chemically defined constituents or groups of constituents of a herbal substance, a herbal preparation or a herbal medicinal product which are of interest for control purposes independent of whether they have any therapeutic activity. Markers serve to calculate the quantity of herbal substance(s) or herbal preparation(s) in the herbal medicinal product if the marker has been quantitatively determined in the herbal substance or herbal preparation.”

Generally, the use of chemical markers is considered as highly reliable and effective quality control measure. But identifying appropriate marker compounds for quality control remains scientifically challenging. India’s regulatory framework depends on heavily on monographs in official pharmacopoeias that specify chemical assays and biological tests, but these methods may not capture the full spectrum of active principles or detect adulteration effectively. Furthermore, the absence of suitable markers or multiple markers for a medicine with multiple active ingredients possesses additional challenges. In the US, although analytical techniques are widely applied, the lack of standardized reference materials across varied botanicals hinders harmonization. EMA mandates the use of markers for content of constituent(s) with known therapeutic activity or unknown therapeutic activity. The choice of markers should be appropriately justified (EMA 2022). Though there is periodic publication of monographs with standardized testing protocols that are issued by EMA's CHMP, complex chemotypes of plants and differences in preparation methods complicate the quality control and batch reproducibility posing challenges for regulatory review processes.

7.7.4 Diverse Regulatory Requirements and Quality Standards

The differences in regulatory requirements and quality control standards have complicated the herbal drug development and marketing compliance by the manufacturers and also impose challenges for the regulators ensuring safety of the medicine. While India relies on traditional knowledge framework embedded in classical textbooks, the mandatory compliance to GMP as mentioned in Schedule T of Drugs and cosmetic Act of 1940 has brought certain uniformity in manufacturing practices. Though the GMP standards are comprehensive, the lack of enforcement has resulted in quality issues. On the other hand, the US adopts a bifurcated system and differentiates dietary supplements from botanical drugs, with markedly different quality expectations. The dietary supplements face relatively modest premarket evaluation, whereas botanical drugs undergo rigorous testing. These differences also pose challenges for regulators and emphasize the need of developing expertise in evaluating heterogeneous dossiers encompassing traditional knowledge and modern scientific data.

7.7.5 Non-compliance with GMPs

Inconsistent product quality is a major problem for Indian herbal drug manufacturers. While GMP adoption is mandatory, the enforcement of such provision is very critical. In the absence of stringent regulation related to sourcing of raw material, storage, processing, or packaging increases the risk of contamination and affects product stability. Moreover, herbal medicines are susceptible to degradation through moisture, heat, and light, requiring packaging standards and stability studies that are sometimes neglected, threatening efficacy and safety.

7.7.6 Scarcity of Validated Analytical Methods and Reference Standards

Availability of validated analytical methods and reference standards for herbal products continues to be a major hurdle. While pharmacopoeial guidelines in India and monographs by the EMA's HMPC provide some standardization, the rapid emergence of new herbal formulations and proprietary blends outpaces availability of validated reference substances. Also, as herbal medicines are quite regional in nature, it is a big scientific challenge to develop standard analytical methods meeting the need of different types of herbal ingredients. The US grapples with a similar challenge in its vast dietary supplement sector, where novel combinations and botanicals are frequently introduced. This gap limits comprehensive quality certification and complicates regulatory surveillance, increasing risk of substandard products entering the market. Further, differences in scientific capabilities across countries bring challenge for uniform adaptation of available techniques.

7.7.7 Adequacy of Pharmacopoeial Standards

The adequacy of pharmacopoeial standards and harmonization is another major challenge. Though India has a rich collection of national pharmacopoeias for ASU drugs; updating and harmonizing these texts with international standards remains a challenging affair. While EU relies on the EMA and the HMPC for harmonized herbal monographs, the US lacks a unified pharmacopoeial framework for dietary supplements. This creates a fragmented and inconsistent regulatory structure. Such structures slow down international trade and mutual recognition, complicating quality assurance on a global scale.

7.7.8 Consumer Awareness and Education

The prominence of misinformation and unverified marketing claims for herbal medicines increases quality concerns and increasing risk. Though pharmacovigilance and adverse drug reaction reporting systems are in place, enforcement constraints demands integrated and cooperative approaches. Consumer awareness and scientific awareness is crucial for safe use of such medicines.

7.8 MITIGATING QUALITY CHALLENGES—THE WAY FORWARD

The intrinsic inconsistencies within botanical raw materials standards, manufacturing capabilities, coupled with differences in the regulatory frameworks and enforcement practices among different countries, has necessitated a coordinated effort to mitigate the quality control issues. In order to achieve quality herbal medicines, the integration of scientific innovation and harmonization of standards are extremely critical. Capacity building and consumer awareness are also essential for safeguarding public health. Implementation of good agricultural practices, standardization of quality specifications, strict quality control, education, and awareness among the farmers/collectors are few of the important steps which need to be practiced. Moving forward, integration of traditional knowledge with modern scientific evidence-based healthcare systems is indispensable. Channelizing efforts in that direction, the Ministry of AYUSH, Govt. of India, has been taking several initiatives to scientifically validate and standardize the drugs with safety and efficacy studies. This is expected to promote an interdisciplinary approach for a more holistic herbal healthcare. However, in addition to this, strong international collaboration and regulatory governance will be critical in assuring the availability of safe, effective, and high-quality herbal medicines globally. These concerted efforts are expected to markedly improve consumer trust and popularity of herbal medicines resulting in an upsurge in market demands. Taken together, all these integrated measures will lead to newer innovations to mitigate public health priorities resulting in furthering the legacy of herbal therapeutics.

REFERENCES

Ahiabor, W. K., S. Darkwah, and E. S. Donkor. 2024. "Microbial Contamination of Herbal Medicines in Africa, 2000-2024: A Systematic Review." *Environ Health Insights* 18:11786302241293345. https://doi.org/10.1177/11786302241293345.

Ayati, Z., M. Ramezani, M. S. Amiri, A. T. Moghadam, H. Rahimi, A. Abdollahzade, A. Sahebkar, and S. A. Emami. 2019. "Ethnobotany, Phytochemistry and Traditional Uses of Curcuma Spp. and Pharmacological Profile of Two Important Species (C. longa and C. zedoaria): A Review." *Current Pharmaceutical Design* 25 (8): 871–935. https://doi.org/10.2174/1381612825666190402163940

AYUSH. 2013. Unani System of Medicine- the science of health and healing: CCRUM. https://ccrum.res.in/writereaddata/UploadFile/Dossier_1325.pdf

Balekundri, A., and Mannur, V. 2020. "Quality Control of the Traditional Herbs and Herbal Products: A Review." *Future Journal of Pharmaceutical Sciences* 6 (1):67. https://doi.org/10.1186/s43094-020-00091-5.

Balu, S., and C. Alagesaboopathi. 1995. "Nomenclatural Confusion of Some Species of Andrographis Wall." *Ancient Science of Life* 14 (3): 136–42.

Barnes, P. M., B. Bloom, and R. L. Nahin. 2008. "Complementary and Alternative Medicine Use Among Adults and Children: United States, 2007." *National Health Statistics Report* (12):1–23.

Basaran, N., D. Pasli, and A. A. Basaran. 2022. "Unpredictable Adverse Effects of Herbal Products." *Food and Chemical Toxicology* 159: 112762. https://doi.org/10.1016/j.fct.2021.112762

Bent, S. 2008. "Herbal medicine in the United States: review of efficacy, safety, and regulation: grand rounds at University of California, San Francisco Medical Center." *Journal of General Internal Medicine* 23 (6):854–9. https://doi.org/10.1007/s11606-008-0632-y.

Betz, J. M., P. N. Brown, and M. C. Roman. 2011. "Accuracy, Precision, and Reliability of Chemical Measurements in Natural Products Research." *Fitoterapia* 82 (1): 44–52. https://doi.org/10.1016/j.fitote.2010.09.011

Bharath, M. N., S. Gupta, G. Vashistha, S. Ahmad, and S. V. Singh. 2023. "Bioprospective Role of Ocimum sanctum and Solanum xanthocarpum Against Emerging Pathogen: Mycobacterium avium Subspecies Paratuberculosis: A Review." *Molecules* 28 (8). https://doi.org/10.3390/molecules28083490

Bouldin, A. S., M. C. Smith, D. D. Garner, S. L. Szeinbach, D. A. Frate, and E. M. Croom. 1999. "Pharmacy and Herbal Medicine in the US." *Social Science & Medicine* 49 (2): 279–89. https://doi.org/10.1016/s0277-9536(99)00118-5

Calapai, G. 2008. "European legislation on herbal medicines: a look into the future." *Drug Saf* 31 (5):428–31. https://doi.org/10.2165/00002018-200831050-00009.

CCRUM. 2025. "Central Council for Research in Unani Medicine." Accessed 10/10/25. https://ccrum.res.in/

Chan, G. K., R. Wu, V. P. Chen, K. Y. Zhu, and Y. Du. 2018. "Quality Control on Herbal Medicine and Its Application." *Evidence-Based Complementary and Alternative Medicine* 2018: 3294398. https://doi.org/10.1155/2018/3294398

Chan, K. 2003. "Some Aspects of Toxic Contaminants in Herbal Medicines." *Chemosphere* 52 (9): 1361–71. https://doi.org/10.1016/S0045-6535(03)00471-5

DCA, India. 1940. Drugs and Cosmetics Act of India. edited by GOI.
Dinakaran, S. K., S. Chelle, and H. Avasarala. 2019. "Profiling and Determination of Phenolic Compounds in Poly Herbal Formulations and Their Comparative Evaluation." *Journal of Traditional and Complementary Medicine* 9 (4): 319–27. https://doi.org/10.1016/j.jtcme.2017.12.001
Dipankar, S. P., M. M. Dani, R. Anirudhan, D. Tripathi, C. Mishra, and S. H. Devi. 2025. "Pharmacological Insights Into Ashwagandha (Withania Somnifera): A Review of Its Immunomodulatory and Neuroprotective Properties." *Cureus* 17 (8): e89856. https://doi.org/10.7759/cureus.89856
EC. 2006. EC Regulation 1924/2006 on nutrition and health claims made on foods edited by European Commission.
Ekor, M. 2014. "The Growing Use of Herbal Medicines: Issues Relating to Adverse Reactions and Challenges in Monitoring Safety." *Frontiers in Pharmacology* 4: 177. https://doi.org/10.3389/fphar.2013.00177
EMA. 2001. DIRECTIVE 2001/83/EC. Available from URL: https://eur-lex.europa.eu/legal-content/EN/TXT/PDF/?uri=CELEX:02001L0083-20210526
Ernst, E. 2004. "Risks of Herbal Medicinal Products." *Pharmacoepidemiology and Drug Safety* 13 (11): 767–71. https://doi.org/10.1002/pds.1014
FICCI. 2025. "AYUSH." FICCI, Accessed April 20, 2025. https://www.ficci.in/sector/ayush.
Fong, H. H. 2002. "Integration of Herbal Medicine into Modern Medical Practices: Issues and Prospects." *Integrative Cancer Therapies* 1 (3): 287–93; discussion 293. https://doi.org/10.1177/153473540200100313
Gatt, A. R., P. Vella Bonanno, and R. Zammit. 2024. "Ethical considerations in the regulation and use of herbal medicines in the European Union." *Frontiers in Medical Technology* 6:1358956. https://doi.org/10.3389/fmedt.2024.1358956.
India, Drugs and Cosmetics Act of. 1940. "Drugs and Cosmetics Act of India." Accessed October 10, https://cdsco.gov.in/opencms/export/sites/CDSCO_WEB/Pdf-documents/acts_rules/2016DrugsandCosmeticsAct1940Rules1945.pdf
Karimi, A., M. Majlesi, and M. Rafieian-Kopaei. 2015. "Herbal Versus Synthetic Drugs; Beliefs and Facts." *Journal of Nephropharmacology* 4 (1): 27–30.
Kumar, G., and R. Patnaik. 2016. "Exploring Neuroprotective Potential of Withania Somnifera Phytochemicals by Inhibition of GluN2B-Containing NMDA Receptors: An in Silico Study." *Medical Hypotheses* 92: 35–43. https://doi.org/10.1016/j.mehy.2016.04.034
Lima, de S., C. Miranda, M. Amoras, T. Fujishima, B. de P. Lima, P. C. Mastroianni, F. F. O. de Sousa, and J. O. da Silva. 2020. "Microbial Contamination in Herbal Medicines: A Serious Health Hazard to Elderly Consumers." *BMC Complementary Medicine and Therapies* 20 (1):17. https://doi.org/10.1186/s12906-019-2723-1.
Luo, L., B. Wang, J. Jiang, M. Fitzgerald, Q. Huang, Z. Yu, H. Li, J. Zhang, J. Wei, C. Yang, H. Zhang, L. Dong, and S. Chen. 2020. "Heavy Metal Contaminations in Herbal Medicines: Determination, Comprehensive Risk Assessments, and Solutions." *Frontiers in Pharmacology* 11:595335. https://doi.org/10.3389/fphar.2020.595335.
Mandal, S. K., M. M. Rehman, A. Katyal, K. Rajvanshi, M. Kannan, M. Garg, S. Murugesan, and P. R. Deepa. 2024. "In Silico Anti-Viral Assessment of Phytoconstituents in a Traditional (Siddha Medicine) Polyherbal Formulation - Targeting Mpro and Pan-Coronavirus Post-Fusion Spike Protein." *Journal of Traditional and Complementary Medicine* 14 (1): 55–69. https://doi.org/10.1016/j.jtcme.2023.07.004
Mehnaz, S., Y. Shamsi, M. W. Akhtar, S. Mohanty, and S. Ahmad. 2024. "Therapeutic Applications of Neem (Azadirachta indica): A Narrative Review." *Advances in Mind Body Medicine* 38 (4): 14–8.
Mikulski, M. A., M. D. Wichman, D. L. Simmons, A. N. Pham, V. Clottey, and L. J. Fuortes. 2017. "Toxic metals in ayurvedic preparations from a public health lead poisoning cluster investigation." *International Journal of Occupational and Environmental Health* 23 (3):187–92. https://doi.org/10.1080/10773525.2018.1447880.

Moshari-Nasirkandi, A., A. Alirezalu, H. Alipour, and J. Amato. 2023. "Screening of 20 Species from Lamiaceae Family Based on Phytochemical Analysis, Antioxidant Activity and HPLC Profiling." *Scientific Reports* 13 (1): 16987. https://doi.org/10.1038/s41598-023-44337-7

Muyumba, N. W., S. C. Mutombo, H. Sheridan, A. Nachtergael, and P. Duez. 2021. "Quality Control of Herbal Drugs and Preparations: The Methods of Analysis, Their Relevance and Applications." *Talanta Open* 4:100070. https://doi.org/10.1016/j.talo.2021.100070.

Naseer, M., and M. Adil. 2025. "Phytochemical Profiling, HPLC Analysis, and Antimicrobial Potential of Curio radicans (L. f.) P.V. Heath." *Scientific Reports* 15 (1): 34753. https://doi.org/10.1038/s41598-025-18323-0

Osman, A., A. G. Chittiboyina, B. Avula, Z. Ali, S. J. Adams, and I. A. Khan. 2023. "Quality Consistency of Herbal Products: Chemical Evaluation." *Progress in the Chemistry of Organic Natural Products* 122: 163–219. https://doi.org/10.1007/978-3-031-26768-0_2

Pelkonen, O., M. Pasanen, J. C. Lindon, K. Chan, L. Zhao, G. Deal, Q. Xu, and T. P. Fan. 2012. "Omics and Its Potential Impact on R&D and Regulation of Complex Herbal Products." *Journal of Ethnopharmacology* 140 (3): 587–93. https://doi.org/10.1016/j.jep.2012.01.035

Pferschy-Wenzig, E. M., and R. Bauer. 2015. "The Relevance of Pharmacognosy in Pharmacological Research on Herbal Medicinal Products." *Epilepsy & Behavior* 52 (Pt B): 344–62. https://doi.org/10.1016/j.yebeh.2015.05.037

Prakash, P., and N. Gupta. 2005. "Therapeutic Uses of Ocimum sanctum Linn (Tulsi) With a Note on Eugenol and Its Pharmacological Actions: A Short Review." *Indian Journal of Physiology and Pharmacology* 49 (2): 125–31.

Rossiter, S. E., M. H. Fletcher, and W. M. Wuest. 2017. "Natural Products as Platforms To Overcome Antibiotic Resistance." *Chemical Reviews* 117 (19):12415–74. https://doi.org/10.1021/acs.chemrev.7b00283.

Sahoo, N., and P. Manchikanti. 2013. "Herbal Drug Regulation and Commercialization: an Indian Industry Perspective." *Journal of Alternative and Complementary Medicine* 19 (12): 957–63. https://doi.org/10.1089/acm.2012.0275

Sahoo, N., P. Manchikanti, and S. Dey. 2010. "Herbal Drugs: Standards and Regulation." *Fitoterapia* 81 (6): 462–71. https://doi.org/10.1016/j.fitote.2010.02.001

Snow, J. E., M. J. Leach, and B. A. Clare. 2017. "Attitudes, Skill and Use of Evidence-Based Practice Among US Western Herbal Medicine Providers: A National Survey." *Journal of Complementary and Integrative Medicine* 14 (1). https://doi.org/10.1515/jcim-2015-0101

Society for the Journal of Indian Medicine and Homoeopathy, Directorate of Indian Medicine and Homoeopathy. 2025. "Siddha." Accessed April 20, 2026. https://jimh.in/siddha.php.

Tallon, M. J., I. Koturbash, and J. L. Blum. 2025. "A Systematic and Ethnobotanical Review of Ashwagandha's (Withania Somnifera) Teratogenic and Abortifacient Potentials." *Phytotherapy Research.* https://doi.org/10.1002/ptr.70079

Turkson, B. K., M. L. K. Mensah, G. H. Sam, A. Y. Mensah, I. K. Amponsah, E. Ekuadzi, G. Komlaga, and E. Achaab. 2020. "Evaluation of the Microbial Load and Heavy Metal Content of Two Polyherbal Antimalarial Products on the Ghanaian Market." *Evidence-Based Complementary and Alternative Medicine* 2020:1014273. https://doi.org/10.1155/2020/1014273.

Validandi, V., S. Kurella, S. Gorain, Y. Sagubandi, A. Rudhra, S. K. D. Annamanedi, N. K. Ramachandrappa, and S. N. Sinha. 2025. "Heavy Metal Contamination in Commonly Consumed Herbal Medicines From Local Markets in Hyderabad, India, and Associated Health Risks." *Food Additives and Contaminants: Part B Surveillance* 18 (2):187–98. https://doi.org/10.1080/19393210.2025.2471065.

Wang, H., Y. Chen, L. Wang, Q. Liu, S. Yang, and C. Wang. 2023. "Advancing Herbal Medicine: Enhancing Product Quality and Safety Through Robust Quality Control Practices." *Frontiers in Pharmacology* 14: 1265178. https://doi.org/10.3389/fphar.2023.1265178

WHO. 2007. *WHO guidelines for assessing quality of herbal medicines with reference to contaminants and residues*. Edited by World Health Organization: WHO.

Welz, A. N., A. Emberger-Klein, and K. Menrad. 2018. "Why People Use Herbal Medicine: Insights from a Focus-Group Study in Germany." *BMC Complementary and Alternative Medicine* 18 (1): 92. https://doi.org/10.1186/s12906-018-2160-6

Wu, K. M., J. Dou, H. Ghantous, S. Chen, A. Bigger, and D. Birnkrant. 2010. "Current Regulatory Perspectives on Genotoxicity Testing for Botanical Drug Product Development in the U.S.A." *Regulatory Toxicology and Pharmacology* 56 (1): 1–3. https://doi.org/10.1016/j.yrtph.2009.09.012

Yan, H., Z. Zhang, L. Lan, and G. Sun. 2025. "A Comprehensive Quality Control System for Chinese Herbal Medicine Based on HPLC Ratio Quantitative Fingerprinting and Spectral Quantized Fingerprinting Combined With Antioxidant Activity: A Case of Siegesbeckiae Herba." *Journal of Chromatography A* 1760: 466313. https://doi.org/10.1016/j.chroma.2025.466313

8 Market Value of Herbal Medicine

Sagar Surendra Deshmukh
and Pankhi Goswami

8.1 INTRODUCTION

Herbal medicine has played an essential role in the traditional healthcare system. Nowadays, with the growing awareness of the side effects of synthetic drugs, the interest in plant-based remedies is reinstated, which can also be termed as "Return to Nature" (Sharma et al. 2008). Herbal medicines are considered safer, cheaper, and more sustainable, and they also contribute significantly to the global economy (Jadhav, Vikhe, and Jadhav 2020; Khan and Ahmad 2019).

According to the World Health Organization (WHO) report on Traditional and Complementary Medicine, 2019, around 80% of the population in the world relies on traditional medicine as their primary healthcare. The WHO defines herbal medicine as finished products made from whole plants or parts of the plants, such as leaves, roots, bark, berries, flowers, and the extracts. These are mostly intended for the therapeutic use of humans (Bareetseng 2022; Kumar et al. 2024). There is a growing demand for herbal medicines in both developing and developed countries, making herbal medicine one of the fastest-growing industries in the world (Hossain, Gera, and Ali 2022). The interests of the countries are mostly derived from consumer expectations for herbal products to be safer with fewer side effects (Mokhtarikhah et al. 2022), the urge to expand coverage of herbal medicine, the rising cost of pharmaceutical products, and reduced budgets for health care (Bareetseng 2022). A forecast by the World Health Organization (WHO) stated that the global herbal market is projected to increase from its current value of USD 62 billion to USD 5 trillion by 2050. This growth demonstrates a rising consumer preference for natural products and the expansion of herbal-based industries (Bhattacharjee et al. 2024)

The herbal medicine market, around the world, is divided into four regions: Europe, the Americas, Asia-Pacific, the Middle East, and Africa. The European region, which includes countries such as Germany, France, Italy, and the UK, is considered to be the largest and fastest-growing region among all. Japan, China, India, South Korea, Indonesia, and Australia constitute the Asia Pacific region and are ranked as the second-largest herbal market (Mazrouei et al. 2022). The reason behind it can be attributed to the growing popularity of Indian medicine and traditional Chinese medicine, due to their natural and side-effect-free treatment, and the adoption by pharmaceutical companies, researchers, and policymakers (Pandey, Rastogi, and Rawat 2013; Patwardhan et al. 2005). The Asia-Pacific and Europe regions dominate the herbal medicine markets by making up 72.36% of global consumption. The United States

DOI: 10.1201/9781003623748-8

has the third-largest herbal medicine market, where these products are primarily used as alternatives for treating conditions such as stroke and arthritis. In contrast, the herbal medicine market in Africa and the Middle East is experiencing the slowest growth (Mazrouei et al. 2022). The use of herbal medicine has increased significantly after COVID-19, as more people are turning to immunity-boosting formulations. This growing demand has opened up new opportunities for farmers, processors, and marketers within the herbal value chain. As a result, herbal medicine is not only seen as a health solution but also as a catalyst for economic development.

8.1.1 Global Market Scenario for Medicinal Plants

The global herbal medicine market, valued at USD 233.08 billion in 2024, is projected to grow at 8.23% CAGR from USD 251.25 billion in 2025 to USD 437 billion by 2032. In 2024, Europe held the largest share of the herbal medicine market, accounting for 44.55% of the total market. In emerging markets, like China and India, the growth is strongest (Hossain, Gera, and Ali 2022). Gradual expansion is also observed in other emerging markets across the Asia/Pacific region, Eastern Europe, Latin America, the Middle East, and Africa (Fortune Business Insight Report 2025) (Figures 8.1–8.3).

The global export value of medicinal plants experienced a significant growth of 31.59% between 2013 and 2024 (Figure 8.1). During this period, key players dominated the global export market, where China emerged as the leading country in 2024 with 21.48% of total global export value; India shared 11.91% of global export value, followed by Canada with 7.54%; and Germany contributed 5.96%. Other major exporters of medicinal plants include the United States, Spain, Egypt, South Korea, Spain, and Poland (Figure 8.4).

FIGURE 8.1 Global export and import value of medicinal plants from the year 2013 to 2024. (Plotted by the author using data from ITC Trade map, 2024.)

FIGURE 8.2 The top exporting countries of medicinal plants in the year 2024. (From ITC Trade Map, 2024.)

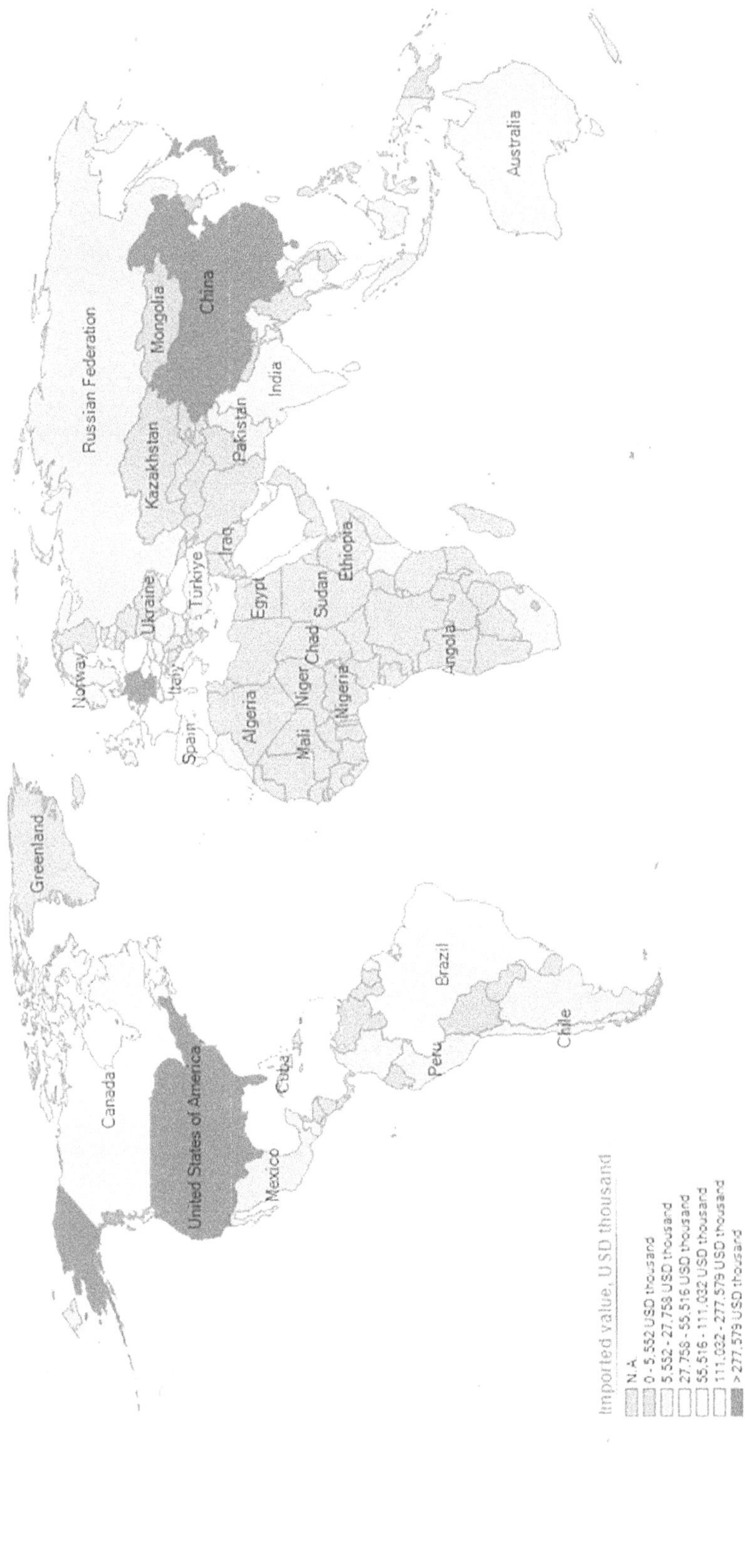

FIGURE 8.3 The top importing countries of medicinal plants in the year 2024. (From ITC Trade Map.)

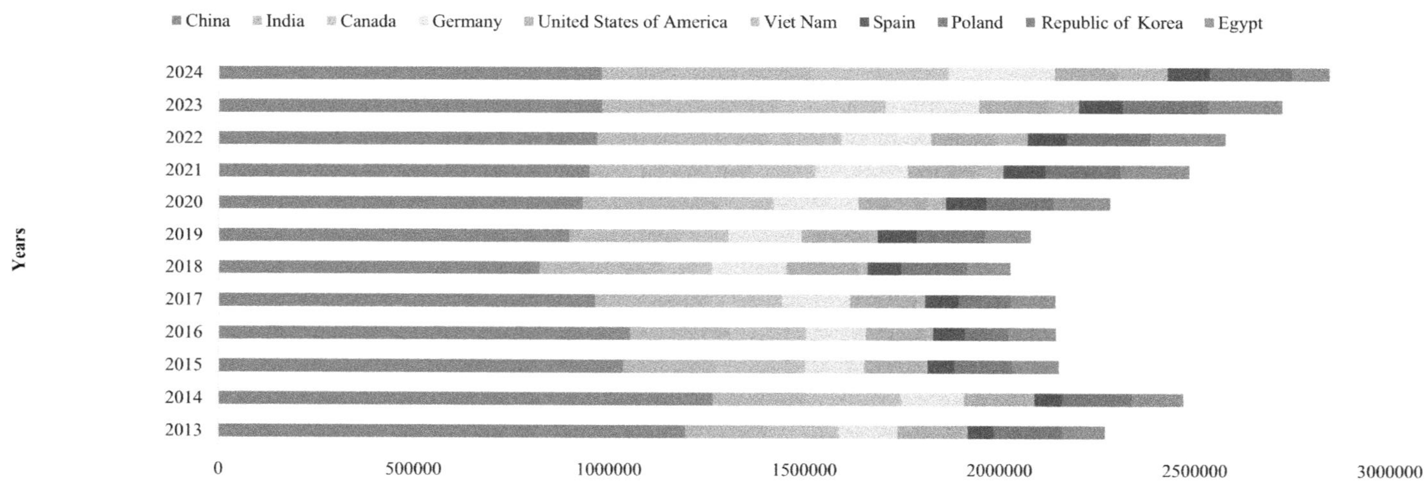

	2013	2014	2015	2016	2017	2018	2019	2020	2021	2022	2023	2024
China	1196641	1265686	1036865	1053143	964461	822422	898270	931661	949966	971020	982466	980136
India	207758	238708	237798	259342	291761	307883	285797	347387	414627	438734	450908	543769
Canada	184911	242694	227114	192908	185565	131849	121397	141659	161364	184236	273862	344342
Germany	148884	161800	151295	151815	175135	190892	187121	217270	237521	230587	239763	272365
United States of America	168206	166935	139989	155193	171226	184139	173354	177211	173515	165118	176681	160709
Viet Nam	12464	13658	22247	17854	19478	23712	20193	46862	71525	80654	78214	126650
Spain	66052	71443	68770	81701	86742	87364	100966	103811	108217	103210	113213	109489
Poland	66269	66941	59758	56835	61828	84050	87680	96909	103682	114596	107161	108873
Republic of Korea	109452	111840	88896	54430	71269	84913	87237	75817	88821	97439	110696	101474
Egypt	109845	130784	118173	120787	115331	110612	117359	144614	175450	192565	189643	96517

Export value (million USD)

FIGURE 8.4 Medicinal plants export value in the top 10 exporting countries during 2013–2024. (From ITC Trade Map.)

The import trends of herbal medicine show a picture where the United States emerged as the largest importer in 2024 with a trade value of 555 million USD, followed by Germany with 486 million USD, accounting for 12.33% and 10.81% of the total import value, respectively. The main market for the USA comprises India and Mexico, contributing 31.62% and 10.6%, respectively (Zamani et al. 2025). The other top importing countries include China, Japan, Hong Kong, Australia, South Korea, Taipei, Italy, France, India, and Korea (Figure 8.5).

8.1.2 Indian Market Scenario for Medicinal Plants

India exports herbal medicine plants to 130 countries, accounting for 11.9% of world exports in this sector, making it the second-largest exporter of herbal medicine plants in the world. The export value of medicinal plants saw a tremendous increase of 295% from 135 million USD in 2010 to 535 million USD in 2024. The expanding growth can be attributed to the increasing cultivation, favourable trade policies, and growing demand for Indian herbal medicine across the world. Psyllium husk and seed, senna leaves and pods, basil, hyssop, sage, rosemary, and sandalwood are some of the most exported herbal medicinal plants from India (Sen et al. 2011; Zamani et al. 2025). The top five countries importing medicinal plants from India in 2024 include the USA, Germany, Italy, China, and the UAE (Figures 8.4 and 8.6).

Medicinal plants are a major source of livelihood for a major section of the Indian population. Approximately 1178 species of medicinal plants are currently in trade, out of which 242 species of herbal medicinal plants exhibit an annual consumption level of more than 100 metric tons/year. In order to promote the conservation, cultivation, trade, and export of the herbal medicine plant sector, the Government of India has taken several measures as follows:

- Setting up the National Medicinal Plants Board (NMPB) in 2000, in the Ministry of AYUSH. NMPB developed an online portal and an Android-based application for trading medicinal plants called "e-charak". This platform helps information exchange and trading among various stakeholders in the medicinal plants sector.
- The NMPB also launched a "Voluntary Certification Scheme for Medicinal Plants Produce (VCSMPP)" on 22nd November 2017. This will encourage the Good Agricultural Practices (GAP) and Good Field Collection Practices (GFCP) of medicinal plants.
- Shellac and Forest Products Export Promotion Council (SHEFEXIL) and Pharmaceutical Export Promotion Council (PHARMEXCIL) are responsible for promoting the export of Herbs and Medicinal Plants. These Export Promotion Councils set up by Department of Commerce and implement various promotional measures to boost export of their products (PIB 2019).
- The Market Access Initiative (MAI) Scheme of the Department of Commerce provides financial assistance to the EPCs for participation and organizing Trade Fairs, Buyer Seller Meets (BSMs), Reverse Buyer Seller Meets (RBSMs), Research & Product Development, Market Studies, etc. (PIB 2019).

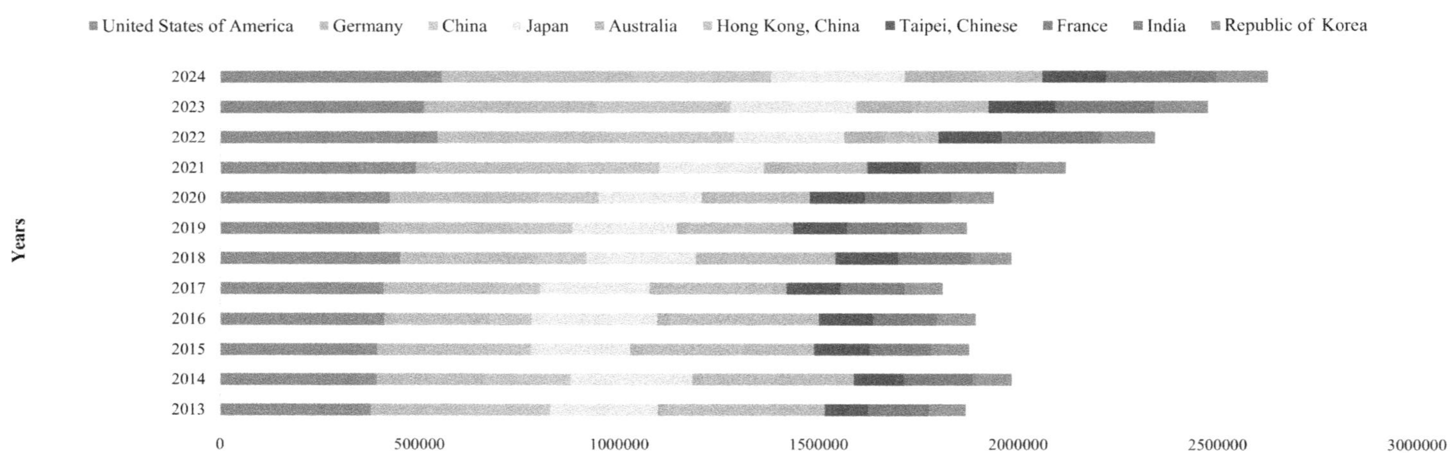

	2013	2014	2015	2016	2017	2018	2019	2020	2021	2022	2023	2024
United States of America	377692	392347	393598	413369	409812	452354	399860	426679	490321	544795	510905	555159
Germany	238536	271037	247173	263350	290126	341077	334159	373488	411942	439924	436117	486350
China	212401	215308	139062	104741	103099	125390	148915	148698	198830	303623	332292	338549
Japan	269115	305342	247345	313334	272931	273463	261573	257859	260365	275170	313741	335427
Australia	25361	27916	35021	31845	35493	37676	27223	34059	50690	96602	141228	182088
Hong Kong, China	393255	376312	427252	373955	307608	311652	263085	236458	209655	140338	190828	162006
Taipei, Chinese	107957	124707	137109	135928	135958	157199	137063	138202	133199	159126	166399	161785
France	97133	104323	84318	91071	99920	102338	104001	113707	134139	132789	135157	143812
India	54831	68575	69258	69742	60813	79267	81324	102184	106613	114981	114385	131184
Republic of Korea	93650	97937	97163	96777	95896	102416	113973	107304	122395	135250	135308	129350

Import value (million USD)

FIGURE 8.5 Medicinal plants import value in the top ten importing countries during 2013–2024. (From ITC Trade Map.)

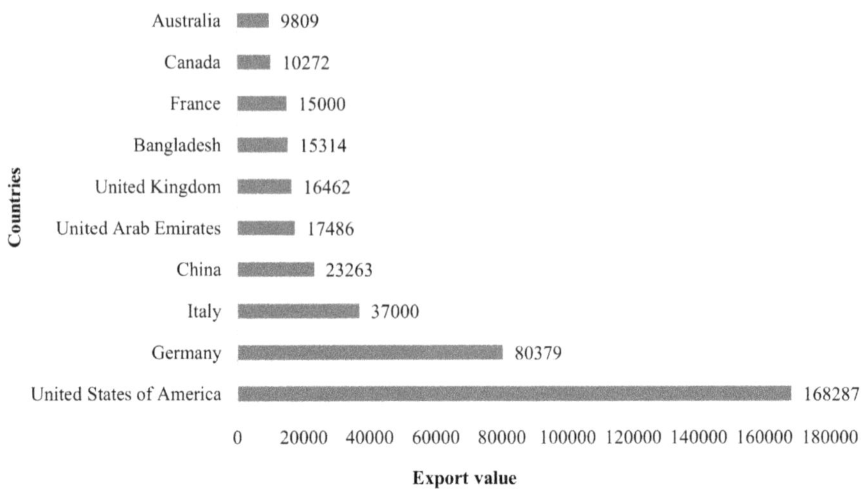

FIGURE 8.6 The top ten countries importing medicinal plants from India in the year 2024. (From ITC Trade Map, 2024.)

- International Cooperation Scheme (ICS) of Ayush Ministry provides financial assistance to exporters for participation in Trade Fairs, organizing International Business Meets & Conferences, and product registration reimbursements.

8.2 MAJOR STAKEHOLDERS IN THE HERBAL MEDICINE SECTOR IN INDIA

The herbal medicine sector involves an exclusive number of key players that include cultivators, processors, farmer cooperatives, pharmaceutical companies, regulatory bodies, and research institutions.

The herbal medicine sector in India is largely dominated by established companies with a strong domestic and global presence (Customer Market Insight 2025). These companies have diversified herbal product portfolios and invest significantly in research and development, as well as marketing. Some of the key Indian companies include

- **Dabur India Limited** – Dabur was founded in Kolkata, West Bengal, by S.K. Burman, an Ayurvedic practitioner, in the year 1884. It is now one of the largest Ayurvedic and Natural Health Care companies with a consumer base in more than 120 countries, with its head office in Ghaziabad. It has 13 manufacturing facilities in different parts of India, along with 8 manufacturing facilities located outside India in countries such as Nepal, UAE, South Africa, Turkey, Egypt, etc. In the year 2023–24, Dabur recorded a turnover of ₹12,404 crore in fiscal year 2023–24, marking a 7.6% growth in consolidated revenue, with the international business experiencing a growth of 16.4% in CC terms. The herbal medicine segment, marketed under the healthcare business, contributed 31.4% to the revenue of Dabur. It also has one of the largest distribution networks in the industry, covering 7.9 million retail outlets and reaching 122,000 villages.

- **Himalaya Wellness Private Limited** – Himalaya Wellness is a multinational company based in Bangalore, Karnataka. It was established by Muhammad Manal in Dehradun in 1930. Its operations are spread across India, the United States, the Middle East, Asia, Europe, and Oceania. Their products are sold in more than 100 countries across the globe. Himalaya Wellness recorded a revenue of ₹4670 crore in the fiscal year 2023–24, resulting in a 14.4% year-on-year growth.
- **Zandu Care Limited** – Zandu Care acquired by Emami Limited in 2008 is renowned brand for Ayurvedic medicines that caters to pain management and general wellness. It was founded by Jugatram S. Vaidya in Mumbai, Maharashtra, in the year 1910. Zandu plays a significant role in the total annual revenue of ₹3578.09 crore generated by Emami, primarily through its broad healthcare and OTC portfolio.
- **Shree Baidyanath Ayurved Bhawan (Pvt) Limited** – Baidyanath was founded by Pandit Ram Narayan Sharma in 1917 in Bihar, currently headquartered in Noida. The company operates 11 factories across India and offers the largest range of Ayurvedic products, with over 700 different formulations. These products are available at more than 100,000 retail outlets all over India. Baidyanath reported a revenue of ₹713.4 crore for the fiscal year 2023–24, reflecting a 4% increase compared to 2022–23, along with a net profit of ₹30.7 crore (Table 8.1).

Farmer cooperatives are also taking up a good shape. Farmer cooperatives like the Retam Farmer Producer Company Ltd. and Krishak Jagriti Manch producer group recently received certification from the Quality Council of India (QCI) for medicinal plants production. The certification will help the two producer organizations to position themselves as key players in the flourishing market by ensuring adherence to quality and sustainability standards (Solidaridad 2024).

One of the most influential regulatory stakeholders in the herbal medicine sector is the Ministry of AYUSH (Ayurveda, Yoga and Naturopathy, Unani, Siddha, and Homoeopathy). AYUSH plays a crucial role in policy development, quality assurance, and promotion of the sector. The ministry supports the industry through various flagship schemes such as

- Ayurgyan Scheme, which promotes education and research in AYUSH
- Ayurswasthya Yojana, which provides financial assistance to eligible individual organizations/institutes for establishing and upgrading their functions and facilities and/or for research and development activities in Ayush

TABLE 8.1
An Overview of Major Ayurvedic Companies in India, Highlighting Their Product Portfolio

Company Name	Product Description	Products Portfolio
Dabur	• Health supplements cater to products that aim to improve overall health, vitality, and immunity • Digestives aid in digestion and relieve gastrointestinal discomfort • Ayurvedic OTCs are preventive and curative herbal remedies for various health issues like respiratory disorders, bone health, etc.	• Health supplements – Chyawanprash, Honey, Glucose • Digestives – Hajmola, Pudin hara, Hingoli • Ayurvedic OTC – Honitus cough syrup, Shilajit, Isabgol, Triphala, Ashwagandha tablets, Tulsi drops
Himalaya	• Pharmaceuticals are mixed herb-based Ayurvedic formulations that treat specific health conditions • General wellness products are single-herb tablets or capsules for physical and mental well-being	• Pharmaceuticals – Liv 52, Septilin, Opthacare eye drops, Gasex, Pilex • General Wellness – Ashwagandha tablets, Giloy Tablets, Tulsi tablets, Cold balm, Neem Tablets, Shatavari Tablets
Zandu care	• Pain management products are topical Ayurvedic products used for muscle and joint pain, inflammation, or headaches • Health supplements include pure herb capsules, powders, juices aimed at achieving optimal health	• Pain Management – Zandu Balm, Pain relief patch • Health Supplements – Shilajit capsules, Triphala capsules, Ashwagandha capsules, Kalamegha capsules, Methi capsules, Vitality booster Juice, Chyawanprash
Baidyanath	• Prescription products are administered under the supervision of a practitioner. These include various herbal Arks (distilled extracts), Bati (tablets), and Kadha (decoction drink) for treating bone issues, respiratory disorders, digestive problems, etc. • Wellness products are to be used for daily personal and lifestyle care and includes herbal and mineral blend churna (powders), bhasma (incinerated powder), Ras rasayana (tablets)	• Prescription products – Tulsi Bati, Guduchi Giloy Bati, Dashmool Ark, Saunf Ark, Kadha • Wellness – Gangadhar Churna, Narayan Churna, Lauh Bhasma, Kashis Bhasma, Ameer Ras, Balark Ras

Source: Compiled by author from annual reports of the companies.

- Ayush Oushadhi Gunvatta Evam Uttpadan Samvardhan Yojana, to provide assistance for equipment upgrades, infrastructure improvements, and regulatory compliance to help beneficiaries meet quality standards and improve their manufacturing capabilities in the Ayush sector
- Conservation development and sustainable management of medicinal plants, to conserve and sustainably manage medicinal plants by promoting habitat protection, quality cultivation, and a strengthened supply chain

- Promotion of International Co-operation- to facilitate international promotion and development of Ayurveda
- The Quality Certification programme, like Ayush Mark and Premium mark, helps set up quality standards

Research institutions are also another major stakeholder in the herbal medicine sector. They contribute significantly by scientific validation, standardization and development of evidence-based herbal medicine formulations (Sen and Chakraborty 2015; Saggar et al. 2022). Some of the prominent research institutions under the Ministry of AYUSH, working for the promotion of herbal medicine, are

- Central Council for Research in Ayurvedic Sciences (CCRAS), New Delhi
- Central Council for Research in Unani Medicine (CCRUM), New Delhi
- Central Council for Research in Siddha (CCRS), Chennai
- Central Council for Research in Yoga and Naturopathy (CCRYN), New Delhi
- Council for Scientific and Industrial Research (CSIR),
- Central Institute for Medicinal and Aromatic Plants (CIMAP), Lucknow
- Central Drug Research Institute (CDRI), Lucknow
- Northeast Institute of Science and Technology (NEIST), Jorhat
- Indian Institute of Integrative Medicine (IIM), Jammu

The CSIR is a notable research institute in India. CSIR and its related laboratories, such as CDRI, IIIM, NEIST, and CIMAP, are involved in the formulation of science-backed herbal medicines for various ailments. Anti-Arthritis" ointment for rheumatoid arthritis and other joint pain, "Fungi-Destruct" multi-action cream for dermal infections developed by NEIST, "Dalzbone" for bone fracture healing, "Joint Fresh" for osteoarthritis developed and marketed by CDRI. IIIM collaborated with Hapico Industries Pvt. Ltd to develop CalPoMag, which is a complete herbal formulation enriched with Calcium, Potassium, Iron, and Magnesium. Likewise, CIMAP is also involved in developing anti-diabetic, pain relievers, cough syrups, nutraceuticals and more.

8.3 MSME AND START-UPS IN THE HERBAL MEDICINE SECTOR IN INDIA

The AYUSH sector of India has a rich history and promising future, offering a plethora of opportunities for entrepreneurs. AYUSH has established itself as a global player, which is evident from its significant exports to key markets like the USA, EU, and UAE. This sector has experienced a huge interest and investment from various global quarters, from Asia to North America to Europe. Initiatives like the AYUSH visa – that allows foreigners to visit India for treatments through Ayurvedic means, as well as endorsements from WHO and G20 has heightened global recognition (Invest India 2025).

MSMEs and start-ups constitute 80% of the sector that has powered its growth story. The number of MSMEs in the AYUSH industry increased from 38,216 in

August 2021 to 53,023 by January 2023, attributing a staggering 40% growth, indicating significant growth and potential. These start-ups are mostly located in Tier-2 and Tier-3 cities with a substantial presence of 55% of women entrepreneurs, which promotes modern entrepreneurial zeal, economic growth, and social development. Ministry of AYUSH has also signed an MOU with the Ministry of Micro, Small, and Medium Enterprises (MSME) in 2018 for the promotion of enterprises related to AYUSH. Since then, various awareness programmes on digital advertisement, e-marketing, along with several National seminars, workshops on packaging, marketing, and export promotion of herbal products were organized by the MSME.

The All India Institute of Ayurveda (AIIA) has set up an incubation centre called Incubation Centre for Innovation and Entrepreneurship (AIIA-ICAINE) to support new start-ups in the Ayush sector. Central Council of Research in Siddha (CCRS) has also set up an Academia-Industry Research Incubation for Value-chain Upgradation (ARIVU) to promote innovation and entrepreneurship through novel ideas and research for commercial use and societal benefit. The central sector scheme AYUSH entrepreneurs also offers several benefits for entrepreneurs, including the setting up of business incubators to provide necessary resources to the entrepreneurs from research facilities to infrastructure and financial help.

Several established start-ups within the AYUSH ecosystem represent the dynamism of the herbal medicine sector as shown in Table 8.2.

TABLE 8.2
Start-Ups in the AYUSH Ecosystem in India

Name of the Company	Company Overview	Founded Year and Location	Annual Revenue (2023–2024)
Life Chart Ayurveda	A company offering scientific herbal medicines for various issues such as piles, fatty liver, constipation, weight loss, and gastric problems	2022, Gurugram	₹1.3Cr ($157K)
Planet Ayurveda	Manufacturer of herbal remedies, natural supplements, and Ayurvedic medicines for various health issues	1990, Chandigarh	₹17.2Cr ($2.43M)
Muditam Ayurveda	A company that offers herbal supplements for diabetes management	2020, Sirmaur	₹1.98Cr ($239K)
Jiva Ayurveda	Jiva Ayurveda deals in herbal medicines for the treatment of diabetes, arthritis, asthma, obesity, hypertension, spondylitis, piles, skin problems, sexual disorders, etc.	1992, Faridabad	₹86.3Cr ($10.4M)
Butterfly Ayurveda	A company that manufactures health supplements, food, and herbal medicines	2014, Gurugram	₹2.88Cr ($388K)
Kerala Ayurveda	A company that offers wellness clinics, hospitals, herbal medicine manufacturing for child health, diabetes, joint care, gastro care, etc.	1945, Aluva	₹93.7Cr ($11.7M)

Source: Tracxn.

8.4 EMERGING TRENDS IN THE HERBAL MEDICINE SECTOR IN INDIA

In recent years, the herbal medicine sector in India has witnessed the emergence of various modern start-ups. These start-ups are redefining the Ayurvedic system by integrating traditional practices with modern scientific advancements. These start-ups focus on plant-based wellness solutions combined with modern nutrition science by offering products like gummies, effervescent tablets, protein powders, health supplements, and more. They focus more on clean label, research-backed formulations, transparency in sourcing, and direct-to-customer models have appealed to the health-conscious young generations. These companies are making herbal medicine more convenient and relevant to modern-day consumers. Table 8.3 and Figure 8.7 depict some top-ranked new-age start-ups and their market share, respectively.

TABLE 8.3
New-Age Start-Ups in the Herbal Medicine Sector

Name of the Company	Company Overview	Founded Year and Location	Annual Revenue (2023–2024)
Kapiva	The company founded by Ameve Sharma and Shrey Badhani offers Ayurvedic healthcare and nutritional products that includes heart care, blood sugar and chronic care, Ayurvedic juices, weight management, and daily wellness products	2016, Bengaluru	₹234Cr ($28.3M)
Ozivia	Founded by Aarti Gill and Mihir Gadani, it is a company that combines modern science with Ayurveda to provide herbal medicine and supplements for skin and hair health, vitamin deficiency, PCOS, and general wellness	2016, Mumbai	₹199Cr ($24.9M)
Plix	Founded by Rishubh Satiya and Akash Zaveri, Plix uses clinically backed plant ingredients to develop herbal supplements gummies, superfood powders and tablets for overall health	2019, Mumbai	₹158Cr ($19.1M)
Wellbeing Nutrition	Founded by Avnish Chabria, deals in the use of plant-based ingredients and nutraceuticals, providing a range of products aimed at enhancing overall well-being	2019, Mumbai	₹72Cr ($8.69M)
Earthful	Founded by sisters Sudha Gogineni and Veda Gogineni, Earthful offers a range of plant-based products that targets to alleviate various health concerns	2020, Hyderabad	₹3.15Cr ($380K)

Source: Tracxn.

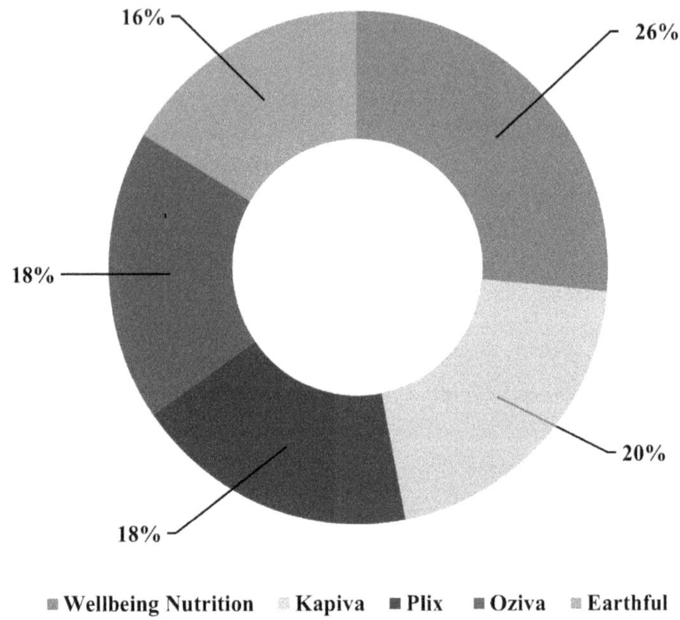

FIGURE 8.7 Market share of new-age start-ups in the herbal medicine sector. (Presented by the author using data from Tracxn).

8.5 REGULATORY STATUS OF HERBAL MEDICINE

The thriving dependency on herbal medicine across the world has entailed the development of a clear, standardized regulatory framework to ensure the safety, efficacy, and quality of plant-based medicines. Herbal medicines are generally considered safe and free from side effects due to their natural origin; however, without adequate quality control and regulatory frameworks, there is a chance of adulteration and misuse (Ekor 2014). Therefore, both international and national bodies have implemented a range of policies to oversee the regulation and promotion of herbal medicine so that they are safe, effective, and of good quality.

8.5.1 Global Regulatory Framework

According to the WHO report on Traditional and Chinese Medicine (2019), the European Region has the most countries with state and national regulations for herbal medicines. This is followed by the African Region, the Region of the Americas, the Eastern Mediterranean Region, the Western Pacific Region, and lastly, the South-East Asia Region.

The European Union established a Committee on Herbal Medicine Products (HMPC) under the European Medicines Agency (EMA) to make the regulation of herbal medicine uniform and to facilitate its marketing (Metta et al. 2020; Xiong et al. 2024). In an attempt to provide a system of complete technical standards and requirements for regulation, the European Pharmacopoeia was developed, which will help create a standard for safety and efficacy in the European Union Market (Metta et al. 2020). In South Africa, herbal medicines, also known as complementary medicine, are governed by the Medicines and Related Substances Act 1965. The South African Health Regulatory Authority is responsible for implementing and enforcing these regulations (Kumar 2017). They ensure that all the herbal medicines meet the required safety, efficacy, and quality standards, while in the United States, the marketing and selling of herbal medicines do not need any evidence for the safety and efficacy of the product, as they are primarily regulated under the Dietary Supplement Health and Education Act (DSHEA) of 1994, which classifies them as dietary supplements rather than medicines (Rojas et al. 2022). The Food and Drug Administration (FDA) also does not do much regarding this, as they do not even consider an herb a medicine. The relatively relaxed regulatory environment has generally benefited manufacturers by reducing the legal barriers they face during production and marketing. However, this situation has also led to tensions between regulatory bodies and producers, particularly regarding the quality control and testing of dietary herbal medicines (Jadhav et al. 2025). Manufacturers often oppose stricter analytical requirements because of the high financial and technical demands they entail. This regulatory gap has raised concerns about product consistency, authenticity, and consumer safety in the herbal supplement market. Kuwait imports herbal medicines from other countries after registering them through the Kuwait Drug and Food Administration (KDFCA). It lacks a unified classification system of Herbal medicine, which leads to Herbal medicines imported into the country being registered under different units of KDFCA according to the regulatory status of the product in their country of origin. As a result, the registration process in Kuwait is disorganized (Ng et al. 2022). Consequently, some health supplements are categorized under the unclassified unit as dietary supplements or classified as functional foods under the food supplements unit, both of which have fewer and less stringent registration requirements (Alostad, Steinke, and Schafheutle 2019). The United Arab Emirates (UAE) and Bahrain import 90% of their herbal medicine, but they have harmonized their regulatory systems due to shared geographic, political, and cultural factors, especially their governance frameworks. Both countries are committed to ensuring the safety, quality, and efficacy of medicines similar to the traditional pharmaceuticals. As a result, they manage herbal medicine regulations in a coordinated and consistent manner (Alostad, Steinke, and Schafheutle 2018; Dubale et al. 2025). In certain Western Pacific countries and Southeast Asia, such as Bhutan, Sri Lanka, and the Maldives, there is a standard regulatory framework for herbal medicine. Certain countries, including Korea, India, Thailand, China, Myanmar, and Indonesia, have their own National Monographs and Pharmacopoeias for herbal medicine. The essential medicine list of countries like China and the Philippines includes herbal medicine as a significant alternative (Sahoo, Manchikanti, and Dey 2010). India has distinct essential medicine lists for traditional herbal medicines, including Ayurveda, Unani, Siddha, and more (Wanjari et al. 2016). In most countries, herbal drugs are available as over-the-counter (OTC) medications, with the exception of Bhutan. In Bhutan, these medicines are only sold with a prescription similar to traditional pharmaceuticals (Sahoo, Manchikanti, and Dey 2010).

8.5.2 Indian Regulatory Framework

In India, herbal medicines are regulated by several key legislations. These include the Indian Central Council of Medicine (IMCC) Act, various research councils like Indian Council of Medical Research (ICMR), Council of Scientific and Industrial Research (CSIR), the Department of AYUSH (Ayurveda, Yoga and Naturopathy, Unani, Siddha, and Homeopathy), and the Drugs and Cosmetics Act of 1940 and its amendments (Singh et al. 2010; Metta et al. 2020; Oyedepo and Palai 2021). According to the amended Drugs and Cosmetics Act of 1940, herbal medicines require adherence to prescribed hygienic conditions under the supervision of qualified individuals, the use of genuine raw materials, and proper labelling of all ingredients utilized. To address challenges in implementing these amendments, the Pharmacopoeia Laboratory for Indian Medicine (PLIM) was established in Ghaziabad by the Ministry of Health and Family Welfare (Vikaas et al. 2017). PLIM includes the Drug Standardization and Testing Unit, a Drug Depot, and Herbarium and Reference Museum (Oyedepo and Palai 2021).

The department of AYUSH has stated that around 600 medicinal plant products are used in traditional Ayurvedic preparations. In the Indian Herbal medicine industry, the quality of herbal medicine has become a rising cause of concern, due to poor standardization practices of the raw materials (Wanjari et al. 2016; Sarmah 2022). The three major issues identified in Indian herbal medicines are substitution, adulteration, and heavy metal contamination. This contamination can occur intentionally or accidentally, starting from the collection of raw materials and continuing through the manufacturing process (Sahoo and Manchikanti 2013; Sarmah 2022). Developed countries mandate chemical analysis to ensure the quality of the product, while these kinds of practices are in a nascent stage in India. The Indian Good Manufacturing Practices do not provide guidelines for marker-based identification (Hossain, Gera, and Ali 2022; Sahoo and Manchikanti 2013; Sarmah 2022). Only 44% of the companies that manufacture herbal medicines in India conduct chemical marker-based studies for their formulations, often at government or private testing laboratories. This process is costly, and many small and medium enterprises lack the necessary resources. Most manufacturers rely on traditional methods for standardizing raw materials, and challenges persist due to the unavailability of reference standards and limited third-party laboratories for testing ingredients (Sahoo and Manchikanti 2013; Sarmah 2022). Marker-based chemical identification of herbal plants is also performed by a very small number of Indian manufacturers, citing a lack of infrastructure, trained personnel, and high operational costs. The Good Manufacturing Practices (GMPs), as mentioned in Schedule T of the Drugs and Cosmetics Rules, 1945, are also not adequately practised due to financial and logistical constraints. Indian states also lack harmonized guidelines for drug registration, licensing, and enforcement, which is one of the major factors that create inconsistencies among manufacturers and make product standardization a difficult task.

India has introduced prominent steps to tackle these challenges. These include the Ayurvedic Pharmacopoeia of India (API), Good Agricultural and Collection

Practices (GACP) guidelines by the National Medicinal Plants Board (NMPB), along with increased support for clinical safety evaluations of the herbal drugs. But these steps must be mediated with awareness among the grassroots level, as implementation at that level still remains finite. To make Indian herbal medicine prominent and credible in the international market, strengthening the infrastructure for herbal drug testing, promoting organic and sustainable cultivation, and harmonizing regulatory protocols among states must be taken up.

8.6 CHALLENGES AND LIMITATIONS IN THE HERBAL MEDICINE MARKET

- **Regulatory challenges** – The regulatory policies around herbal medicines keep evolving continuously. In certain countries like the USA, herbal medicines are considered as dietary supplements or food (Singh et al. 2010), so they do not undergo rigorous testing like traditional medicines, ergo this raises safety concerns for health authorities as well as the general public, whereas in the European Union, Russia, and the Middle East, the regulations regarding herbal medicines are much stringent to maintain safety and efficacy. These regulatory inconsistencies between agencies and pharmacovigilance often hinder the export and import of herbal medicines among countries. In India, though the primary regulatory framework for herbal medicine remains the Drugs and Cosmetics Act of 1940, with AYUSH as the regulatory authority, the ever-changing guidelines around corporate governance, tax laws, sourcing of materials, waste generation, and quality of finished products may have an impact on the business activities of herbal medicine (Saggar et al. 2022)
- **Supply chain challenges** – Supply chain limitations of the herbal medicine market continue to be one of the prominent challenges today. The availability of herbal ingredients is dependent on nature, growing seasons, and harvesting seasons. Limited availability of raw materials for production and single vendor sourcing create vulnerabilities in the supply chain. These issues can lead to delays, supply shortages, and quality concerns. Therefore, it has become important for companies dealing with herbal medicines to move cultivation closer to their manufacturing units by involving local farmers. Certain businesses are also changing their inventory models from "Just in Time" to "Just in Case" by maintaining safety stock of critical raw materials (Thillaivanan and Samraj 2014; Moloughney 2022)
- **Economic Uncertainty** – The price of medicinal herbs has increased over the recent years due to economic factors like international conflicts and market uncertainty. The companies reduced the margin at first to keep prices low for the herbal products, but recent trends have shown an increase in the price of the finished herbal product. Inflation can impact the consumption habits of consumers and reduce the profit margin percentage of companies in this sector (Moloughney 2022)

8.7 CONCLUSION

The Herbal medicine industry is experiencing significant growth worldwide, driven by customer preference for natural remedies, which are considered free of side effects. The market has emerged as a dynamic and critical factor for healthcare, with an expected global market value to reach USD 5 trillion by 2050.

India, known for its rich biodiversity and traditional knowledge system of Ayurveda, plays a strategic role in the global market of herbal medicine. Although India is the second largest exporter of herbal medicinal plants, the industry still faces significant challenges related to standardization, quality control and regulatory challenges. It is necessary to handle these challenges through technological innovation, better enforcement of Good Manufacturing Practices (GMP) and capacity building at the grassroots levels.

Regulatory frameworks around the world differ significantly. Some regions practice strict quality standards, while others adopt a more lenient approach like the USA. Markets in regions such as the European Union, North America, Southeast Asia, and the Middle East are increasingly seeking certified, plant-based alternatives to synthetic pharmaceuticals. However, India is still behind in obtaining international certifications like EU-GMP, and WHO-GMP for many of its herbal products. For India to take full benefit of international opportunities, it is crucial to align domestic practice with global benchmarks. Improving the traceability and certification of herbal products are essential to achieve these benchmarks.

The herbal medicine sector has great potential for improving healthcare and wellness, as well as a source of livelihood for millions of cultivators and entrepreneurs. With strong policy support, innovation, and sustainable practices, it can significantly contribute to creating a healthier, more inclusive, and economically resilient future.

REFERENCES

Alostad, A. H., D. T. Steinke, and E. I. Schafheutle. 2018. "International Comparison of Five Herbal Medicine Registration Systems to Inform Regulation Development: United Kingdom, Germany, United States of America, United Arab Emirates and Kingdom of Bahrain." *Pharmaceutical Medicine* 32 (1): 39–49. https://doi.org/10.1007/s40290-018-0223-0

Alostad, A. H., D. T. Steinke, and E. I. Schafheutle. 2019. "A Qualitative Exploration of Bahrain and Kuwait Herbal Medicine Registration Systems: Policy Implementation and Readiness to Change." *Journal of Pharmaceutical Policy and Practice* 12 (1): 32. https://doi.org/10.1186/s40545-019-0189-7

Bareetseng, S. 2022. "The Worldwide Herbal Market: Trends and Opportunities." *Journal of Biomedical Research & Environmental Sciences* 3 (5): 575–84. https://doi.org/10.37871/jbres1500

Bhattacharjee, B., K., Sandhanam., S., Ghose, and D., Barman,. 2024. Market Overview of Herbal Medicines for Lifestyle Diseases. Role of Herbal Medicine. Springer Nature. 10.1007/978-981-99-7703-1_30.

CSIR – Indian Institute of Integrative Medicine. 2025. "The National Startup Festival 2025 organised by CSIR-IIIM, Jammu, empowered 45+ Startups, promoting Biotech, AgriTech & Health Innovation." CSIR-IMM. Accessed May 5, 2025. https://iiim.

res.in/the-national-startup-festival-2025-organised-by-csir-iiim-jammu-empowered-45-startups-promoting-biotech-agritech-health-innovation/

Customer Market Insight. 2025. India Herbal Products Market Size, Trends, Share 2033." Customer Market Insight. Accessed January 23, 2025. https://www.custommarketinsights.com/report/india-herbal-products-market/

Dubale, S., R. E. Usure, Y. T. Mekasha, G. Hasen, F. Hafiz, D. Kebebe, and S. Suleman. 2025. "Traditional Herbal Medicine Legislative and Regulatory Framework: A Cross-Sectional Quantitative Study and Archival Review Perspectives." *Frontiers in Pharmacology* 16. https://doi.org/10.3389/fphar.2025.1475297

Ekor, M. 2014. "The Growing Use of Herbal Medicines: Issues Relating to Adverse Reactions and Challenges in Monitoring Safety." *Frontiers in Pharmacology* 4. https://doi.org/10.3389/fphar.2013.00177

Fortune Business Insight. 2025. "Herbal Medicine Market Size, Share & Industry Analysis, By Form (Powder, Liquid & Gel, and Tablets & Capsules), By Application (Pharmaceutical & Nutraceutical, Food & Beverages, and Personal Care & Beauty Products), and Regional Forecast, 20252032." Fortune Business Insight. Accessed January 22, 2025. https://www.fortunebusinessinsights.com/herbal-medicine-market-106320

Hossain, C. M., M. Gera, and K. A. Ali. 2022. "Current Status and Challenges of Herbal Drug Development and Regulatory Aspect: A Global Perspective." *Asian Journal of Pharmaceutical and Clinical Research* 15 (2): 31–41.

International Trade Centre. 2024. "Trade Map - List of Exporters for the Selected Product (Plants and Parts of Plants, Incl. Seeds and Fruits, of a Kind Used Primarily in Perfumery)." International Trade Centre. Accessed May 30, 2025. https://www.trademap.org/Country_SelProduct_TS.aspx?nvpm=1%7c%7c%7c%7c%7c1211%7c%7c%7c4%7c1%7c1%7c2%7c2%7c1%7c2%7c1%7c1%7c1

Invest India. 2025. "AYUSH". Invest India. Accessed May 24, 2025. https://www.investindia.gov.in/sector/ayush

Jadhav, C. A., D. N. Vikhe, and R. S. Jadhav. 2020. "Global and Domestic Market of Herbal Medicines: A Review." *Research Journal of Science and Technology* 12 (4): 327–30. 10.5958/2349-2988.2020.00049.2

Jadhav, R., J. Das, S. Rajyaguru, S. Kyada, V. Kale, P. K. Teja, D. O. Bhutia, and S. K. Chauthe. 2025. "Regulations, Current Development, and Future Prospects of Phytopharmaceuticals, a New Class of Herbal Medicines in India." *Discover Pharmaceutical Sciences* 1 (1): 6. https://doi.org/10.1007/s44395-025-00006-4

Khan, M. S. A., and I. Ahmad. 2019. "Herbal Medicine: Current Trends and Future Prospects." In *New look to Phytomedicine*, 3–13. Elsevier.

Kumar, D., K. Keshav, N. Mehra, A. Ojha, and K. Kishan. 2024. "Herbal Medicine: Current Trends and Future Prospects." *International Journal of Pharmaceutical Sciences Review and Research* 84 (1). https://doi.org/10.47583/ijpsrr.2024.v84i01.019

Kumar, V. 2017. "Herbal Medicines: Overview on Regulations in India and South Africa." *World Journal of Pharmaceutical Research* 6: 690–8. https://doi.org/10.20959/wjpr20178-9091

Mazrouei, N. A., A. Z. A. Meslamani, R. Alajeel, G. Alghadban, N. Ansari, M. A. Kaabi, A. Sadeq, R. Ibrahim, and O. M. Ibrahim. 2022. "The Patterns of Herbal Medicine Use in the United Arab Emirates; A National Study." *Pharmacy Practice (Granada)* 20 (3): 2698. https://doi.org/10.18549/PharmPract.2022.3.2698

Metta, A.M., Kingumahanthi, N.L., Kalidindi, V.R., Juturi, R.K. and Boddu, V., 2021. "Scope for Harmonisation of Herbal Medicine Regulations." *International Journal of Pharmaceutical Sciences and Research* 12(4): 2012–2020. http://dx.doi.org/10.13040/IJPSR.0975-8232.12(4).2012-20

Ministry of AYUSH. n.d. "Ministry." Ministry of AYUSH, Government of India. Accessed January 22, 2025. https://ayush.gov.in/#!/aboutus

Mokhtarikhah, G., E. Mohammad-Taghi, and A. Mahdi. 2022. "Agro-Morphological and Phytochemical Studies of Spearmint Landraces (Mentha spicata L.) in Iran." *Industrial Crops and Products* 176: 114367. https://doi.org/10.1016/j.indcrop.2021.114367

Moloughney, S. 2022. "A Growing Herbal Products Market Still Faces Formidable Threats, Challenges." Nutraceuticals World, July 18, 2022. https://www.nutraceuticalsworld.com/a-growing-herbal-products-market-still-faces-formidable-threats-challenges/

Ng, J. Y., M. Kim, and A. Suri. 2022. "Exploration of Facilitators and Barriers to the Regulatory Frameworks of Dietary and Herbal Supplements: a Scoping Review." *Journal of Pharmaceutical Policy and Practice* 15 (1):55. https://doi.org/10.1186/s40545-022-00447-7

Oyedepo, T. A., and Santwana Palai. 2021. "Chapter 6 - Herbal Remedies, Toxicity, and Regulations." In *Preparation of Phytopharmaceuticals for the Management of Disorders*, edited by Chukwuebuka Egbuna, Abhay Prakash Mishra, and Megh R. Goyal, 89–127. Academic Press.

Pandey, M. M., Subha Rastogi, and A. K. S. Rawat. 2013. "Indian Traditional Ayurvedic System of Medicine and Nutritional Supplementation." *Evidence-Based Complementary and Alternative Medicine* 2013 (1): 376327. https://doi.org/10.1155/2013/376327

Patwardhan, B., D. Warude, P. Pushpangadan, and N. Bhatt. 2005. "Ayurveda and Traditional Chinese Medicine: A Comparative Overview." *Evidence-Based Complementary and Alternative Medicine* 2 (4): 629537. https://doi.org/10.1093/ecam/neh140

Press Information Bureau. 2019. "*Export of Herbal Medicines.*" Press Information Bureau, Ministry of Commerce & Industry, Government of India. Accessed May 24, 2025. https://www.pib.gov.in/Pressreleaseshare.aspx?PRID=1594929

Rojas, P., H. Jung-Cook, E. Ruiz-Sánchez, I. S. Rojas-Tomé, C. Rojas, A. M. López-Ramírez, and A. A. Reséndiz-Albor. 2022. "Historical Aspects of Herbal Use and Comparison of Current Regulations of Herbal Products Between Mexico, Canada and the United States of America." *International Journal of Environmental Research and Public Health* 19 (23): 15690. https://doi.org/10.3390/ijerph192315690

Saggar, S., P. A., Mir, N. Kumar, A. Chawla, J. Uppal, and A. Kaur. 2022. "Traditional and Herbal Medicines: Opportunities and Challenges." *Pharmacognosy Research* 14(2): 107–114. 10.5530/pres.14.2.15

Sahoo, N., and P. Manchikanti. 2013. "Herbal Drug Regulation and Commercialization: An Indian Industry Perspective." *The Journal of Alternative and Complementary Medicine* 19 (12): 957–63. https://doi.org/10.1089/acm.2012.0275

Sahoo, N., P. Manchikanti, and S. Dey. 2010. "Herbal Drugs: Standards and Regulation." *Fitoterapia* 81 (6): 462–71. https://doi.org/10.1016/j.fitote.2010.02.001

Sarmah, D. 2022. "Indian Herbal Drug Industry: Prospects and Current Scenario." *Current Trends in Pharmaceutical Research* 9 (1): 162–78.

Sen, S., and R. Chakraborty. 2015. "Toward the Integration and Advancement of Herbal Medicine: A Focus on Traditional Indian Medicine." *Botanics: Targets and Therapy* 5 (null): 33–44. https://doi.org/10.2147/BTAT.S66308

Sen, S., R. Chakraborty, and B. De. 2011. "Challenges and Opportunities in the Advancement of Herbal Medicine: India's Position and Role in a Global Context." *Journal of Herbal Medicine* 1 (3): 67–75. https://doi.org/10.1016/j.hermed.2011.11.001

Sharma, A., C. Shanker, L. K. Tyagi, M. Singh, and C. V. Rao. 2008. "Herbal Medicine for Market Potential in India: an Overview." *Academic Journal of Plant Sciences* 1 (2): 26–36.

Singh, A., V. A. Saharan, V. Kharb, and B. Anil. 2010. "Current Status of Regulations for Herbal Medicines in Europe, United States and India." *Journal of Natural Conscientia* 2 (3): 406–22.

Solidaridad. 2024. "Medicinal Plant Farmer Groups in India Receive Top Sustainability and Quality Certification." Solidaridad Network. Accessed May 24, 2025. https://www.solidaridadnetwork.org/news/medicinal-plant-farmer-groups-in-india-receive-top-sustainability-and-quality-certification/

Thillaivanan, S., and K. Samraj. 2014. "Challenges, Constraints and Opportunities in Herbal Medicines-A Review." *International Journal of Herbal Medicine* 2 (1): 21–4.

Vikaas, B., Y. Sunita, and C. Manjusha. 2017. "A Comprehension Study on Regulation of Herbal Drugs in USA, European Union and India." *International Journal of Drug Regulatory Affairs* 5 (4). https://doi.org/10.22270/ijdra.v5i4.205

Wanjari, A.S., S. Magar, S. Chapalgaokar, N.B. Chouragade, and D.S. Wanjari. 2016. "Evaluation and Standardization of Herbal Formulation." *Research Journal of Pharmacognosy and Phytochemistry* 8(3): 133–140.

World Customs Organization. "What is the Harmonized System (HS)?" World Customs Organization. Accessed May 30, 2025. https://www.wcoomd.org/en/topics/nomenclature/overview/what-is-the-harmonized-system.aspx

World Health Organization. 2019. *WHO Global Report on Traditional and Complementary Medicine 2019.* World Health Organization.

Xiong, R., H. Zhang, Y. Li, G. Zeng, Y. Liu, and Y. Xu. 2024. "Enlightenment of EU Herbal Medicinal Products Regulation on the Registration of Traditional Chinese Medicines in the Guangdong-Hong Kong-Macao Greater Bay Area." *Journal of Holistic Integrative Pharmacy* 5 (4): 344–53. https://doi.org/10.1016/j.jhip.2024.11.005

Zamani, S., M. Fathi, M.-T. Ebadi, and Á. Máthé. 2025. "Global Trade of Medicinal and Aromatic Plants. A Review." *Journal of Agriculture and Food Research* 21: 101910. https://doi.org/10.1016/j.jafr.2025.101910

9 Current Perspectives of Patenting in Herbal Drug Sector in India

Padmavati Manchikanti and Laxmi Ashok Kumar Dubey

9.1 INTRODUCTION

Herbal medicines are used by about 80% of the world's population, predominantly in developing countries for primary healthcare (Kamboj 2000). The global herbal medicine market is valued at \$30–60 billion, with India's market at about \$1 billion and exports of crude extracts at \$80 million. India's herbal drug sector is vast, complex and embedded in traditional knowledge. This encompasses several herbal industries, small units and individual healers. It is supported by a complex trade network that channels herbal raw material from diverse supply sources to end users. The marketing of medicinal plant produce is primarily through informal market places and wholesale markets. Traditional Knowledge (TK) is expressed through stories, rituals, various other forms and is central to Indian systems of medicine, including Ayurveda, Unani, Siddha, and Yoga. The associated knowledge with respect to use of bioresources is distinct as it is practised at the household level. India's biodiversity is huge with 7–8% of recorded species and 16 of the world's 26 agroclimatic zones. Despite this, India's exports and standardization efforts need to be enhanced with greater publication of monographs and larger exports.

'AYUSH' encompasses traditional systems of medicine such as Ayurveda, Yoga, Unani, Siddha, and Homeopathy. There is increased R&D and capacity building in these areas. The AYUSH sector in India is emerging as a prominent domain in terms of revenue and employment. It has attracted interest from both policymakers and patent practitioners. There is an increasing awareness, demand for therapies and government initiatives to promote traditional medicine systems. The Ministry of AYUSH has been promoting international cooperation in international exchange of experts and officers, incentives to AYUSH entrepreneurs and institutions, support for international market development, translation and publication of AYUSH literature, establishment of AYUSH information cells and health centers and support to foreign nationals to undertake related courses in India (Ministry of AYUSH 2025).

The AYUSH healthcare delivery system in India is bifurcated into public and private sectors. India's AYUSH sector benefits from a rich heritage of ancient healing practices and a vast pool of skilled practitioners. The country's cost-effective approach to AYUSH treatments makes it a preferred destination for individuals seeking alternative and holistic healthcare solutions. As part of the MoU signed

DOI: 10.1201/9781003623748-9

between the Ministry of AYUSH and the Indian Council of Medical Research (ICMR) in 2023, the activities initiated include the AYUSH-ICMR advanced center for Integrated Health Research, initiation of the multi-Centre randomised controlled trials for iron deficiency anemia and research capacity building. Several efforts are being undertaken to identify collaboration opportunities to explore the potential of AYUSH with countries such as Vietnam, Malaysia and other countries. In the year 2024, an MOU was signed between the Pharmacopoeia Commission for Indian medicine and homeopathy and the Indian Pharmacopoeia Commission to promote and facilitate the 'One Herb, One Standard' initiative.

9.2 GROWTH OF INDIAN HERBAL DRUG SECTOR AND INTERNATIONAL RECOGNITION

The India-WHO AYUSH (Ayurveda, Yoga & Naturopathy, Unani, Siddha, and Homeopathy) cluster refers to a formal partnership and collaborative platform established between the Government of India (through the Ministry of AYUSH) and the World Health Organization (WHO). It aims to advance, standardize, and promote traditional medicine systems globally. The establishment of the WHO Global Centre for Traditional Medicine, in 2022 and the inclusion of traditional medicine chapter under the ICD-11 (ICD refers to International Classification of Diseases) is a significant development (World Health Organization 2022). Another milestone is the recognition of Indian Ayurvedic manufacturing units with Certificates of Pharmaceutical Product (COPP) from the Central Drugs Standard Control Organisation (CDSCO). A COPP is a WHO-recommended certificate that proves a medicine is manufactured under proper quality and regulatory standards in the exporting country. Ayurvedic units in India have been obtaining COPPs since 2014, after joint inspections by CDSCO and AYUSH authorities (Central Drugs Standard Control Organization n.d.). This certificate is very vital for export because many foreign regulators require it before allowing sales in their markets. The WHO recognition and COPP system enable India's traditional medicines greater trust worldwide, improve export opportunities and support cooperation with other countries in healthcare. COPPs provide product-level evidence of regulatory compliance. The practical use of COPPs enhances confidence among foreign regulators, buyers and research partners.

The National AYUSH Mission's objective is to promote AYUSH medical systems through cost effective services and foster the enforcement of quality control of AYUSH drugs (National AYUSH Mission n.d.). Regulatory frameworks for herbal products in India have evolved in response to both domestic demand and international expectations. The Drugs and Cosmetics Act 1940, and its subsequent amendments, provide the backbone for herbal drug regulation, while Schedule T specifically addresses Good Manufacturing Practices (GMP) for Ayurveda, Siddha, and Unani (ASU) pharmaceuticals (Schedule 1940, Schedule T). These guidelines have been updated to align more closely with WHO recommendations, aiming to ensure that herbal medicines produced in India meet global standards for safety, quality, and efficacy.

Despite these advancements, quality control and standardization are recurring concerns, as the efficacy of herbal medicines can be compromised by variations in plant sourcing, cultivation, and processing. Clinical validation is another area where India lags; most herbal formulations have not undergone rigorous, large-scale clinical trials, making it difficult to substantiate efficacy claims to the satisfaction of international regulatory bodies. This, in turn, hampers the export potential of Indian herbal products, even as domestic demand continues to grow (Indian Brand Equity Foundation 2025).

9.3 GLOBAL ADOPTION OF TKDL: HOW PATENT OFFICES WORLDWIDE ARE LEVERAGING INDIA'S TRADITIONAL KNOWLEDGE DATABASE

India's Traditional Knowledge Digital Library (TKDL) stands as a pioneering global initiative to protect traditional medicinal knowledge from misappropriation and bio-piracy (Traditional Knowledge Digital Library n.d.). The Indian government has implemented several measures by way of TKDL access agreements with several patent offices. This serves as a safeguard against biopiracy by providing patent examiners worldwide with recorded traditional medicine information. Enhancing patent legislation, optimizing regulatory frameworks and promoting research-driven innovations in the herbal sector are important considerations for India to progress towards its aspiration of becoming a global leader in traditional and herbal medicine exports (Sen, Chakraborty, and De 2011). In 2001, the Council of Scientific and Industrial Research (CSIR) in collaboration with the Ministry of Health and Family Welfare initiated and developed the TKDL for the systematic documentation and digitization of the traditional medicinal knowledge of Ayurveda, Unani, Siddha, Sowa Rigpa, and Yoga. Using classification codes the information is catalogued into a format accessible and comprehensible to international patent examiners. This was developed in response to erroneous patent grants on Indian knowledge, such as the turmeric and neem patents. The TKDL overcomes significant language and format barriers by translating and structuring ancient texts, originally written in Sanskrit, Hindi, Arabic, Persian, Urdu, and Tamil, into five major international languages: English, Japanese, French, German, and Spanish. As of 2025, TKDL contains over 454,000 formulations and practices, making it the world's largest and most sophisticated prior-art database on traditional medicine. The database uses an innovative Traditional Knowledge Resource Classification (TKRC) system which is compatible with the International Patent Classification (IPC). This allows patent examiners to efficiently search for relevant prior art.

Globally, the TKDL is recognized as a highly effective defensive tool against the misappropriation of Indian traditional knowledge. Patent offices in sixteen countries including the European Patent Office, United States Patent & Trademark Office, Japan Patent Office, UK Patent Office, Canadian Intellectual Property Office, German Patent Office, IP Australia, Chile Patent Office, Russian Patent Office, Malaysian Patent Office, and others have signed non-disclosure access agreements with India, permitting their examiners to use the TKDL for search and examination purposes (TKDL Access Agreement n.d.). These agreements include several safeguards to ensure that the proprietary information in the TKDL is not disclosed to third parties except as necessary for citation in patent proceedings.

The TKDL has enabled patent examiners worldwide to reject or revoke hundreds of wrongful patent applications that attempted to claim inventions already documented in Indian traditional knowledge (Chaudhary and Singh 2012). As of 2023, more than 324 patent applications have been set aside, withdrawn, or amended based on TKDL evidence. The TKDL team also actively submits third-party observations and pre-grant oppositions at various international patent offices in an effort to protect India's traditional knowledge.

The success of the TKDL has led several countries and organizations, including the World Intellectual Property Organization (WIPO), expressing interest in replicating the TKDL model for their own traditional knowledge systems. The database is now considered a global benchmark for the digital protection of traditional knowledge and has been cited in international fora as a best practice for preventing the erroneous grant of patents on indigenous resources (World Intellectual Property Organization 2011).

In August 2022, the Indian government approved the widening of access to the TKDL database, allowing not just patent offices but also researchers, academic institutions, industry, and other stakeholders to use the database through a phased, paid subscription model (Press Information Bureau 2022). This initiative is expected to further stimulate research and innovation based on India's traditional knowledge, while maintaining the primary mandate of defensive protection against bio-piracy. The relevance of the AYUSH was highlighted during the COVID-19 pandemic, which saw widespread use of Indian traditional medicines for immune support, symptom relief, and antiviral activity.

While its primary mandate is to prevent of wrongful appropriation of Indian traditional knowledge, the TKDL is also poised to inspire innovation and the development of safer, more effective solutions for a healthier society. The integration of scientific validation, modern processing techniques, and international collaboration can enhance market credibility and create additional economic opportunities for the Indian herbal industry.

9.4 HERBAL DRUG PATENTS - PATENT LEGISLATION AND GUIDELINES

The development, manufacturing and marketing of herbal drugs in India is governed by a unique intersection of intellectual property laws, traditional knowledge protection, and regulatory oversight. At the heart of this framework lies the Patents Act, 1970, which establishes the criteria for what constitutes a patentable invention in India. According to Section 2(1)(j) of the Act, an invention must be a new product or process involving an inventive step and capable of industrial application to qualify for patent protection. This means that, for herbal drugs, only those formulations, processes, or uses that are truly novel, have an inventive step and industrially applicable are eligible for patenting.

The Section 3(p) of the Act is particularly significant, as it excludes from patentability any invention that is essentially traditional knowledge or merely an aggregation of known properties. This provision is designed to prevent the privatization of community-held wisdom and to ensure that only derived inventions are eligible for patent protection (Indian Patents Act 1970, sec. 3(p)). The Indian Patent Office has implemented guidelines that require applicants to demonstrate novelty, inventive step, and industrial applicability, especially for AYUSH-related inventions. Landmark legal cases, such as the revocation of the turmeric and neem patents, have underscored the importance of robust prior art searches and the role of the TKDL in safeguarding Indian interests (Sheridan 2005; World Intellectual Property Organization n.d.). The courts have made it clear that simply combining traditional components does not make an invention patentable; there must be a significant transformation or inventive step involved.

Government initiatives, such as the National Mission on Medicinal Plants and the promotion of organic farming, are steps in the right direction, but sustained effort is needed to bridge the gap between traditional knowledge and commercialization. The high demand and lack of market transparency have raised concerns about the quality and authenticity of raw drugs (Pujari, Shrikanth, and Naik 2021).

9.5 GUIDELINES FOR EXAMINATION OF PATENT APPLICATIONS RELATED TO TK AT INDIAN PATENT OFFICE

India has developed a robust legal framework in recognition of the value and vulnerability of TK, particularly in the context of herbal drugs. In order to enable IPR protection to traditional systems of Indian medicines, a Committee was constituted by the Department of Promotion of Industries and Internal Trade (DPIIT). Under this a sub-committee has been constituted to suggest amendments to the existing TK guidelines for processing patent applications. The Indian Patent Office had issued guidelines for the examination of AYUSH-related inventions, requiring that patent applications demonstrate novelty, inventive step, and industrial applicability (Guidelines for processing of patent applications relating to traditional knowledge and biological material n.d.). Combinations of known ingredients must show evidence of synergistic effects to be considered inventive, and all applications are screened to determine if they fall under the category of traditional knowledge before being examined. In recognition of increasing patent activity in this domain, the Indian Patent Office issued on January 31, 2025 the Draft Guidelines for Processing Patent Applications of AYUSH Systems and Related Inventions. These guidelines complement existing guidelines and focus on the patentability of AYUSH products, devices, and nutraceuticals, and underscore that such inventions must still comply with Sections 2(1)(j) (invention definition), 3(p) (traditional knowledge exclusion), and 3(e) (mere admixture exclusion) of the Patents Act, 1970.

The 2025 draft guidelines identify several categories of potentially patentable AYUSH inventions:

a. *Products and devices*: These include new formulations or delivery systems (e.g., patches, inhalers), as well as devices used uniquely in AYUSH practices. Devices unlike traditional herbal mixture that are patentable if they show novelty and inventive step.
b. *Food recipes and nutraceuticals*: Although nutritional supplements are derived from herbs, they may be patentable if they demonstrate a synergistic effect or novel formulation beyond traditional knowledge

c. *Procedures and processes*: Novel extraction, purification, or processing methods of herbal components if distinguished from traditional processes can qualify as patentable inventions

The guidelines articulate six guiding principles to assess novelty and inventiveness (Table 9.1).

Extracts and isolates that are inherently present in known AYUSH formulations are not novel. However, novel extraction processes can be patentable if they meet inventive step criteria. Known combinations are presumed obvious unless they deliver unexpected technical benefits due to specific ratios or synergistic effects. Optimization

TABLE 9.1
Guiding Principles for TK Inventions with Respect to Patent Applications at Indian Patent Office

Guiding Principle	Aspect Covered	Not Patentable	Patentable
1.	Extracts/alkaloids and/or isolation of active ingredients of plants	Claims that relate to extracts/alkaloids and/or isolation of active ingredients of plants, which are naturally/inherently present in plants, such claims cannot be considered as novel and/or inventive when use of such plants is pre-known in AYUSH systems.	Processes for obtaining above-mentioned extracts/isolates may be considered patentable subject to the requirements of novelty and inventive step
2.	Combination of ingredients from plants/minerals/animal origin/existing formulations already known for the treatment of a disease as a part of Traditional Knowledge	In case combination of ingredients from plants/minerals/animal origin/existing formulations already known for the treatment of a disease as a part of Traditional Knowledge, then it is obvious that a combination product comprising these known ingredients with further ingredients from plants/minerals/animal origin/existing formulations with the same known therapeutic effect would be more effective than each of the ingredient when applied separately (additive effect).	Specific ratios leading to unexpected technical effects of such combinations may be considered to establish non-obviousness.
3.	Ingredients are already known for the treatment of a disease (in a combination product)	In case an ingredient is already known for the treatment of a disease, then it creates a presumption of obviousness that a combination product comprising this known active ingredient would be effective for the treatment of same disease.	Unexpected technical effects of such combinations may be considered to establish non-obviousness

(Continued)

TABLE 9.1 *(Continued)*
Guiding Principles for TK Inventions with Respect to Patent Applications at Indian Patent Office

Guiding Principle	Aspect Covered	Not Patentable	Patentable
4.	Optimum or workable range of ingredients	Discovering the optimum or workable ranges of traditionally known ingredients by routine experimentation is not inventive.	–
5.	Multiple ingredients with same therapeutic activity	In case multiple ingredients are known to have the same therapeutic activity as per traditional knowledge, taking one component out of them cannot be considered as inventive.	–
6.	Equipment or devices used in AYUSH systems	–	If the subject matter of the claims relates to inventions regarding equipment/device used in AYUSH systems, then such inventions may be patentable if novel and inventive over the prior art

Source: Draft Guidelines for processing patent applications of AYUSH systems and related inventions, Indian Patent Office.

of known ingredients through routine experimentation is not considered inventive. Isolating a single component from a known mixture is non-inventive unless the isolate shows unanticipated therapeutic effect. Devices and equipment specifically designed for AYUSH applications are eligible for patenting provided they are novel and non-obvious. These principles offer substantive guidance to patent examiners.

The Draft Guidelines also reaffirm that any invention involving biological resources derived from India must comply with the Biological Diversity Act, 2002, as amended in April 2024. Under Section 6(1) of this Act, prior approval from the National Biodiversity Authority (NBA) is mandatory if the invention is based on Indian biological material. Failure to disclose this properly may result in patent refusal or opposition.

Among the five research councils under the Ministry of AYUSH, some have successfully pursued patents. There are several patents filed in the case of two research councils, namely, the Central Council of Research in Ayurvedic Sciences (CCRAS) and the Central Council for Research in Unani Medicine (CCRUM). Assignee analysis of the patent applications filed reveals that some inventions are a result of the collaborative effort (Table 9.2). A 2020 patent landscaping study identified 1,560 AYUSH patents filed worldwide between 2015–2020, with most filings in Yoga, Homeopathy, and Ayurveda (Singh and Purushotham 2020).

TABLE 9.2
Patent Applications Filed by Research Councils Under the Ministry of AYUSH

S.No	Organisation under AYUSH	Joint Applicant(s)	Application Number and Title of the Invention
1.	Central Council for Research in Ayurvedic Sciences (CCRAS)	Maharashtra Arogya Mandal	2589/MUM/2013-A composition for hemoglobinopathy and method of preparation thereof
		Lovely Professional University	202511049597-Development and standardisation of alternative preparation techniques of netra putapaka formulation for opthalmic use
		Mangalayatan University, Aligarh, PGI Dental Sciences and Hospital, Lucknow, Baba Mastnath University, Haryana, IIMT University, Meerut ITM College of Pharmacy and Research, Gorakhpur, Moradabad Institute of Technology, Moradabad, Madhyanchal Professional University, Bhopal	202311064840-Formulation and characterisation of PLGA nanoparticles of Crataegus oxycantha extract for the treatment of obstructive sleep apnea syndrome
		Maharana Pratap University of Agriculture and Technology, Udaipur, VBS Purvanchal Unviersity, Jaunpur, Manipal College of Pharmaceutical Sciences, Karnataka, Rungta College of Pharmaceutical Sciences and Research, Bhilai, Madhyanchal Professional University, Bhopal, Oriental University, Indore, GVM College of Pharmacy, Sonipat, Babasaheb Bhimrao Ambedkar University, Lucknow, Mahavir Institute of Pharmacy, Nashik	202311061132-Formulation and characterisation of dendrimer nanogel containing essential oil of *Eugenia uniflora* to evaluate their antifungal activity
		Anurag University, Medchal-Malkajgiri, Gujrat Technical University, Surat, Dr Chunibhai Vallabhbhai Patel College of Pharmacy, Surat, Future Institute of Pharmacy, Faridpur, Swami Vivekanand Sansthas Institute of Pharmacy, Nashik	202341051786- A Formulation of paclitaxel-loaded, micro-sponge-based hydrogel for the treatment of Kaposi's sarcoma and method thereof

(Continued)

TABLE 9.2 *(Continued)*
Patent Applications Filed by Research Councils Under the Ministry of AYUSH

S.No	Organisation under AYUSH	Joint Applicant(s)	Application Number and Title of the Invention
		Rakshpal Bahadur College of Pharmacy, Bareilly, JMCT Institute of Pharmacy, Nashik, IIMT University, Meerut, MIT Campus, Moradabad, Krishna Pharmacy College, Bijnor, J.S. Singh Institute of Pharmacy, Sitapur, Siddhartha Institute of Pharmacy, Dehradun,	202311049599- Development and characterisation of *Phyllanthus niruri* extract based polymeric nanoparticles for the management of hyperglycemia
		Institute of Pharmaceutical Research, GLA University, NH-19, Mathura, Modern International College of Pharmacy, Jasana, KIET Group of Institutions, Ghaziabad, NGSM Institute of Pharmaceutical Sciences (NGSMIPS), Department of Pharmacology, Mangaluru, India, Department of Fashion Technology, Jamal Mohamed College, Tiruchirappalli, Periyar College of Pharmaceutical Sciences, Tiruchirappalli,	202311048128- Innovative active bioactive dressing for treatment of diabetic wounds
2.	Central Council for Research in Unani Medicine (CCRUM)	Jiwaji University, Gwalior	1827/DEL/2009- Propolis extract a potent antioxidant and hepatoprotective agent and the process for preparation thereof
		Regional Research Center, Panalwan Chuahara, Allahabad	201811011609-A novel herbal composition and a process for preparation thereof effective against bronchial asthma (Zeequn Nafas)

Source: Author's own compilation

The regulation of herbal drugs in India falls under the Drugs and Cosmetics Act, 1940, with the Department of AYUSH responsible for enforcing Good Manufacturing Practices (GMP), quality control, and standardization. The National Medicinal Plants Board (NMPB) plays a vital role in promoting the cultivation, conservation, and certification of medicinal plants, further supporting the integrity and sustainability of India's herbal sector (Katiyar et al. 2023). Despite this the lack of harmonized global regulation means that herbal medicines are often marketed and sold internationally with varying degrees of oversight. The World Health Organization (WHO) has called for countries to adopt national quality specifications, GMP standards, and licensing schemes for the manufacturing, import, and marketing of herbal medicines to ensure safety and efficacy (WHO guidelines on good manufacturing practices (GMP) for herbal medicines). In Europe, for example, the European Directive on Traditional Herbal Medicinal Products (Directive 2004/24/EC) requires all herbal medicinal products to obtain marketing authorization and comply with GMP, serving as a model for quality assurance and regulatory harmonization (European Medicines Agency n.d.).

International discussions have highlighted the need for both "defensive" and "positive" protection of TK. Defensive protection aims to prevent the unauthorized patenting or misappropriation of TK (for example, through disclosure of origin requirements in patent law), while positive protection enables TK holders to acquire and enforce their own IP rights over their knowledge (Gopalakrishnan 2002). Despite progress in international negotiations, the modalities for protecting TK are still evolving, and there is not yet a universally agreed international regime.

9.6 APPLICATION OF THE BIOLOGICAL DIVERSITY ACT 2002

The Biological Diversity Act, 2002 (BDA), enacted to fulfil India's obligations under the 1992 Convention on Biological Diversity (CBD), aims to conserve the country's rich biodiversity, ensure the sustainable utilisation of biological resources, and guarantee fair and equitable benefit sharing arising from their use. It becomes especially relevant in the context of herbal medicines, traditional knowledge, and patenting of biological inventions.

The BDA establishes a three-tier regulatory structure; the National Biodiversity Authority (NBA), the apex body at the national level. It is responsible for granting approvals for access to biological resources and associated traditional knowledge, especially for foreign individuals, companies, or for any transfer of results of research outside India. At the next level are State Biodiversity Boards (SBBs). These boards operate at the state level, regulating access to biological resources by Indian citizens and ensuring benefit sharing within the state. At the local level are the Biodiversity Management Committees (BMCs). These are local bodies that document biological resources and traditional knowledge, and maintain People's Biodiversity Registers (PBRs)

An important provision of the Act is Section 6, which mandates that no person or organization, whether Indian or foreign, can apply for any intellectual property right including patents based on research or information derived from Indian biological resources without prior approval from the NBA. A no objection must be obtained NBA for an invention derived from bioresources of India. Even if the patent application has been examined it will not be officially sealed for grant until the NBA approval is available. This mechanism is designed to balance the encouragement of innovation with the need to ensure that the benefits of India's biological resources are shared equitably. Moreover, the Patents Act, 1970 complements the BDA by requiring all patent applicants to disclose the source and geographical origin of any biological material used in their invention. This dual requirement of disclosure and prior approval is intended to prevent the misuse or misappropriation of India's biological wealth.

For the herbal drug and traditional medicine sector, the BDA has significant implications. Any company or researcher seeking to use Indian medicinal plants or associated traditional knowledge for commercial purposes must secure prior approval from the NBA and agree to benefit-sharing arrangements. The Act thus serves as both a protective and enabling instrument, supporting conservation, sustainable use, and equitable development (Lakshmanan 2020).

Section 18(4) of the Biological Diversity Act, 2002, grants the National Biodiversity Authority (NBA) a unique and powerful mandate; the authority to take *suo-moto* action to oppose the grant of intellectual property rights in any country outside India on any biological resource obtained from India or knowledge associated with such biological resource derived from India. This provision is a critical safeguard against biopiracy and the unauthorized patenting of India's biological resources and traditional knowledge by foreign inventors or companies.

The NBA actively monitors both Indian and international patent publications to detect applications that may involve Indian biological resources or associated knowledge. This is accomplished through systematic keyword searches and manual verification of patent documents. The NBA tracks patent filings in major foreign patent offices, including the European Patent Office (EPO), United States Patent and Trademark Office (USPTO), Canada, China, Japan, WIPO, Australia, and Taiwan. The NBA has taken action in over 74 cases globally under Section 18(4), including filing oppositions at the EPO (23 cases), USPTO (32 cases), Canada (10 cases), Australia (5 cases), and other jurisdictions (4 cases) (National Biodiversity Authority 2020). The NBA's role in monitoring, detection and opposition of patents has set an important precedent for other biodiversity-rich countries.

9.7 PATENTING ACTIVITY IN INDIAN HERBAL DRUG SECTOR-EMERGING TRENDS

During the past decade, there was a consistent increase in filings of herbal drug patents by Indian organizations and inventors. Globally, Indian companies have been exceptionally active, particularly in the US and EU economies, demonstrating a clear strategic shift towards global protection of India's herbal innovations. While India has taken defensive actions such as the development of the TKDL to counter wrong patent awards overseas, the efficacy of benefit-sharing regimes and prior informed consent requirements is yet to be completely realised. There is a trend towards giving more emphasis to global commercialization and strategic portfolio management due to competitive global markets.

Analysis of patenting activity for the duration of 2001 to 2010, indicates that Indian inventors and organizations steadily increased patent filings for herbal drugs with a peak in 2007. Of the 159 granted patents, 49% are of individual inventors (mainly herbal practitioners), 29% to research institutes (led by CSIR) and 22% to companies. Most patents addressed diabetes, skin diseases, cancer, liver conditions, and immune disorders like HIV/AIDS. Many covered new herbal compositions, extraction methods, and improved dosage forms such as 'Asavas' and 'Aristhas'. More individual inventors filed in the EU than in the US. Most international patents covered new compositions and extraction methods for diabetes and liver disorders (Sahoo, Padmavati, and Dey 2011).

9.7.1 Data Collection and Screening

In this study the analysis was performed on the prosecution trends of herbal drug patent applications filed by Indian manufacturers certified under the COPP. The primary aim was to identify companies that are both active in patenting as well compliant with recognised manufacturing standards. The first stage of the search strategy involved identifying the target companies for analysis. Two primary criteria were applied: the company must hold a WHO Certificate of Pharmaceutical Product (CoPP) Good Manufacturing Practices (GMP) certification, and it must have an active patent portfolio in the herbal drug domain. First, a general search was conducted through publicly available online resources and industry directories to compile a list of herbal pharmaceutical manufacturers holding WHO-CoPP or GMP certification. Once identified, each company name was used as an assignee search term in Orbit 3.0 to retrieve its associated patent filings. Patent data collection was carried out using a combination of online searches, database queries, and official records from the Indian Patent Office. The Orbit 3.0 database (www.orbit.com) was chosen for its advanced search capabilities and wide coverage of patent records from multiple jurisdictions, including access to both application and prosecution histories. The platform provides detailed metadata such as the earliest filing year, application number, publication number, assignee name, inventor details, International Patent Classification (IPC) codes, and territorial jurisdictions. These features make Orbit 3.0 well-suited for compiling a robust dataset of patents and for conducting prosecution history analysis, which forms the core of this study.

In addition to keyword restrictions, International Patent Classification (IPC) codes relevant to herbal drug formulations and manufacturing processes were reviewed and applied as filters. This step reduced irrelevant entries and improved the precision of the dataset. The classification filter ensured that the selected patents specifically addressed herbal or botanical-based pharmaceutical compositions, excluding unrelated chemical or synthetic drug patents.

This dataset was then restricted to patents with Indian assignees with 18 companies identified, reflecting the study's focus on domestic herbal drug manufacturers. For each selected patent, prosecution details were obtained from the Indian Patent Office's online database (www.ipindia.gov.in). This included reviewing the complete prosecution history from the First Examination Report (FER) to the applicant's responses and any subsequent examination steps.

The prosecution history analysis aimed to identify common grounds for objections and rejections, amendments made during prosecution, and the role of regulatory compliance (such as NBA approvals) in the grant process. This dual-layered approach combining bibliographic patent data from Orbit with official Indian Patent Office records allowed for a comprehensive analysis of both the legal and procedural challenges faced by herbal drug patent applicants.

9.7.2 Legal Status of the Patents and Extent of Claim Amendments

In the current study the focus was on identifying companies engaged in herbal drug manufacturing that also hold CoPP GMP certification. Despite holding GMP certification and having a strong market presence with unique products, many companies do not pursue patent protection. Several factors could explain this trend, such as restrictions to inventions particularly Section 3 of the Patents Act, which poses significant barriers to the patenting of herbal drugs. This legal framework has made it especially challenging for herbal drug manufacturers to secure patent rights.

A total of 278 patent applications was identified. The analysis reveals a concerning trend. About 54% of these patents are no longer active. Additionally, 42% of the applications were voluntarily withdrawn by the applicants, often after publication or examination. Many of these withdrawals seem to be strategic decisions to discontinue the prosecution process. Only 18% of the patents have been granted so far. Meanwhile, 13% of the applications are still pending due to various reasons, including ongoing examination or delays in obtaining approvals from the National Biodiversity Authority (NBA) (Figure 9.1).

A detailed prosecution analysis of the granted patents was conducted with respect to only those companies who have patents in force (which includes only nine companies). During the prosecution phase many patents have undergone claim deletion and only after deletion this patent grant has been permitted. Out of 18% granted patents, 87.5% patent applications had claim deletions or claims merged with the previous claim and remaining 12.5% had amended claims in view of the non-compliance under Section 10 of the Patents Act 1970 (with respect to disclosure norms) (Figure 9.2.). A detailed analysis of the claims with respect to each company before and after grant is depicted in Figure 9.3.

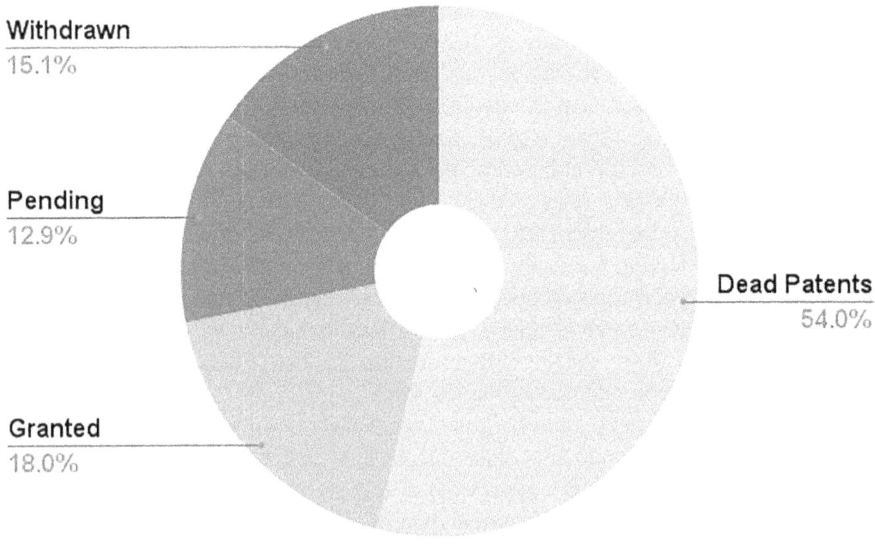

FIGURE 9.1 Analysis of the legal status of the granted patents.

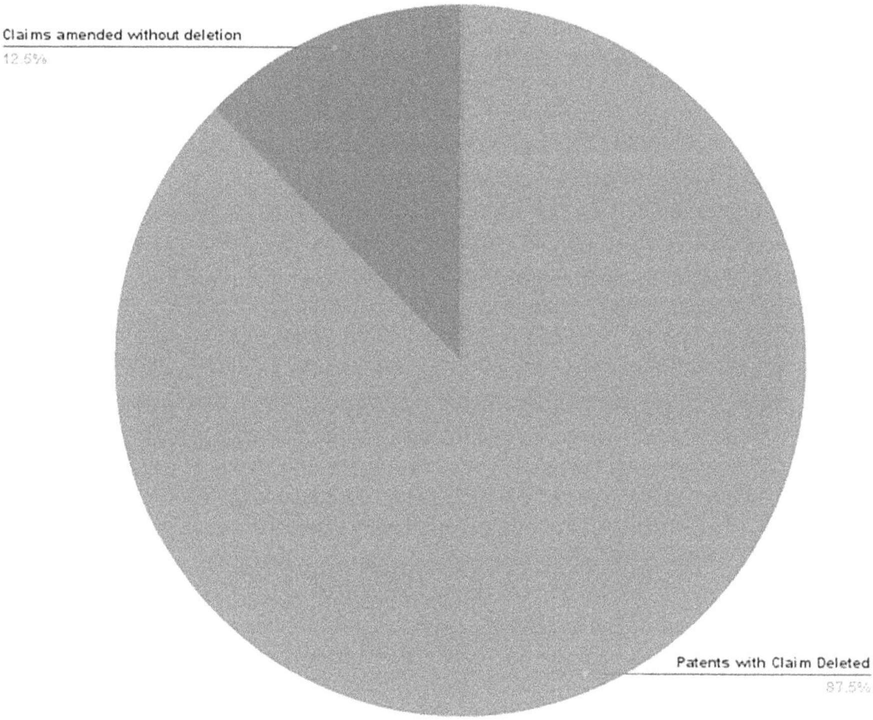

FIGURE 9.2 Extent of claim amendments.

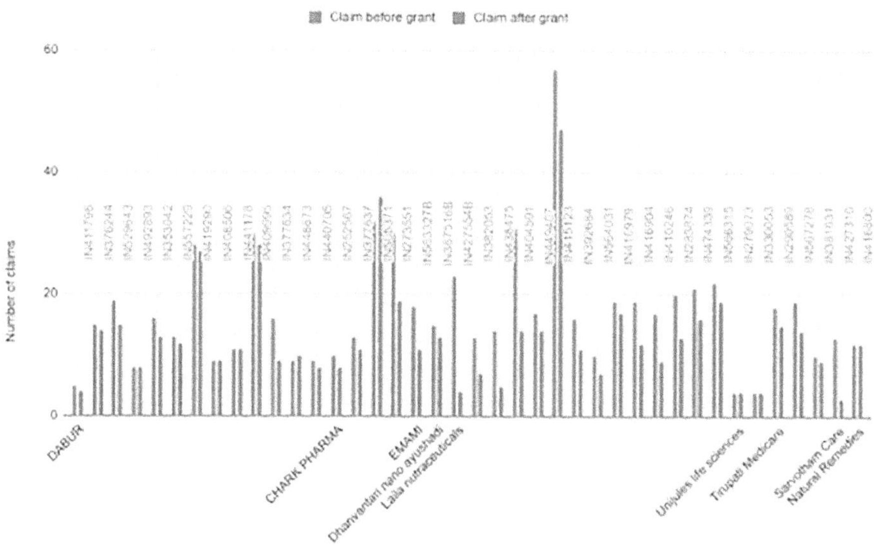

FIGURE 9.3 Company wise claim analysis before and after grant of patent.

9.7.3 Analysis of Restrictions under Section 3 of the Patents Act 1970

An analysis of the patent prosecution reveals that Section 3 objections, under the Indian Patent Acts 1970 which pertain to subject matter eligibility, are raised in approximately 87.5% of granted patents (Figure 1.4). This indicates that Section 3 remains a significant barrier, particularly in the domain of herbal and composition-based inventions. Interestingly, the remaining patents that did not attract Section 3 objections primarily relate to non-herbal formulations or involve claims where traditional knowledge or natural origin was not central to the invention. In contrast, every patent involving herbal compositions encountered some form of restriction under Section 3. This highlights a consistent trend wherein the Indian Patent Office (IPO) is particularly stringent when assessing herbal or plant-based inventions, likely due to concerns over traditional knowledge misuse and lack of inventive step.

To further analyse the nature and scope of restriction under Section 3, categorization of specific sub-clauses of Section 3 that were invoked in patent applications provides clarity on the types of subject matter exclusions that are most relevant in herbal patent filings. This analysis reveals a significant pattern in the application of various sub-sections under Section 3 of the Indian Patent Act. These sub-sections act as major hurdles for patent applicants, especially in domains like herbal and pharmaceutical inventions where the line between traditional knowledge, discovery, and innovation is often blurred.

One of the least-invoked subsections is Section 3(a), cited in about 1.3% of the cases. It deals with inventions that are considered frivolous or contrary to well-established natural laws. While rare, this ground becomes relevant when the claimed invention appears to defy scientific logic-something particularly scrutinized when applicants propose miraculous or unproven effects of natural products. A more frequently cited objection comes under Section 3(c), used in around 7.5% of cases. This section excludes the mere discovery of a scientific principle or the formulation of an abstract theory. In the herbal context, this often arises when applicants simply isolate a compound from a plant and claim novelty, without demonstrating a concrete technical application beyond the known traditional use.

The next most common objection was under Section 3(d), which accounted for around 20% of the cases. This provision bars patents for the mere discovery of a new form or new use of a known substance unless it results in enhanced efficacy. While this clause is often discussed in the context of pharmaceutical patents, it also creates serious challenges for herbal patents. For example, if an applicant claims a new therapeutic use for a known herb without robust scientific backing demonstrating increased efficacy, the application is likely to face rejection under this clause. Sub-section 3(e) has been the cited objection, in approximately 25% of the cases. This clause deals with claims where the invention is merely an aggregation of known substances without any synergistic effect. In the context of herbal product patenting, this is a particularly relevant barrier. Many applicants tend to combine multiple herbs in a composition, assuming that the mixture itself will qualify as an invention. However, without concrete evidence of a synergistic interaction between the components, such claims are routinely rejected. This makes 3(e) a critical point of concern for drafters working on herbal compositions, as failure to demonstrate synergy can significantly narrow the scope of protection or even lead to claim deletions.

Other significant subsection in the context of herbal medicine is Section 3(p), which accounted for 22.5% of objections. This clause prohibits the patenting of inventions that are traditional knowledge or are an aggregation or duplication of known properties of traditionally known components. In India, where Ayurvedic and other traditional medicine systems are well-documented, patent examiners closely scrutinize herbal patent applications for overlaps with such knowledge. Even if the formulation is not directly copied from a traditional text, the examiner may still reject it if the therapeutic use is widely known in folklore or common practice.

Other sub-sections frequently cited by examiners are Sections 3(i) and 3(j). Section 3(i), referenced in about 16.3% of the cases, prohibits patents on methods of treatment for humans or animals, whether medical, surgical, curative, or preventive. This creates challenges in herbal medicine, where the healing method is often central. Hence, claims must emphasize the composition, not its therapeutic use. Section 3(j), cited in 7.5% of cases, excludes patents on plants and their parts (like seeds, roots, or bark), unless transformed into a novel formulation. These objections underline the need to focus on product innovation rather than traditional use or raw plant materials.

This data shows that Sections 3(e), 3(d), and 3(p) are the major hurdles in patenting, especially in the herbal and pharmaceutical fields (Figure 9.4). Applicants are often required to prove therapeutic efficacy under Section 3(d) and synergistic effect under Section 3(e). Since herbal patents frequently involve combining different herbs, the 25% objection rate under Section 3(e) is expected. However, this highlights how crucial these sections are in prosecution. Therefore, it's important for drafters to structure the application carefully to avoid claim deletions, as that would limit the scope of protection.

In recent years, a notable shift has been observed in the way Indian patent examiners assess herbal patent applications. A new and increasingly challenging trend is the simultaneous invocation of multiple restrictive provisions under Section 3 of the Indian Patents Act. While individual sub-sections such as Section 3(d), 3(e), or 3(p) have historically posed hurdles, their combined application is now emerging as a formidable barrier to patent grant, particularly in the domain of traditional or herbal-based inventions (Figure 9.5).

Each of these sub-sections carries a distinct rationale for refusal. Section 3(d) targets incremental innovations lacking enhanced efficacy; Section 3(e) focuses on mere admixtures without synergy; Section 3(p) addresses traditional knowledge and publicly known substances. When cited individually, these provisions already require the applicant to submit robust technical and scientific justifications. However, when cited together in combinations, they raise the bar significantly.

From the analysis of granted patents and prosecution history, it is evident that the most commonly encountered combination is Section 3(d) read with Section 3(e).

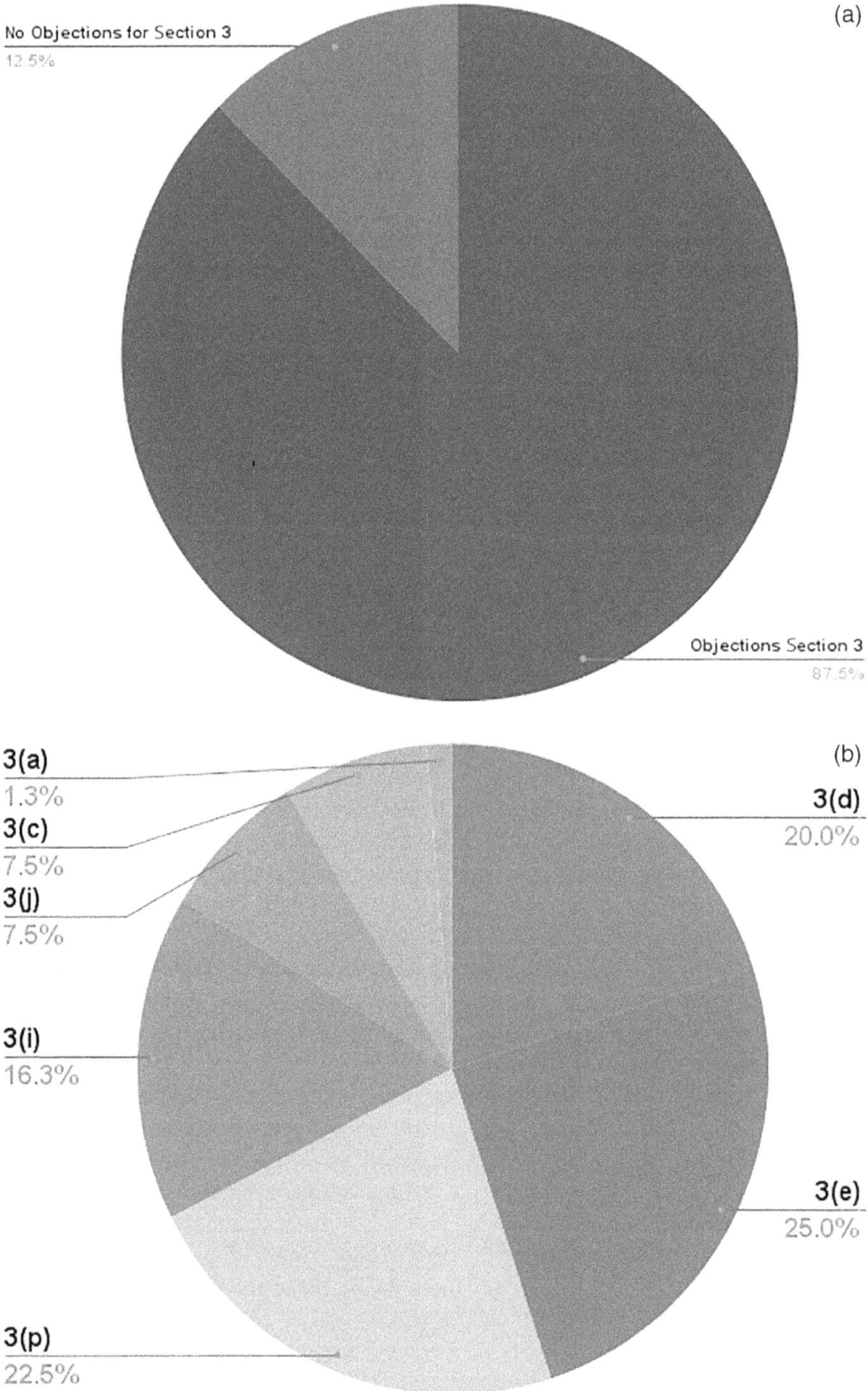

FIGURE 9.4 Analysis of the (a) rejections under Section 3 and (b) its subsections.

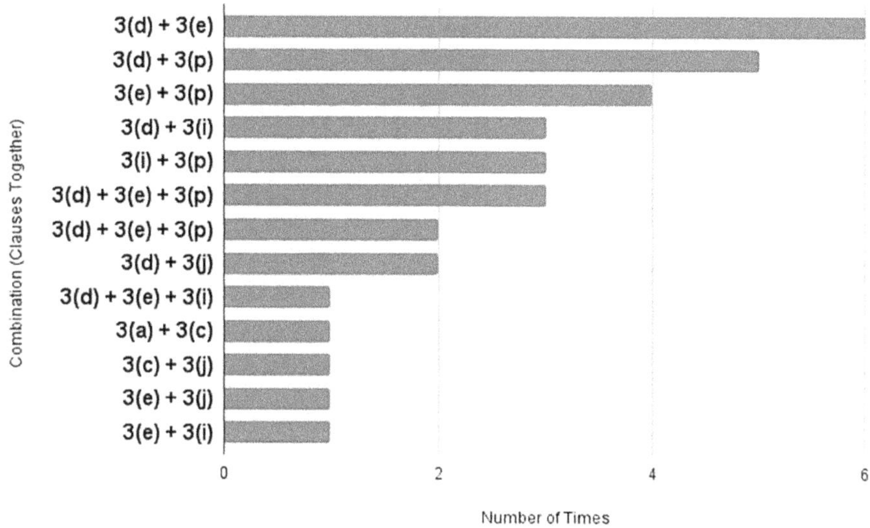

FIGURE 9.5 Analysis of the combination clause cited with respect to Section 3 of the Patents Act 1970.

This pair is frequently used to question both the novelty and inventiveness of compositions, especially in the context of herbal formulations. The challenge becomes even greater when this pair is further accompanied by Section 3(p), which questions the novelty by referencing traditional or folklore-based knowledge.

The most complex objections were found in cases where all three Section 3(d), 3(e), and 3(p) were jointly applied. In such scenarios, applicants often struggle to successfully overcome the rejections, as they needed to prove not just enhanced efficacy and synergistic effect, but also establish that the invention was not part of traditional knowledge and had sufficient inventive merit.

This growing use of layered objections significantly increases the burden on applicants. It necessitates a more sophisticated approach to claim drafting, technical disclosure, and legal argumentation. Importantly, to address such objections, applicants are often compelled to amend or delete claims during the prosecution process. Claim deletion, either before or after the grant of the patent, can directly impact the scope, enforceability, and commercial value of the patent.

9.7.4 Patentability Criteria Analysis in Herbal Patent Applications

During the process of patent examination in India, applications often encounter two primary hurdles: restrictions and objections. Restrictions are imposed under Section 3 of the Indian Patent Act, which outlines categories of subject matter deemed ineligible for patent protection. When a claimed invention falls under any of these categories, it is considered restrictive subject matter and cannot be granted protection unless suitably amended to remove the ineligible portion.

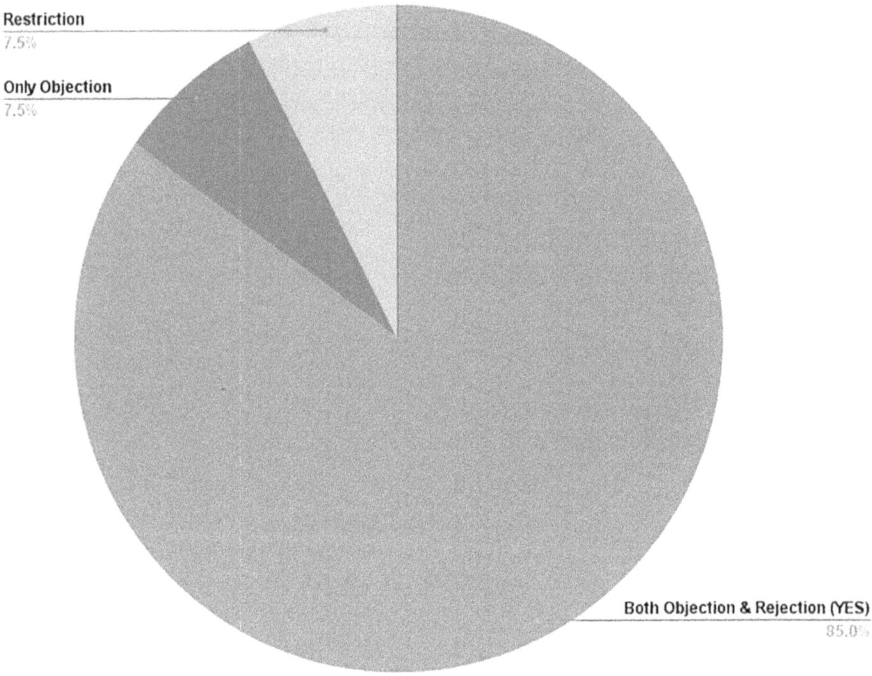

FIGURE 9.6 Analysis of the percentage of objections and restriction in patent applications.

In contrast, objections refer to other deficiencies identified by the examiner that do not necessarily relate to subject matter eligibility. These may include lack of novelty, inventive step, inadequate disclosure, unclear claims, or issues with formal requirements. While objections can often be overcome through clarifications, amendments, or supporting evidence, restrictions under Section 3 require substantive changes to the nature or scope of the invention itself. The data analysis reveals that Section 3 restrictions are a predominant factor influencing grant outcomes. In the sample examined 85% of patents faced both restrictions under Section 3 and other objections. 7.5% faced only restrictions under Section 3 without additional objections. 7.5% faced only objections without Section 3 restrictions (Figure 9.6).

The company wise graph for objection and restriction faced by patent prosecution to get clear understanding of position of companies (Figure 9.7).

9.7.5 Analysis of NBA Approval for the Patent Applications

One of the most critical requirements in the process of herbal drug patenting is obtaining National Biodiversity Authority (NBA) approval. Without this approval, a patent application involving biological resources cannot be granted. This requirement ensures that the use of biological resources is regulated, preventing undue exploitation and safeguarding resources that may require special protection, while also considering their availability.

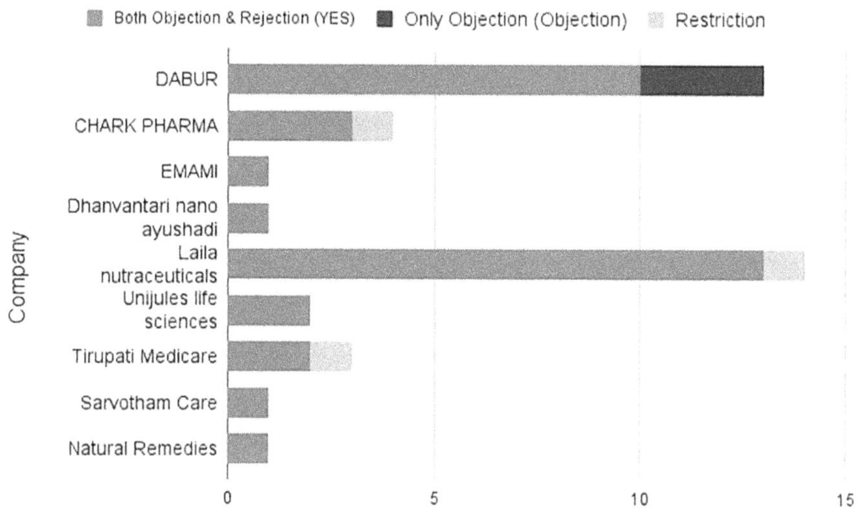

FIGURE 9.7 Company wise analysis of objection and restrictions.

In this study it was observed that 57.5% of the patent applications encountered the requirement for NBA approval. However, not all applicants complied by submitting the approval. A section of the applicants responded to the First Examination Report (FER) with explanations stating that the approval was not applicable to their invention. Out of the 57.5% who were required to submit the approval, only 37.5% actually obtained and filed the NBA approval, while the remaining 20% did not file it, instead providing written justifications during prosecution (Figure 9.8).

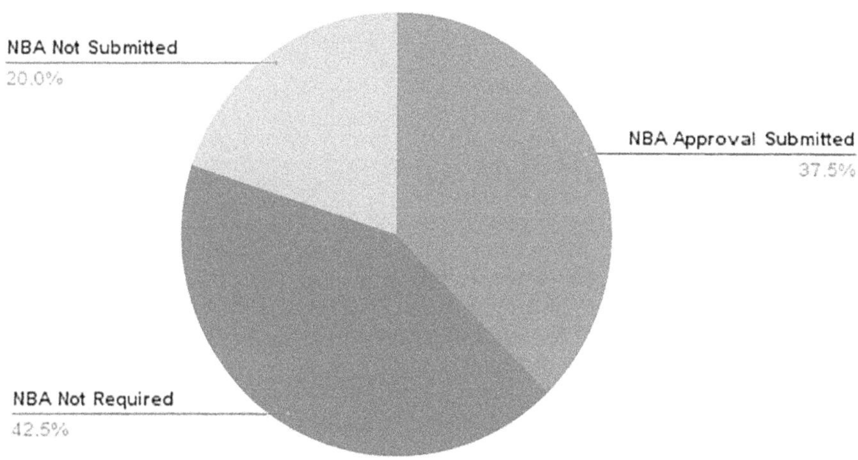

FIGURE 9.8 Analysis of NBA approval.

9.8 CONCLUSION AND WAY FORWARD

The present study provides a comprehensive analysis of the patenting activity of herbal drug companies that are also COPP certified. The collaborations with WHO, including the establishment of the WHO Global Centre for Traditional Medicine, and the introduction of certification frameworks like COPP, reflect India's growing stature in global herbal medicine trade. The documentation and protection mechanisms exemplified by the Traditional Knowledge Digital Library (TKDL) demonstrate India's proactive stance in combating biopiracy and defending indigenous knowledge in international intellectual property regimes. The analysis of patent prosecution reveals the specific challenges these companies face in securing intellectual property rights for their herbal innovations.

The study also reveals the complexity of patent prosecution in the herbal drug domain, where patentability is rigorously scrutinized under multiple provisions of the Indian Patent Act, especially Section 3 and its sub-clauses. The patent system balances grant of patent rights to inventions (new formulations, extraction methods, and devices) and also the protection of community traditional knowledge from privatization. The high frequency of patent objections and claim deletions illustrates the challenges faced by herbal drug innovators. Regulatory fragmentation, lack of harmonized standards, and challenges in meeting global GMP requirements disproportionately affect smaller manufacturers and continue to hinder export potential. The mandatory compliance with the Biological Diversity Act, especially NBA approval, adds an essential layer to benefit-sharing but also introduces procedural complexities in patent filing.

Securing patent rights offers immense potential to companies to leverage global trade in traditional medicines. However, realizing these potential demands greater integration of scientific validation, standardized quality control, and streamlined intellectual property strategies. Expansion of access to critical resources like the TKDL to wider stakeholders, continued governmental impetus towards harmonizing regulations, and fostering innovation through collaborative R&D are crucial for sustaining growth. Safeguarding India's traditional knowledge is undeniable. However, fostering breakthrough innovations in this sector will facilitate promotion of India's invaluable ethnomedical heritage.

REFERENCES

Central Drugs Standard Control Organization (CDSCO). n.d. "COPPs Granted by CDSCO for Ayurvedic Products." Accessed July 30, 2025. https://cdsco.gov.in/opencms/opencms/en/Approval_new/COPPs-Ayurvedic-Products/

Chaudhary, A., and N. Singh. 2012. "Intellectual Property Rights and Patents in Perspective of Ayurveda." *AYU: An International Quarterly Journal of Research in Ayurveda* 33 (1): 20–6. https://doi.org/10.4103/0974-8520.100298

European Medicines Agency. n.d. "Herbal Medicinal Products." Accessed July 30, 2025. https://www.ema.europa.eu/en/human-regulatory-overview/herbal-medicinal-products

Gopalakrishnan, N. S. 2002. "Protection of Traditional Knowledge: The Need for a Sui Generis Law in India." *Journal of World Intellectual Property* 5: 725.

Guidelines for Processing of Patent Applications of AYUSH Systems and Related Inventions. 2025. Accessed July 30, 2025. https://ipindia.gov.in/frontend/pdf/patents/guidelines/Guidelines%20for%20Examination%20of%20Ayush%20Related%20Inventions.pdf

Guidelines for Processing of Patent Applications Relating to Traditional Knowledge and Biological Material. n.d. Accessed July 30, 2025. https://ipindia.gov.in/writereaddata/Portal/IPOGuidelinesManuals/1_39_1_5-tk-guidelines.pdf

Indian Brand Equity Foundation (IBEF). 2025. "AYUSH Industry Report." May 2025. https://www.ibef.org/industry/AYUSH

Indian Patents Act, 1970. n.d. Section 3(p): "An Invention Which, in Effect, is Traditional Knowledge or Which is an Aggregation or Duplication of Known Properties of Traditionally Known Component or Components." Chapter II, *Inventions Not Patentable.* Accessed July 30, 2025. https://ipindia.gov.in/writereaddata/Portal/IPOAct/1_113_1_The_Patents_Act__1970___incorporating_all_amendments_till_1-08-2024.pdf

Kamboj, Ved P. 2000. "Herbal Medicine." *Current Science* 78 (1): 35–39. https://www.jstor.org/stable/24103844

Katiyar, C. K., A. K. Chakrabarty, S. K. Dubey, P. K. Pandey, T. Manash, and A. Narwaria. 2023. "Enhancing Quality of AYUSH Products: Strategies and Efforts of the Ministry of AYUSH and AYUSH Industry." *Journal of Research in Ayurvedic Sciences* 7 (Suppl 1): S73–81. https://doi.org/10.4103/jras.jras_92_23

Lakshmanan, P. K. 2020. "Regulated Use of Biological Resources and Traditional Knowledge for Sustainable Development: The Experience from India." *OIDA International Journal of Sustainable Development* 13 (5): 59–70. https://doi.org/10.2139/ssrn.5283585

Ministry of AYUSH. 2025. "Promotion of International Cooperation." Accessed July 28, 2025. https://ayush.gov.in/#!/poicooparetion

National AYUSH Mission. n.d. "National AYUSH Mission." Accessed July 30, 2025. http://namayush.gov.in

National Biodiversity Authority. 2020. "Annual Report 2019–20." http://nbaindia.org/uploaded/pdf/AR2019-20.pdf

Press Information Bureau. 2022. "Cabinet Approves Widening Access of the Traditional Knowledge Digital Library (TKDL) Database to Users, Besides Patent Offices." News release, August 17, 2022. https://www.pib.gov.in/PressReleseDetailm.aspx?PRID=1852530

Pujari, V., P. Shrikanth, and S. Naik. 2021. "A Review on Trends of Herb Market in India." *AYUSHDHARA*. https://ayushdhara.in/index.php/ayushdhara/article/view/855/688

Sahoo, N., M. Padmavati, and S. H. Dey. 2011. "Herbal Drug Patenting in India: IP Potential." *Journal of Ethnopharmacology* 137 (1): 289–97. https://doi.org/10.1016/j.jep.2011.05.022

Schedule, T., and Drugs and Cosmetics Act, 1940. n.d. Accessed July 30, 2025. https://www.wbhealth.gov.in/ayush/download/ISM/Schedule-T.pdf

Sen, S., R. Chakraborty, and B. De. 2011. "Challenges and Opportunities in the Advancement of Herbal Medicine: India's Position and Role in a Global Context." *Journal of Herbal Medicine* 1 (3–4): 67–75. https://doi.org/10.1016/j.hermed.2011.11.001

Sheridan, C. 2005. "EPO Neem Patent Revocation Revives Biopiracy Debate." *Nature Biotechnology* 23 (5): 511–12. https://doi.org/10.1038/nbt0505-511

Singh, K., and H. Purushotham. 2020. "Worldwide Patent Landscaping of AYUSH Technologies." *Journal of Intellectual Property Rights* 25 (5): 265–75. https://doi.org/10.56042/jipr.v25i5.34455

TKDL Access Agreement. n.d. Accessed July 30, 2025. https://www.tkdl.res.in/tkdl/langdefault/common/terms.pdf

Traditional Knowledge Digital Library (TKDL). n.d. Accessed July 30, 2025. https://tkdl.res.in/tkdl/langdefault/common/Home.asp?GL=Eng

"WHO Guidelines on Good Manufacturing Practices (GMP) for Herbal Medicines." Accessed July 30, 2025. https://iris.who.int/bitstream/handle/10665/43672/9789241547161_eng.pdf

World Health Organization. 2022. "WHO Establishes the Global Centre for Traditional Medicine in India: Maximizing Potential of Traditional Medicines through Modern Science and Technology." News release, March 25, 2022. https://www.who.int/news/item/25-03-2022-who-establishes-the-global-centre-for-traditional-medicine-in-india

World Intellectual Property Organization (WIPO). n.d. "The Use of Turmeric in Wound Healing." Accessed July 30, 2025. https://www.wipo.int/export/sites/www/tk/en/docs/use-of-turmeric-in-wound-healing-e.pdf

World Intellectual Property Organization (WIPO). 2011. "International Conference on the Utilization of the Traditional Knowledge Digital Library as a Model for Protection of Traditional Knowledge, March 22–24, 2011, New Delhi, India." https://www.wipo.int/meetings/en/2011/wipo_tkdl_del_11/

10 Innovation and Entrepreneurship in Herbal Drug Development

Dimple Chauhan and Trishna Bal

10.1 INTRODUCTION

10.1.1 Innovation and Entrepreneurship of HDD

a. *The Techno-Entrepreneurial Landscape of Herbal Therapeutics*: Recognition of the hidden potential of herbal medicines is reshaping global healthcare, marking a shift towards evidence-based integration. This transformation is driven by technological advances that overcome limitations of traditional formulations. The convergence of science and entrepreneurship has created a dynamic HDD landscape, enabling innovative delivery systems and strategic business models that support the development and commercial success of reliable herbal therapeutics.
b. *The Need for Innovation in Herbal Drug Development*: Despite their long-standing use, herbal remedies often lack the precision and reliability required by modern healthcare. The complexity of herbal extracts necessitates advanced drug delivery approaches to address issues such as poor solubility, low bioavailability, inconsistent standardisation, limited targeting, and palatability. HDD resolves these challenges, unlocking nature's pharmacopoeia to deliver more effective, reliable, and clinically acceptable herbal therapies.

10.1.1.1 Technological Innovations Driving Advancements in HDD

The use of cutting-edge technologies to increase the efficacy and accessibility of herbal remedies is what distinguishes herbal drug development.

a. *Nanotechnology-Based Delivery Systems – Transforming Bioavailability and Specific Targeting*

   Nanotechnology has ushered in a new era of HDD by enabling precise molecular and cellular drug delivery. Biodegradable polymeric nanoparticles, such as chitosan or PLGA, enhance solubility and protect hydrophobic herbal drugs from gastrointestinal degradation (Chowdhury et al. 2024).

DOI: 10.1201/9781003623748-10

Lipid-based carriers like SLNs and NLCs improve bioavailability and tissue targeting, while metallic nanoparticles offer antimicrobial and anticancer potential. Novel nanocarrier systems present significant therapeutic and commercial opportunities.

b. *Controlled Release through Microencapsulation and Microparticulate Systems*

To improve the pharmacokinetic characteristics of herbal medications and reduce the frequency of dosing, microencapsulation offers a strong basis for achieving controlled and prolonged release (Neekhra et al., 2022). To improve patient comfort and sleep quality, a company could, for instance, use cellulose-based polymers to create microencapsulated formulations of valerian root extract that would allow its sedative components to release gradually throughout the night. Making pH-sensitive microparticles that release herbal anti-inflammatory drugs only in the colon, minimising stomach discomfort and increasing local efficacy for ailments like ulcerative colitis, could be another business opportunity. The development of scalable and economical manufacturing processes, as well as the optimisation of microencapsulation techniques and materials to achieve desired release profiles for particular herbal products, represents the entrepreneurial potential here.

c. *Lipid-Based Delivery Systems: Enhancing Absorption of Lipophilic Actives*

Lipid-based delivery systems overcome the poor oral bioavailability of lipophilic herbal compounds. Liposomes enhance intestinal absorption of extracts such as *Ginkgo biloba* by encapsulating both hydrophilic and lipophilic drugs (Soman et al., 2020). Innovations like patented phytosomes improve efficacy and reduce dosage requirements (Cicero, Tartagni, and Borghi 2012), offering strong commercial and therapeutic advantages.

d. *Hydrogels and Mucoadhesive Systems: Enabling Localised and Prolonged Action*

Hydrogels offer versatile platforms for herbal drug delivery, particularly for localised therapy. Bioadhesive chamomile extract-loaded hydrogels can treat oral mucositis by adhering to mucosa and promoting healing (Rana et al. 2020). Thermosensitive and mucoadhesive hydrogels enable sustained herbal release for wound healing, vaginal, and nasal applications, supporting targeted therapeutic outcomes.

e. *Transdermal delivery*

For herbal medications that are prone to first-pass metabolism or induce gastrointestinal distress, transdermal delivery presents an alluring alternative. Companies are advancing this field by developing sophisticated transdermal patches with penetration enhancers or special carrier systems like ethosomes, which are extremely flexible lipid vesicles that can penetrate the skin barrier with ease (Alkilani et al. 2022; Goyanes et al. 2015). An entrepreneurial initiative could centre around designing a transdermal patch with a standardised extract of Boswellia serrata for consistent relief from arthritic pain, effectively bypassing the digestive system and maintaining stable therapeutic levels. Another potential idea is to develop

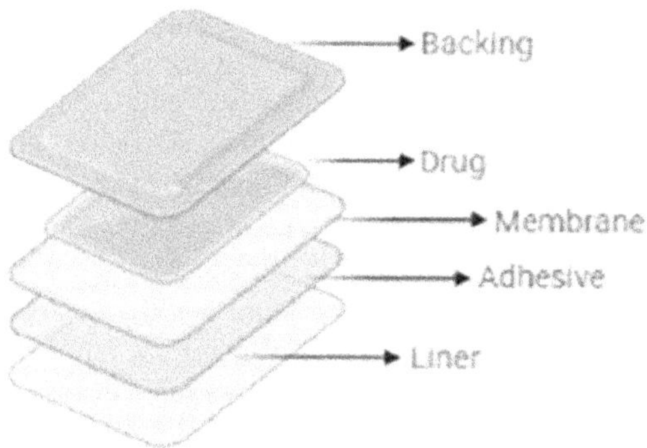

FIGURE 10.1 Schematic layered structure of a transdermal patch showing backing layer, drug reservoir, membrane, adhesive, and liner.

microneedle-based transdermal systems for larger herbal compounds that typically have low skin permeability. Transdermal drug delivery systems, shown in Figure 10.1, are a way to administer drugs through the skin with localised therapeutic/Skin patches effects.

f. *3D printing*

The use of 3D printing in HDDS is still in its early phases, but presents significant possibilities for crafting personalised herbal treatments (Hewlings and Kalman 2017). Various methods as shown in Figure 10.2 are used for 3D printing of drug delivery systems using natural products are ideally two types: [i] nozzle-based deposition systems and [ii] ink-based printing system. Under the nozzle-based deposition system, there are two

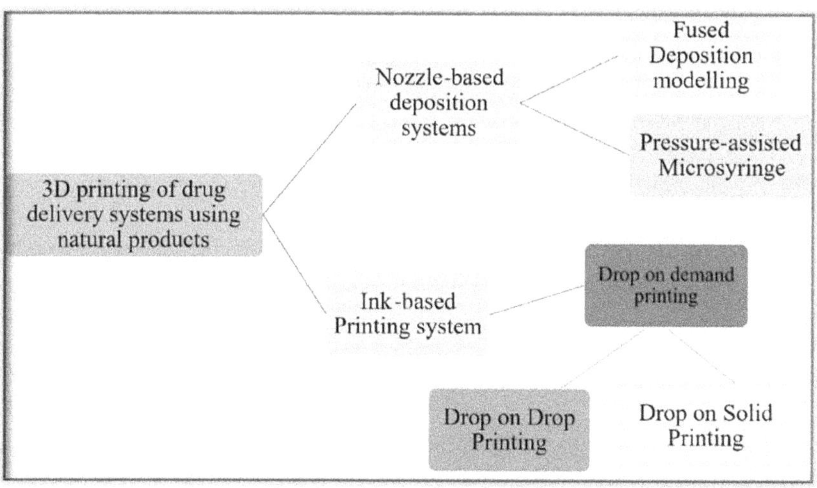

FIGURE 10.2 Classification of 3D printing techniques for drug delivery systems using natural products, including nozzle-based deposition and ink-based printing systems.

categories, namely, fused deposition modelling and pressure-assisted microsyringe, whereas under the ink-based printing system, drop on demand printing prevails, which is further categorised into two types, namely, drop on drop printing and drop on solid printing. Envision a pharmacy utilising 3D printing technology to create tailored herbal formulations that align with a patient's individual requirements and metabolic profile, adjusting the dosages and release characteristics of various herbal components within a single polypill. An entrepreneurial venture could focus on producing specialised 3D printers and biocompatible materials refined for herbal drug formulations. Another pathway is establishing software platforms that empower healthcare professionals to design and fabricate customised herbal medications, paving the way for new opportunities for precision medicine in phytotherapy.

10.2 TECHNOLOGICAL PROGRESS IN HDDS IS SPURRING A VARIETY OF ENTREPRENEURIAL VENTURES

a. *Nanomaterial Start-Ups*: Companies concentrating on the development and marketing of nanoparticle-based HDDS for targeted delivery and improved bioavailability.
b. *Controlled Release Developers*: Businesses that specialise in microencapsulation and other controlled release methodologies for herbal extracts, offering formulation services to herbal medicine producers.
c. *Lipid-Based Delivery Ventures*: Enterprises committed to creating and promoting liposomal and phytosomal formulations of numerous herbal drugs known for their enhanced absorption rates.
d. *Hydrogel and Biomaterials Innovators*: Companies focusing on developing new hydrogel delivery systems for localised and sustained release of herbal treatments.
e. *Transdermal Technology Providers*: Firms that design advanced transdermal patches and delivery systems for herbal medications, emphasising improved skin penetration.
f. *3D Printing Suppliers for Herbal Pharmacies*: Companies providing 3D printing equipment, materials, and software for the personalised preparation of herbal medications.
g. *Analytical and Quality Control Firms*: Given the increasing complexity of HDDS, there is a rising demand for specialised analytical services to ensure the quality, stability, and standardisation of these sophisticated formulations.
h. *Regulatory Consulting Businesses*: With the intricate regulatory landscape surrounding new HDDS, specialised consulting firms can assist companies in navigating the approval process.

These entrepreneurial activities are not only propelling technological innovation but also establishing novel business models and market segments within the herbal medicine industry.

10.2.1 Challenges and Strategies for Techno-Entrepreneurial Success

a. *Ensuring Quality Consistency and Standardisation* – The natural variability of herbal extracts requires strong quality control methods and standardisation procedures for advanced delivery systems. Entrepreneurs need to invest in advanced analytical techniques and establish dependable sourcing practices.
b. *Understanding Complex Regulatory Frameworks* – The regulations governing herbal drugs and innovative drug delivery systems can be unclear and significantly differ across regions. Entrepreneurs should engage proactively with regulatory bodies and develop a thorough understanding of the approval processes.
c. *Scaling Manufacturing While Keeping Costs Down* – Moving from small-scale lab development to large-scale, cost-effective production of advanced herbal drug delivery systems demands specialised knowledge in pharmaceutical engineering and supply chain management.
d. *Producing Strong Clinical Evidence* – Showing the efficacy and safety of herbal drugs delivered through new systems necessitates well-conceived pre-clinical and clinical trials, which can be both financially and time-intensive.
e. *Protecting Intellectual Property and Encouraging Innovation* – Obtaining patents for new formulations and delivery technologies is critical for sustaining a competitive edge. Entrepreneurs must create comprehensive intellectual property strategies.

10.2.2 Case Studies of Entrepreneurial Achievements in Herbal Drug Development

To further illustrate the vibrant relationship between innovation and entrepreneurship in herbal drug delivery systems, let us review some potential and in some cases reflective of actual trends case studies.

Case Study 1 – *Employing Nanoparticles for Improved Curcumin Bioavailability*

a. *Company Profile* – Numerous groups develop nanoparticle curcumin formulations; for instance, Thera Values Corporation's patented Super Curcumin uses micronation and emulsification to enhance bioavailability (Anand et al. 2008).
b. *Innovation* – Firms such as Thera Value enhance the solubility and absorption of curcumin using advanced formulations. Super-Curcumin shows higher plasma levels than native curcumin (Derry et al. 2017). Nanoparticle-based curcumin delivery is also being explored for inflammatory diseases and cancer therapy (Hosseinzadeh et al. 2015).
c. *Entrepreneurial Approach* – Companies commercialising bioavailable curcumin target the nutraceutical market with premium supplements, emphasising scientific validation, enhanced absorption, and health benefits, while pursuing pharmaceutical partnerships and building strong, efficacy-driven brands.

d. *Techno-Entrepreneurial Aspect* – The success of these businesses relies on the innovative application of formulation technologies to overcome bioavailability issues, rigorous scientific validation of their products, and effective marketing to health-conscious consumers and potentially the pharmaceutical market.

***Case Study 2** – Controlled Release Formulations in Herbal Medicine*

a. *Company Overview* – Indian S.P.A. is well-known for its sophisticated botanical extracts and formulations, including some with modified release characteristics designed to enhance therapeutic effects and improve patient adherence (Indian Corporate Website).
b. *Innovation* – Firms such as Indian companies invest in research to control herbal drug release using microencapsulation or polymer matrices, optimising pharmacokinetics for targeted or sustained delivery of compounds like silymarin and Ginkgo biloba.
c. *Entrepreneurial Strategy* – These companies supply scientifically validated herbal extracts and formulations to pharmaceutical and nutraceutical manufacturers, enabling differentiated products with enhanced efficacy and patient convenience, supported by robust scientific evidence.
d. *Techno-Entrepreneurial Aspect* – Achieving success in this field depends on expertise in the extraction and purification of botanicals, advanced formulation technologies for controlled release, stringent quality control, and the capability to provide scientific proof regarding the benefits of their ingredients and formulations.

***Case Study 3** – Transdermal Delivery of Herbal Actives*

a. *Company Overview* – Transdermal delivery of herbal medicines is an emerging area, mainly explored by research groups and small enterprises. Studies have investigated patch-based transdermal delivery of capsaicin for pain management (Othman and Farooqui 2015; Torchilin, 2014).
b. *Innovation* – Research focuses on overcoming the skin barrier using penetration enhancers, innovative patches, and microneedles to deliver herbal compounds effectively, enabling local or systemic therapy while avoiding oral administration-related complications.
c. *Entrepreneurial Strategy* – New ventures may target niche applications, such as analgesic patches containing pain-relieving herbal extracts. Their success depends on demonstrated transdermal efficacy, ease of use, and proven therapeutic benefits supported by clinical trials or strong user feedback.
d. T*echno-Entrepreneurial Aspect* – Innovation is centred around the creation of safe and effective transdermal delivery systems for specific herbal compounds, combined with a focus on user-friendly product design and evident therapeutic benefits in targeted applications.

***Case Study 4** – Personalised Herbal Formulations via Advanced Compounding*

a. *Company Overview* – The application of 3D printing in personalised herbal medicines is still nascent, with limited adoption. Companies like FabRx are developing pharmaceutical 3D-printing technologies that could support future personalised herbal formulations.

b. *Innovation* – The concept involves using 3D printing to create personalised herbal dosage forms with customised extract combinations and release profiles. This requires advanced software, compatible printing materials, and a strong understanding of herbal pharmacokinetics.
c. *Entrepreneurial Strategy* – Initial adoption may take place in specialised compounding pharmacies that partner with practitioners prescribing personalised herbal treatments. The value proposition is highly customised therapy, enhanced patient adherence, and the likelihood of improved treatment results.
d. *Techno-Entrepreneurial Aspect* – Success is reliant on developing dependable and precise 3D printing technologies for herbal formulations, intuitive software for designing personalised medications, and ensuring safety and efficacy through adequate quality control and clinical evidence (El-Atroush, Magdy, and Werner 2015).

10.3 THE FUTURE OF HERBAL THERAPEUTICS THROUGH INNOVATION AND ENTREPRENEURSHIP

The evolution of herbal medicine is closely tied to entrepreneurship and advances in drug delivery systems. Modern HDD address the limitations of traditional formulations, enabling safer, more effective, and patient-friendly therapies. Continued scientific and technological progress will drive innovative herbal products, supporting their integration into mainstream healthcare and addressing global health challenges through science–entrepreneurship synergy.

10.3.1 Innovations Transforming Herbal Therapeutics

Herbal medicine is evolving from traditional remedies into advanced pharmaceutical products through technological innovations such as.

- Nanotechnology, phytotomies, and nano-emulsions that improve bioavailability, targeted delivery, and stability of herbal compounds.
- Artificial intelligence (AI) and machine learning (ML) are applied in phytochemical analysis, predictive modelling, personalised formulation, and faster clinical trials.
- Novel extraction methods using green chemistry and biotechnological techniques like CRISPR for plant genome editing.

10.3.2 The Significance of Herbal Medicine and Innovation in HDD

Herbal medicines have long been integral to global healthcare, offering accessible and affordable therapeutic options with fewer side effects than synthetic drugs (Liu et al. 2015). However, their variability, poor solubility, and limited bioavailability necessitate innovation in the development of herbal drugs. Advanced technologies such as nanotechnology, microencapsulation, and lipid-based delivery systems are being used to enhance stability, absorption, and therapeutic reliability (Chowdhury et al. 2024).

i. *Changing the Research Paradigm in HDD*

The research environment in HDD is changing quickly due to advances in several scientific domains and a growing understanding of the unrealised potential of herbal remedies when paired with contemporary drug delivery techniques. Several key areas highlight this paradigm shift in research.

ii. *Focusing on Boosting Bioavailability*

Extensive research targets improving the bioavailability of poorly absorbed herbal compounds using enzyme inhibitors, lipid-based carriers such as phytosomes, and nano-formulations including nanoparticles and nanoliposomes (Mak, Wong, and Pichika 2024; Soman et al. 2020). Nano-formulations of curcumin have shown enhanced absorption and therapeutic efficacy against inflammatory diseases (Alkilani et al. 2022).

iii. *Targeted Medication Delivery Systems*

Another crucial field of research is the development of targeted drug delivery systems for herbal remedies. This reduces systemic exposure and any adverse effects by modifying drug carriers to deliver herbal medicines precisely to the site of action. The techniques employ stimuli-responsive systems that release their payload in response to physiological conditions, e.g., pH, temperature at the disease site, as well as ligands that have an affinity for receptors on target cells or tissues (Hasenclever et al. 2017).

iv. *Quality Control and Standardisation*

Advances in analytical techniques are greatly improving the standardisation and quality control of herbal drug composition. Assuring batch-to-batch consistency and quality, sophisticated techniques such as gas chromatography-mass spectrometry (GC-MS) and high-performance liquid chromatography (HPLC) are employed to identify and measure the active ingredients in herbal extracts and formulations. Furthermore, a technique for confirming herbal ingredients and avoiding adulteration is being researched (Palmberg 2004).

v. *Integration of "Omics" Technologies*

The integration of omics technologies, including genomics, proteomics, and metabolomics, enhances understanding of herbal medicine mechanisms and biological interactions. Metabolomic studies identify key bioactive compounds and assess how delivery systems influence their metabolism and distribution, enabling more rational and efficient herbal drug delivery strategies.

vi. *Machine Learning and Artificial Intelligence*

Artificial intelligence and machine learning are emerging tools in herbal drug development, enabling large-scale data analysis, prediction of pharmacokinetic and pharmacodynamic properties, and advanced delivery system design. Current HDD innovation emphasises bioavailability, targeted delivery, standardisation, omics integration, and AI/ML approaches, improving reliability, efficacy, and techno-entrepreneurial opportunities. Figure 10.3 highlights key drivers and challenges.

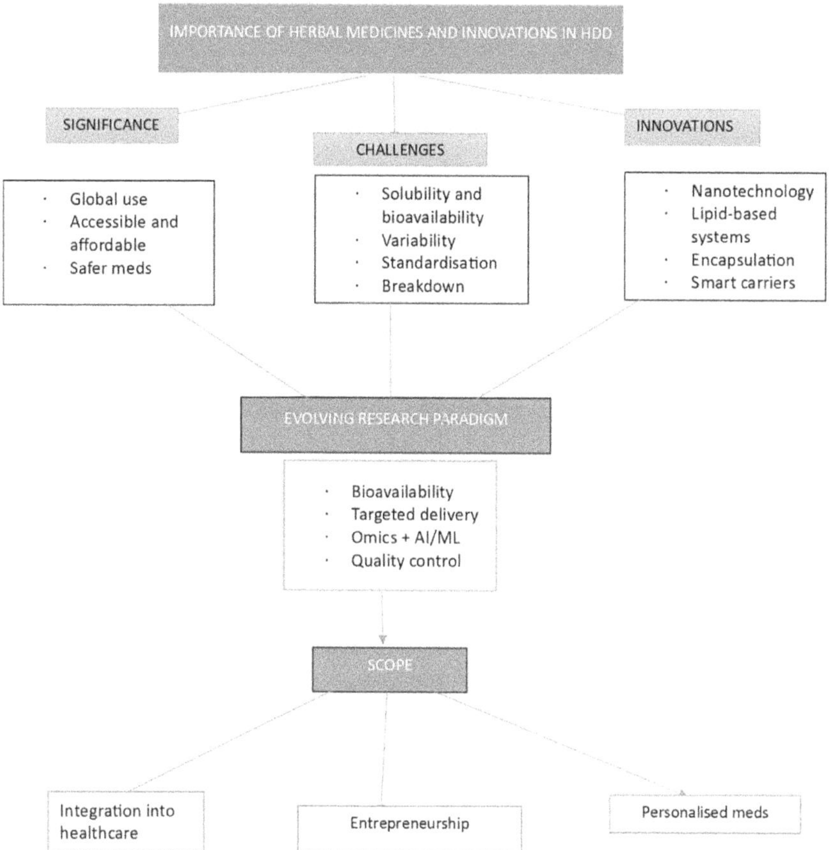

FIGURE 10.3 Importance of herbal medicines and innovations in herbal drug development (HDD), highlighting significance, challenges, technological advances, evolving research paradigms, and future scope.

10.4 CHANGING MARKET IN HDD

Driven by changing consumer preferences, scientific validation, and the integration of traditional knowledge with modern technology, the global herbal drug development (HDD) market is undergoing rapid transformation. According to Fortune Business Insights, the market is expected to grow from USD 233.08 billion in 2024 to USD 437 billion by 2032, at a CAGR of 8.23%. Rising demand for natural and organic products has increased the use of herbal remedies, functional foods, and natural personal care items (Parveen et al. 2015). The popularity of CBD products and clean beauty brands highlights this shift.

Integration of traditional systems such as Ayurveda and Traditional Chinese Medicine with modern research has enabled standardised, safer, and more effective herbal formulations (Ekor 2014). E-commerce platforms have further expanded global access and supported niche and direct-to-consumer brands (Sharma et al. 2006).

Despite challenges related to quality, adulteration, and clinical validation, companies are strengthening quality control and research investments. Innovations in drug delivery, nanotechnology, sustainability, and ethical sourcing are expected to drive continued market growth.

10.5 ENTREPRENEURSHIP IN HDD

10.5.1 Market Trends and Consumer Demand

Entrepreneurial activity in the HDD sector is rising due to health-conscious consumer demand and strong market growth, with the global herbal market projected to reach USD 437 billion by 2032, creating opportunities for new ventures.

10.5.2 Factors Influencing Entrepreneurship in HDD

a. *Rising Consumer Preference for Organic and Natural Products* – Consumer preference for organic and natural alternatives to synthetic medicines is increasing due to safety concerns and holistic wellness goals (Kessler, et al. 2001; Yuan et al. 2016). The success of adaptogenic herbs such as Rhodiola and Ashwagandha reflects this trend, with entrepreneurs developing premium herbal formulations for diverse wellness needs (Mukherjee et al. 2022).
b. *Integration of Traditional Medicine with Modern Science* – Scientific validation of traditional systems like Ayurveda and TCM is creating entrepreneurial opportunities in HDD (Shahriari et al. 2023). Entrepreneurs develop standardised, evidence-based formulations using modern extraction and delivery methods, such as nanotechnology-enhanced Curcuma longa, supported by clinical research for competitive advantage (Dolgui et al. 2018; Kilbourne et al. 2018; Longhi and Rochhia 2015).
c. *Proliferation of E-commerce and Direct-to-Consumer Paradigms* – Online retail platforms have lowered entry barriers in the HDD industry (Ritchie 2007). Entrepreneurs use e-commerce, direct consumer interaction, and subscription models to market niche botanicals, handmade herbal products, and curated wellness packages globally (Straus, 2002).
d. *Growing Focus on Preventive Healthcare* – Rising healthcare costs and awareness of reactive medicine drive consumer interest in preventive healthcare (Ritchie 2007). Entrepreneurs offer herbal products for general health, immunity, and lifestyle imbalances, exemplified by *Sambucus nigra* and *Silybum marianum* formulations (Grunwald, 2009; Klein-Junior et al. 2021).
e. *Demand for Transparency and Quality Assurance* – Modern consumers value the origin, quality, and ethical sourcing of herbal ingredients (Thakkar et al. 2020). Entrepreneurs build trust and loyalty through transparent supply chains, strict quality controls, certifications, and third-party testing for purity and potency (Rastogi and Yadav 2012; Schork 2019).

The entrepreneurial environment in the HDD market is vibrant. By responding strategically to trends and consumer demand for natural, effective, transparent,

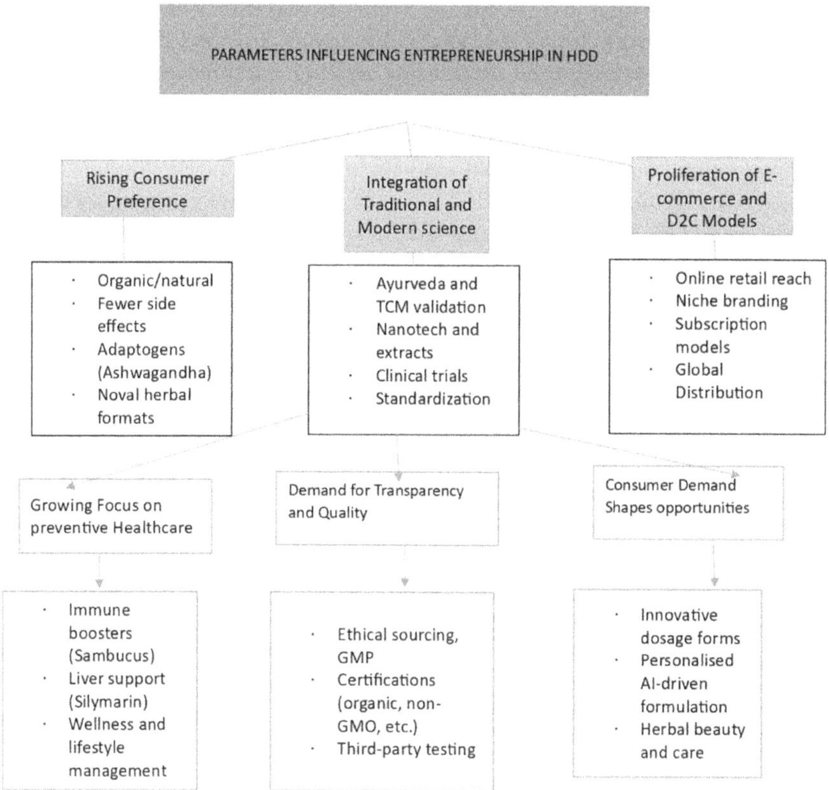

FIGURE 10.4 Key parameters influencing entrepreneurship in herbal drug development (HDD), including consumer preference, scientific integration, e-commerce growth, quality transparency, preventive healthcare, and emerging market opportunities.

and scientifically validated herbal products, entrepreneurs can build successful firms (Kushwah et al. 2023).The following Figure 10.4 shows the parameters influencing HDD.

Details of Parameters discussed below:

a. *Growing Consumer Desire for Natural Alternatives* – Consumers increasingly prefer plant-based remedies over synthetic medicines due to perceived safety and holistic benefits (Kellerman, 2023). The global CBD market has expanded rapidly, driven by perceived therapeutic effects, with entrepreneurs developing diverse CBD-infused products such as oils, tinctures, foods, and topical formulations (Taft et al. 2024).
b. *Increasing Prevalence of Chronic Ailments* – The rising prevalence of chronic diseases such as diabetes, arthritis, and cardiovascular disorders increases demand for long-term management, where herbal medicines

are viewed as safer complementary therapies (Chaudhry and Walsh 1995; Morris and Avorn 2003). Extracts of *Curcuma longa* and *Zingiber officinale* are increasingly used for arthritis, driving entrepreneurial activity in their formulation and marketing (Bull and Willard 1993).

c. *Integration of Traditional Medicinal Systems* – Market growth reflects integration of Ayurveda, TCM, and Unani into healthcare (Straus 2002), remedies (Mukherjee et al. 2022) standardised by entrepreneurs, India's Patanjali and Himalaya (Erdem and Chandra 2003; Senthil Kumar and Gunasundari 2018).
d. *Growing Senior Population* – The growing global elderly population drives herbal medicine demand, seeking gentler therapies for chronic, age-related conditions such as arthralgia and cognitive impairment, boosting interest in remedies like *Ginkgo biloba* and *Bacopa monnieri* (Shankar and Balasubramanian 2009; Steinhoff 1999).
e. *Growth of Distribution Channel* – The growing availability of herbal products through a variety of channels, such as online platforms, pharmacies, supermarkets, and niche stores, is boosting market accretion and valuation (Kessler et al. 2001). E-commerce has facilitated entrepreneurship by allowing smaller herbal companies to reach a larger consumer base (Steinhoff 1999).
f. *Increasing Investment in Research and Development* – Rising research on herbal efficacy and safety boosts credibility and market growth (Mangold and Faulds 2009). Entrepreneurs using scientific validation, bioactive isolation, and novel delivery systems gain competitive advantage and value (Kaplan and Haenlein 2010).
g. *Growing Demand for Cosmetics and Personal Care* – Herbal extracts and essential oils in personal care and cosmetics boost market value (Nurhandayani, Syarief, and Najib 2019). Entrepreneurs benefit from growing demand for natural beauty products (Qiraouani Boucetta et al. 2015; Yuan et al. 2016), with a strong market for argan oil, aloe vera, and other herbal hair and skin care ingredients (Kamsu-Foguem and Foguem 2014).

10.6 ONLINE PLATFORMS FOR PROMOTING MEDICINAL PLANTS

The rise of the digital era has transformed marketing strategies in a variety of industries, including the HDD industry. Entrepreneurs can market medicinal flora and their products through virtual platforms in a way that is unique and effective, reaching a worldwide audience, increasing transparency, and facilitating direct interaction with customers (Durach et al. 2021). The growing internet penetration, the growing consumer desire for natural health solutions (and the urgent need for readily available information on these resources (Easterlin 2009) are all factors that are pushing the development of virtual platforms for marketing medicinal plants.

10.6.1 For the Purpose of Marketing Medicinal Plants, Several Virtual Platform Archetypes Are Beginning to Appear, Each Offering Unique Benefits for Entrepreneurs

a. *E-commerce Marketplaces Specialising in Botanicals* – Online herbal marketplaces provide entrepreneurs with centralised platforms to display and sell products to niche audiences (Rowley, 2002). Features like customer reviews, secure payments, and search tools (Kumar and Ramasamy, 2024; Kwahk and Ge 2012) build trust. Platforms like Etsy and specialised medicinal plant sites offer curated selections and expert guidance (Van't Riet et al. 2001).
b. *Online Shops and Direct-to-Consumer websites* – Entrepreneurs sell medicinal plants via branded websites, controlling branding and customer experience (Hasenclever et al. 2017). Highlighting origin, uses, and research builds connections, supported by platforms like Shopify and WooCommerce (Drucker and Maciariello 2014).
c. *Online educational portals and informational websites* – Virtual platforms offering detailed information on medicinal flora, traditional uses, scientific validation, and cultivation act as effective marketing tools (Attele, Wu, and Yuan 1999; Sharkie 2003). Though non-transactional, they build trust, attract niche audiences, establish credibility, and indirectly drive sales via e-commerce links or affiliate marketing.
d. *Influencer Marketing and Social Media Platforms* – Social media platforms like Instagram, Facebook, and Pinterest enable visual promotion of medicinal flora (Kunle, Egharevba, and Ahmadu 2012; Liang, Xie, and Chan 2004). Through engaging content, community building, and influencer collaborations, entrepreneurs expand reach and market herbal products effectively (Chang, El-Rayes, and Shi 2022; Uppoor 2001).
e. *Telehealth Platforms and Virtual Consultations* – Virtual platforms enable remote consultations and customised herbal treatments, connecting entrepreneurs, practitioners, and clients beyond geographical limits (Allen and Cullis 2013). Telehealth allows assessment, prescription, and monitoring, with integrated e-commerce supporting personalised herbal formulation purchases.
f. *Blockchain-Based Platforms for Transparency and Traceability* – Blockchain-based virtual platforms provide transparent, verifiable data on medicinal plant origin, processing, and quality, enhancing trust and differentiation (Rehman et al. (2017); Sun et al. 2010), for example via QR codes tracing products from farm to consumer.

10.6.2 Benefits of Using Online Platforms to Promote Medicinal Plants

Using online platforms to promote medicinal plants offers numerous advantages, ranging from global market access and public awareness to conservation and education. Digital tools such as e-commerce, social media, and virtual herbal gardens have revolutionised how medicinal plants are marketed and appreciated worldwide like.

i. *Global Reach* – Entrepreneurs may reach clients all over the world using virtual platforms since they are not limited by geographical boundaries (Kaplan and Haenlein 2010; Rowley 2002)
ii. *Cost-Effectiveness* – Virtual marketing may be more cost-effective than traditional brick-and-mortar stores and traditional advertising, particularly for new businesses with limited funding (Bull and Willard 1993).
iii. *Direct Customer Engagement* – Virtual platforms allow businesses to interact directly with customers through online communities, assessments, and feedback mechanisms, which helps entrepreneurs establish relationships and customise their products (Savant and Kareppa 2022).
iv. *Improved Transparency and Information Dissemination* – Entrepreneurs may offer comprehensive facts about their medicinal flora, such as its origin, growing techniques, chemical makeup, and scientific data supporting its uses (Kwahk and Ge 2012).
v. *Data Analytics and Targeted Marketing* – Virtual platforms provide useful data on consumer behaviour and preferences, allowing entrepreneurs to improve their marketing plans and target consumer groups (Gertsch 2009; Kaplan and Haenlein 2010; Mangold and Faulds 2009; Rowley 2002).

In conclusion, virtual platforms have become essential tools for entrepreneurs in the HDD sector, providing unparalleled opportunities to market medicinal flora and their products effectively. By utilising e-commerce marketplaces, D2C websites, informational portals, social media, telehealth platforms, and new technologies like blockchain, business owners can reach a worldwide audience, establish trust via transparency, and encourage direct interaction with customers, which will ultimately promote the growth and sustainability of their businesses in the expanding herbal industry.

10.7 ENTREPRENEURSHIP IN HDD RECOGNISING AND FILLING MARKET GAPS

The very foundation of successful entrepreneurship lies in the wise identification and effective resolution of unsatisfied needs or underserved groups in a specific industry. Despite its enormous growth trajectory and inherent potential, the HDD industry has several major market gaps that astute business owners may strategically focus on for innovation, business creation, and meaningful healthcare contributions (Allen and Cullis 2013; Gilani 2005). It is imperative to have a thorough understanding of the changing demands of consumers, the dynamic landscape of technological breakthroughs, the complexity of regulatory frameworks, and the inherent constraints of current herbal goods and services to identify and seize these opportunities (Ali et al. 2005; Sireesha and Haritha 2024).

10.7.1 The HDD Market Still Has Several Critical Market Gaps, Each of Which Provides Distinct and Promising Opportunities for Entrepreneurial Endeavours

a. *Lack of Standardised and Clinically Validated Herbal Products* – Despite extensive traditional use, many medicinal floras lack rigorously standardised, clinically validated formulations meeting modern scientific standards (Bandaranayake 2006; Kartal 2007). Batch variability limits medical acceptance. Entrepreneurs can address this through robust preclinical and clinical research, as shown by Bionomic's trial-proven herbal medicines (Narang et al. 2022; Rastogi and Yadav 2012).
b. *Limited Availability of High-Quality, Traceable, and Authenticated Herbal Ingredients* – Quality, purity, and authenticity concerns hinder HDD industry growth (Parveen et al. 2015). Misidentification, adulteration, and inconsistent sourcing reduce trust and pose health risks. Entrepreneurs building traceable supply chains, using tools like blockchain, ensure quality, enhance transparency, and differentiate products, strengthening consumer confidence (Grunwald 2009; Nurhandayani, Syarief, and Najib 2019).

10.7.2 Examples of Entrepreneurial Businesses Using a Strategic Approach to Fill Market Gaps

- Businesses That Concentrate on Clinically Validated Herbal Extracts: Some cutting-edge firms are concentrating their efforts on the rigorous scientific validation of certain herbal extracts for health disorders, carrying out thorough preclinical and clinical studies to give healthcare practitioners evidence-based solutions.
- Blockchain-Enabled Herbal Supply Chain Platforms: By utilising the transparency and immutability of blockchain technology, new businesses are developing platforms that monitor the entire provenance of medicinal plants and offer customers verifiable data about their origin, quality, and ethical sourcing.
- Businesses that use nanotechnology to deliver herbal medicines: Entrepreneurs are employing cutting-edge nanotechnology to create unique formulas that greatly improve the bioavailability, targeted delivery, and sustained release of herbal compounds that have been used historically, hence enhancing their therapeutic benefits.

With novel solutions based on scientific rigour, ethical behaviour, and a thorough understanding of consumer needs, entrepreneurs in the HDD industry may strategically identify and creatively address these major market gaps, helping to establish lucrative and sustainable firms as well as playing a vital role in promoting the legitimacy, effectiveness, and responsible integration of herbal medicine into the worldwide healthcare industry.

10.8 ADVANCING INNOVATION IN HERBAL DRUG DEVELOPMENT

There is a growing global demand for herbal medicines, creating a need for innovative approaches in herbal drug development. Although traditional medicinal systems are invaluable, they often lack uniformity, reproducibility, and scientific validation. Variations in the quality and composition of herbal materials across geographical regions, climates, and processing methods necessitate the development of advanced standardisation techniques. Modern analytical tools such as metabolomics, genomics, and cheminformatics enable better characterisation and standardisation of herbal drugs, ensuring batch-to-batch consistency and accurate dosing (Sharma et al. 2006).

Advanced analytical techniques, including NMR and mass spectrometry (MS), enhance the isolation and characterisation of bioactive compounds from complex plant sources (Ganesan, 2004). High-throughput X-ray crystallography further supports the identification and optimisation of lead compounds from medicinal plants (Shyur, Liu, and Chien 2013). Metabolomics tools such as GC–MS, LC–MS, and NMR are widely used to evaluate plant composition, improve crop varieties, ensure the safety of botanical supplements, and identify health-related phytochemicals (Kumar 2023).

Innovations in drug delivery systems, particularly nanotechnology-based carriers such as phytosomes, liposomes, nanoparticles, and solid lipid nanoparticles, improve bioavailability, site specificity, and therapeutic efficacy, including enhanced BBB penetration (Azwanida, 2015; Gangwal and Lavecchia 2025). Artificial intelligence further accelerates herbal drug discovery through target prediction and virtual screening (Usman, Nakagawa, and Cheng 2023). Advanced extraction techniques, such as supercritical fluid, microwave-assisted, and ultrasound-assisted extraction, improve yield, efficiency, and the preservation of thermolabile compounds (Eisenberg et al. 1998).

10.8.1 Entrepreneurship in Herbal Drug Development

Entrepreneurship in herbal drug development involves establishing businesses that cultivate, process, and market herbal products and medicines, capitalising on rising consumer demand for natural health solutions and government support for the AYUSH sector. Key steps include market research, sourcing quality raw materials, developing formulations, navigating legal and regulatory requirements, and implementing effective marketing strategies to reach target consumers seeking preventive and personalised care.

Entrepreneurs can focus on various segments of the medicinal and aromatic plants sector, including

- ***Cultivation:*** Growing medicinal herbs and plants.
- ***Processing and Manufacturing:*** Developing formulations from raw materials.

- ***Supply Chain and Distribution:*** Connecting farmers, processors, and the pharmaceutical industry.
- ***Ancillary Services:*** Providing technical guidance, consultancy, and testing facilities.

The following Table 10.1 gives the details of various segments of the medicinal and aromatic plants sector and their activities.

TABLE 10.1
Various Segments of the Medicinal and Aromatic Plants Sector and Their Activities

Segment	Stage/Step	Key Activities
Cultivation	Site Selection and Planning	Assess climate, soil, drainage, and sunlight. choose species per agro-climatic condition; plan layout and irrigation system
	Land Preparation	Clear weeds/debris; plough/till for aeration; add organic manure/compost; test soil pH and nutrients, adjust with lime/fertilisers
	Propagation Method	Seed sowing (direct/nursery); vegetative propagation (cuttings, rhizomes, root division, tissue culture).
	Planting/Transplanting	Sow seeds at proper depth/spacing; transplant healthy seedlings; provide initial irrigation
	Irrigation and Nutrient Management	Water as needed; apply organic/recommended fertilisers; mulch to conserve moisture and control weeds
	Weed, Pest, and Disease Control	Manual weeding, eco-friendly herbicides; Integrated Pest Management (IPM) with beneficial insects/biopesticides; regular monitoring
	Growth Monitoring and Cultural Practices	Pruning, thinning, staking; maintain plant density and airflow; record growth data
	Harvesting	Harvest at correct maturity for max active constituents (leaves before flowering, roots/rhizomes at season end, flowers at bloom)
	Post-Harvest Handling	Clean and sort; dry in shade or controlled drying; store in airtight containers
	Quality Testing and Packaging	Check moisture, active compounds, contaminants; pack, label for market or processing
Processing and Manufacturing	Raw Material Reception and Inspection	Verify species and plant part; check for impurities; record supplier and harvest data
	Cleaning and Sorting	Remove stones/dust; sort by size, colour, and plant part
	Drying (Curing)	Shade drying for sensitive compounds; oven drying with controlled heat; freeze drying for high-value extracts

(Continued)

TABLE 10.1 *(Continued)*
Various Segments of the Medicinal and Aromatic Plants Sector and Their Activities

Segment	Stage/Step	Key Activities
	Grinding / Pulverisation	Crush/mill dried material; sieve for uniform particle size
	Extraction of Active Compounds	Use solvents (water, ethanol, hydro-alcoholic, CO_2); techniques: maceration, percolation, Soxhlet, ultrasonic, microwave; concentrate extracts
	Purification and Standardisation	Filter/clarify; adjust extract concentration for consistency
	Formulation Development	Solid: tablets, capsules, powders; semi-solid: ointments, gels, creams; liquid: syrups, tinctures, teas; add excipients
	Quality Control and Testing	Physicochemical tests (moisture, ash, pH); microbiological safety; active marker assay (HPLC, GC-MS); stability studies
	Packaging and Labelling	Airtight, light-resistant containers; labels with botanical name, batch, expiry, dosage, and approvals
	Storage and Distribution	Cool, dry storage, protected from sunlight; maintain GMP and traceability systems
Supply Chain and Distribution	Cultivation and Harvesting (Farm Level)	Farmers/herbal cooperatives follow GACP; harvest at optimum maturity; initial drying on farms
	Primary Collection and Aggregation	Farmers deliver dried/semi-dried herbs to mandis, grading, batch numbering, moisture, and contamination checks
	Processing Units/ Primary Processing	Cleaning, drying, grinding, initial crude extraction; pack bulk raw material
	Secondary Processing and Manufacturing	Bioactive extraction; formulation into products; quality control, GMP compliance
	Storage and Warehousing	Climate-controlled facilities; inventory/batch traceability.
	Distribution and Logistics	Wholesalers, herbal distributors, cold chain for sensitive extracts; exports with phytosanitary compliance
	Retail and End Users	Pharmacies, herbal shops, supermarkets, online; end users include pharma, nutraceuticals, cosmetics, and consumers

10.8.2 Supply Chain & Distribution

The supply chain of herbal drug delivery systems integrates traditional and modern pharmaceutical practices, spanning raw material sourcing to patient use. As shown in Figure 10.5, suppliers provide active ingredients and excipients, which are converted into finished products through manufacturing. These products are stored with real-time inventory tracking, distributed via logistics networks to

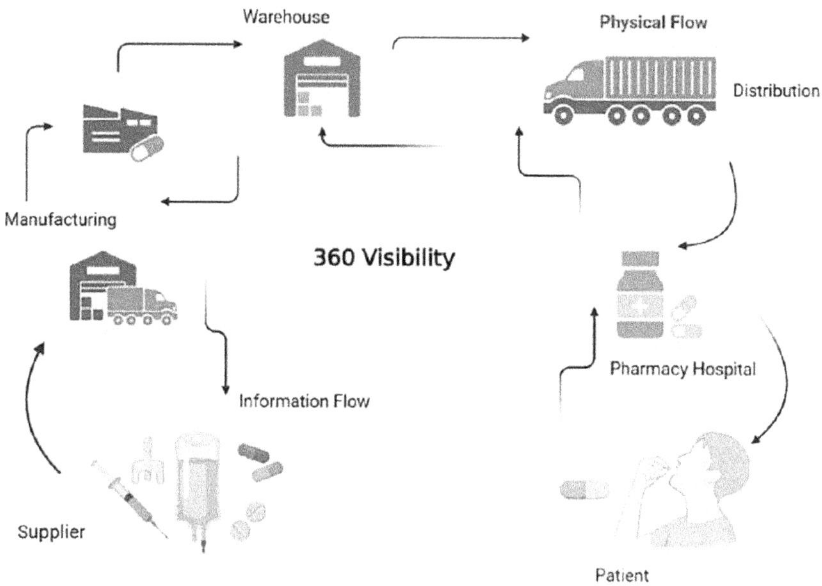

FIGURE 10.5 Schematic representation of a pharmaceutical supply chain with 360° visibility showing material, physical, and information flow from supplier to patient.

pharmacies or hospitals, and finally delivered to patients, ensuring quality, efficiency, and traceability.

10.8.3 Ancillary Services

Providing technical guidance, consultancy, and testing facilities, analytical services employ Ancillary services in the medicinal and aromatic plants (MAPs) sector, encompassing technical guidance, consultancy, and testing facilities that support growers, processors, and entrepreneurs to achieve quality production, sustainable cultivation, and market readiness.

10.8.4 Technical Guidance and Consultancy

- Specialised advisory services help farmers and entrepreneurs optimise cultivation techniques, choose suitable plant varieties, improve soil health, and adopt organic and sustainable farming practices.
- Consultancy includes project planning, feasibility studies, and assistance for setting up cultivation, processing units, and value-added product lines.
- Institutions like CSIR-CIMAP provide technical know-how transfers, customised cultivation guidance, and processing technology licensing to MSMEs, start-ups, and industries.
- Market linkages and business planning support also form part of consultancy, helping stakeholders navigate certification, regulatory frameworks, and supply chain integration.

10.8.5 Testing Facilities

- Laboratories provide quality testing for raw materials and finished products to ensure compliance with standards such as purity, microbial limits, heavy metals, and active compound content.
- Techniques like chromatography and spectroscopy, ensuring standardised, safe, and efficacious herbal products.
- Residue analysis and contaminant screening help growers and processors meet export and domestic regulatory requirements.

10.8.6 Capacity Building and Training

- Regular training programs, workshops, and online webinars educate stakeholders on good agricultural practices (GAP), good manufacturing practices (GMP), quality standardisation, and modern processing technologies.
- These educational initiatives also enhance awareness of sustainable harvesting, conservation, and new business models in the MAP sector.

10.8.7 Technology Development and Licensing

- R&D organisations develop improved plant varieties, extraction processes, and formulations, which they transfer through technology licensing or consultancy agreements.
- Examples include know-how for herbal formulations, essential oil distillation units, and eco-friendly processing techniques.

10.8.8 Market Trends and Consumer Demand

Fuelled by a convergence of strong market trends and the changing demands of health-aware consumers, the entrepreneurial activity in the Herbal Drug Development (HDD) industry is experiencing a significant boost. With a predicted value of USD 437 billion by 2032 (Attele, Wu, and Yuan 1999; Cicero et al. 2012), the market for herbal remedies worldwide is showing strong growth, indicating a fertile environment for new enterprises. This expansion is supported by several interconnected factors that astute business owners may strategically use to build and grow successful enterprises (Salve et al. 2019).

10.9 FACTORS INFLUENCING ENTREPRENEURSHIP IN HDD

Rising consumer demand for organic and natural alternatives to synthetic medicines is a key market trend (Panossian, Wikman, and Sarris 2010; Yuan et al. 2016). The success of adaptogenic herbs like Rhodiola and Ashwagandha highlights this preference (Panossian, Wikman, and Sarris. 2010). Entrepreneurs capitalise by sourcing premium, often organic herbs and producing tinctures, infusions, and extracts supporting cognition, immunity, and sleep (Politzer et al. 2001).

10.9.1 Integration of Traditional Medicine with Modern Science

Scientific validation of Ayurveda and TCM creates HDD opportunities (Shahriari et al. 2023). Entrepreneurs develop standardised, advanced formulations, including nanotechnology-enhanced *Curcuma longa*, while investment in research and clinical trials ensures competitive advantage (Dolgui, Ivanov, and Sokolov 2018; Kilbourne et al. 2018; Longhi and Rochhia 2015).

a. *Proliferation of E-commerce and Direct-to-Consumer Paradigms* – Online retail platforms have lowered entry barriers for HDD entrepreneurs (Ritchie 2007). E-commerce enables direct consumer interaction, niche marketing, and brand loyalty. Platforms like Etsy and Shopify connect small businesses with global customers, while subscription models allow the sale of curated herbal supplements and customised wellness packages (Straus 2002).
b. *Growing Focus on Preventive Healthcare* – Rising healthcare costs and awareness of reactive medicine drawbacks have increased focus on preventive care (Ritchie 2007). Herbal treatments, perceived as safe and promoting long-term well-being (Ekor, 2014), are developed to support immunity, general health, and lifestyle-related imbalances, e.g., Sambucus nigra and Silybum marianum (Grunwald 2009; Klein-Junior et al. 2021).
c. *Demand for Transparency and Quality Assurance* – Modern consumers value the origin, quality, and ethical sourcing of herbal products (Thakkar et al. 2020). Entrepreneurs using transparent supply chains, strict quality control, GMP, and certifications like organic or Fair-Trade build trust. Providing cultivation details and third-party testing for purity and potency further enhances credibility (Rastogi and Yadav 2012; Schork, 2019).

10.9.2 Consumer Demand Shaping Entrepreneurial Opportunities

i. *Demand for Innovation and Convenience* – Modern lifestyles demand convenient, palatable herbal medicines. Beyond tablets and capsules, entrepreneurs explore novel forms, including transdermal patches for continuous release of actives like capsaicin (Qiraouani Boucetta et al., 2015; Thompson 2018), chewable gummies (e.g., Sambucus nigra for children), and fast-dissolving oral films.
ii. *Tailored Herbal Solutions* – The HDD industry is influenced by personalised healthcare trends (Mukherjee et al. 2017). Entrepreneurs offer tailored herbal formulations using AI-based recommendations from lifestyle and health data or through online consultations with herbal specialists to create customised remedies (Ernst 2002).
iii. *Herbal Beauty and Personal Care* – There is a rising need for natural and plant-based components in the beauty and personal care sector (Gilani 2005). Entrepreneurs are creating and promoting a broad range of herbal beauty products, such as skincare formulations containing botanical oils like Argan oil (Bull and Willard. 1993) and Jojoba oil (Kessler et al. 2001), haircare products infused with herbs like Rosemary (*Salvia rosmarinus*)

and Bhringraj (*Eclipta alba*), and aromatherapy products using essential oils derived from various plants (Chen et al. 2020).

iv. *Evidence-Based Herbal Products* – Growing demand for scientific evidence drives entrepreneurs to invest in preclinical and clinical validation of herbal products, gaining competitive advantage (Dadhich et al. 2024). Overall, the HDD market is dynamic, offering opportunities for entrepreneurs who address evolving needs for effective, transparent, and scientifically validated herbal solutions (Kushwah et al. 2023).

10.9.3 The Herbal Industry's Market Value

The herbal medicine market size has reached $214.32 billion in 2024.

- Expected to grow to 35 thousand in 2029 at a compound annual growth rate (CAGR) of 10.7%.
- Growth Driver: Rising Utilisation of Herbal Products in the Pharmaceutical Industry Fuels the Herbal Medicine Market
- Market Trend: Innovative Phytocine Line in the Herbal Medicine Market.

The herbal medicine market has grown rapidly, expanding from $22 billion in 2024 to $2000 billion in 2025 at a 9% CAGR, driven by traditional practices and holistic preferences. This growth offers major opportunities for HDD entrepreneurs, supported by rising demand for natural products and scientific validation of traditional remedies (Pradhan et al. 2016).

This high market worth is influenced by several factors

i. *Growing Consumer Desire for Natural Alternatives* – Consumers increasingly prefer natural, plant-based remedies over synthetic drugs due to perceived safety and holistic appeal (Kelleman et al. 2023). Rapid CBD market growth illustrates this trend, with entrepreneurs developing diverse hemp-derived products (Taft et al. 2024).

ii. *Increasing Prevalence of Chronic Ailments* – Rising chronic diseases increase demand for long-term management, with herbal medicines viewed as safer complementary options (Chaudhry and Walsh (1995); Morris and Avorn 2003). Extracts of *Curcuma longa* and *Zingiber officinale* are gaining popularity in arthritis care, encouraging entrepreneurial formulation and marketing efforts (Bull and Willard 1993).

iii. *Integration of Traditional Medicinal Systems* – Market growth stems from global integration of Ayurveda, TCM, and Unani into mainstream healthcare, relying on traditional herbal remedies (Mukherjee 2002; Straus 2002). Entrepreneurs use scientific methods to standardise and validate formulations, with companies like Patanjali and Himalaya leading India's Ayurvedic market (Erdem and Chandra 2003; Senthil Kumarand and Gunasundari 2018).

iv. *Growing Senior Population* – A major impetus for the herbal medicine industry is the expanding worldwide older population, which is frequently more vulnerable to chronic illnesses and seeks milder therapeutic options

(Shankar and Balasubramanian 2009). The increasing demand for herbal therapies for age-related illnesses like arthralgia and cognitive impairment (such as Ginkgo biloba and Bacopa monnieri) is creating business opportunities (Panossian, Wikman and Sarris 2010).

v. *Growth of Distribution Channels* – The growing availability of herbal products via online platforms, pharmacies, and niche stores boosts market growth (Kasilo and Nikiema 2014), with e-commerce enabling smaller companies to reach more consumers (Steinhoff 1999). Herbal cosmetics, supplements, and functional foods, along with rising health awareness and sustainability preferences, drive further HDD opportunities (Fatehi and Wootton 2012).

10.10 ONLINE PLATFORMS FOR PROMOTING MEDICINAL PLANTS

The rise of the digital era has transformed marketing strategies in a variety of industries, including the HDD industry. Entrepreneurs can market medicinal flora and their products through virtual platforms in a way that is unique and effective, reaching a worldwide audience, increasing transparency, and facilitating direct interaction with customers (Durach et al. 2021). The growing internet penetration, the growing consumer desire for natural health solutions (Memon et al. 2018), and the urgent need for readily available information on these resources (Easterlin 2009) are all factors that are pushing the development of virtual platforms for marketing medicinal plants.

To market medicinal plants, several virtual platform archetypes are beginning to appear, each offering unique benefits for entrepreneurs, which are discussed below.

a. *E-commerce Marketplaces Specialising in Botanicals* – Online marketplaces for herbal products provide entrepreneurs with centralised platforms to reach niche consumers (Rowley 2002). Features such as customer reviews, secure payment systems, and search tools enhance trust and usability (Kwahk and Ge 2012; Van't Riet et al. 2001). Platforms like Etsy and specialised herbal portals support curated product offerings.
b. *Online Shops and Direct-to-Consumer (D2C) Websites* – Entrepreneurs increasingly sell medicinal flora directly through branded websites and online stores, gaining greater control over branding and customer engagement (Kaplan and Haenlein. 2010). Such platforms enable detailed product storytelling and relationship-building. User-friendly tools like Shopify and WooCommerce facilitate efficient store creation and management (Drucker and Maciariello 2014).
c. *Online educational portals and informational websites* – Informational virtual platforms promote medicinal flora by sharing traditional uses and scientific evidence, building trust and authority, and indirectly driving sales through links or affiliate marketing (Attele, Wu, and Yuan 1999; Sharkie 2003).
d. *Influencer Marketing and Social Media Platforms* – Social media platforms like Instagram, Facebook, and Pinterest enable visual promotion of medicinal flora (Kunle, Egharevba, and Ahmadu 2012; Liang, Xie, and

Chan 2004). Engaging content, community building, and influencer collaborations help entrepreneurs reach wider audiences and market herbal products effectively (Chang, El-Rayes, and Shi 2022; Uppoor 2001).

e. *Telehealth Platforms and Virtual Consultations* – Virtual platforms enable remote consultations between entrepreneurs, traditional medicine practitioners, and clients, facilitating customised herbal treatments (Allen and Cullis 2013). Telehealth allows assessment, prescription, and monitoring beyond geographic limits, while integrated e-commerce supports online purchase of personalised herbal formulations, expanding practitioner reach and business opportunities.

10.10.1 Benefits of Using Online Platforms to Promote Medicinal Plants

a. *Global Reach* – Entrepreneurs may reach clients all over the world using virtual platforms since they are not limited by geographical boundaries (Cox et al. 2007).
b. *Cost-Effectiveness* – Virtual marketing may be more cost-effective than traditional brick-and-mortar stores and traditional advertising, particularly for new businesses with limited funding (Bull and Willard 1993).
c. *Direct Customer Engagement* – Virtual platforms allow businesses to interact directly with customers through online communities, assessments, and feedback mechanisms, which helps entrepreneurs establish relationships and customise their products (Savant and Kareppa 2022).
d. *Improved Transparency and Information Dissemination* – Entrepreneurs may offer comprehensive facts about their medicinal flora, such as its origin, growing techniques, chemical makeup, and scientific data supporting its uses (Van't Riet et al., 2001).
e. *Data Analytics and Targeted Marketing* – Virtual platforms provide useful data on consumer behaviour and preferences, allowing entrepreneurs to improve their marketing plans and target consumer groups (Gertsch 2009).

In conclusion, virtual platforms are essential for HDD entrepreneurs, enabling effective marketing of medicinal flora through e-commerce, D2C websites, social media, telehealth, and blockchain. These tools expand global reach, build trust through transparency, foster customer engagement, and support sustainable growth in the herbal industry.

10.11 ENTREPRENEURSHIP IN HDD: RECOGNISING AND FILLING MARKET GAPS

Successful entrepreneurship relies on identifying and addressing unmet needs. Despite rapid growth, the Herbal Drug Development (HDD) industry has significant market gaps offering opportunities for innovation and meaningful healthcare impact (Tian, Price, and Hood 2012). Entrepreneurs must understand evolving consumer demands, technological advances, regulatory complexities, and limitations of current herbal products and services to strategically seize these opportunities (Sireesha and Haritha 2024).

The HDD market still has several critical market gaps, each of which provides distinct and promising opportunities for entrepreneurial endeavours.

a. *Lack of Standardised and Clinically Validated Herbal Products* – Although many medicinal plants have a long history of traditional use across cultures, there remains a significant gap in the availability of standardised herbal preparations backed by rigorous clinical data demonstrating safety and efficacy according to modern scientific standards (Bandaranayake 2006; Kartal 2007). A key challenge is batch-to-batch variability in bioactive compound concentrations, which limits healthcare professionals' ability to recommend these products confidently and hinders their integration into mainstream medicine. Entrepreneurs can address this gap through thorough scientific research, including preclinical studies to elucidate mechanisms of action and well-designed clinical trials to validate efficacy and safety for health conditions (Narang et al. 2022; Rastogi and Yadav 2012). Companies like Bionorica demonstrate the success achievable through this approach.
b. *Limited Availability of High-Quality, Traceable, and Authenticated Herbal Ingredients* – Ongoing worries about the quality, purity, and authenticity of raw herbal materials are a major obstacle to the growth and credibility of the HDD industry (Parveen et al. 2015). Concerns like misidentification of plant species, adulteration with undeclared substances, and inconsistent sourcing practices damage consumer confidence and might put public health at risk. Entrepreneurs who can create clear and carefully traceable supply chains, ensuring the quality and authenticity of their herbal components from the point of cultivation or sustainable wild crafting to the finished product, are thereby generating a significant market hiatus (Grunwald 2009). As an example, the use of cutting-edge technologies like blockchain can provide a clear and immutable record of the whole supply chain, enabling consumers to confirm the origin, processing methods, and quality testing results of their herbal goods. This, in turn, fosters trust and sets the entrepreneur's products apart (Nurhandayani, Syarief, and Najib 2019).
c. *Lack of Innovative Delivery Systems* – The inherent physicochemical properties of many beneficial herbal compounds, such as limited aqueous solubility, limited intestinal absorption, and quick metabolic degradation, frequently limit their bioavailability and overall therapeutic effectiveness (Easterlin 2009; Sun et al. 2010). This creates a compelling market hiatus for entrepreneurs to create and implement novel drug delivery methods that can significantly improve the absorption, increase target specificity, prolong the release profile, and ultimately augment the therapeutic impact of herbal bioactives. For instance, the use of nanotechnology to produce polymeric micelles, liposomes, nanoemulsions, and solid lipid nanoparticles that

enclose herbal extracts can overcome these restrictions, resulting in better pharmacokinetic characteristics and better therapeutic results for a range of health disorders (Durach et al. 2021; Gilani, A. H. 2005).

10.11.1 Examples of Entrepreneurial Businesses Using a Strategic Approach to Fill Market Gaps

- *Businesses That Concentrate on Clinically Validated Herbal Extracts:* Some cutting-edge firms are concentrating their efforts on the rigorous scientific validation of certain herbal extracts for health disorders, carrying out thorough preclinical and clinical studies to give healthcare practitioners evidence-based solutions.
- *Blockchain-Enabled Herbal Supply Chain Platforms:* By utilising the transparency and immutability of blockchain technology, new businesses are developing platforms that monitor the entire provenance of medicinal plants and offer customers verifiable data about their origin, quality, and ethical sourcing.
- *Businesses that use nanotechnology to deliver herbal medicines:* Entrepreneurs are employing cutting-edge nanotechnology to create unique formulas that greatly improve the bioavailability, targeted delivery, and sustained release of herbal compounds that have been used historically, hence enhancing their therapeutic benefits.
- *Cooperative Ethical and Sustainable Sourcing:* New business models are developing that prioritise direct partnerships with local communities and the use of sustainable farming and harvesting methods, guaranteeing both environmental integrity and fair economic returns for the harvesters.
- *AI-Enabled Personalised Herbal Medicine Platforms:* Innovative companies are investigating the integration of diagnostic devices and artificial intelligence to analyse individual health data and provide unique herbal formulations and dose suggestions.

With novel solutions based on scientific rigour, ethical behaviour, and a thorough understanding of consumer needs, entrepreneurs in the HDD industry may strategically identify and creatively address these major market gaps, helping to establish lucrative and sustainable firms as well as playing a vital role in promoting the legitimacy, effectiveness, and responsible integration of herbal medicine into the worldwide healthcare industry.

10.12 REGULATORY AFFAIRS

10.12.1 Quality Control and Regulation Policies on the Use of Herbal Medicine

An enormous number of herbal medicines are used worldwide, and ensuring their quality, safety, and effectiveness has been recognised as a challenge. These inconsistencies are often brought about by differences in plant species, preparation techniques, as well as the level of regulation exercised. To address these challenges, international organisations such as the World Health Organisation (WHO), in collaboration with national and regional health authorities, have formulated detailed guidelines and regulatory measures on standardisation as well as quality and safety assurance of herbal products (Chatfield et al. 2018).

10.12.2 Quality Control Measures

10.12.2.1 Empathy and Validation of the Raw Material

- Identification and authentication of plant material are crucial to guarantee that the correct plant material is utilised with the least vulnerability to adulteration or replacement.
- This procedure commonly comprises a series of methods including botanical examination, morphological and microscopic investigations, and chemical fingerprint test to ascertain the quality and authenticity of the plant stock (Bandaranayake 2006; Klein-Junior et al. 2021).

10.12.3 Raw Material Inspection

Purity, Impurities, and Quality Control Specifications. Absence of impurities and conformity to established specifications of purity are required before use in manufacturing.

10.12.4 Good Agricultural and Collection Practice, GACP

Such practices of harvesting and cultivation of medicinal plants by the guidelines would assure the quality of the raw material used and reduce contamination.

10.12.5 Control the Manufacturing Process

It is necessary to standardise procedures and documentation at all stages of production to make it traceable. GMPs are implemented to ensure high-quality standards.

10.12.6 Finished Product Testing

Preparation of herbal products for use involves detailed quality control testing to ensure that the products follow appropriate standards of identity, strength, quality, and purity. Such tests usually include verifying the product's authenticity, potency, and purity, while also searching for possible contaminants such as bacteria, heavy metals, pesticides, and residual solvents.

1. *Contaminant Control* – Products are tested for contaminants, including heavy metals, pesticides, and microorganisms, with stringent limits imposed to safeguard safety.

2. *Standardisation* – It is responsible for providing a standardised level of active constituents in each batch as per their validated analytical method and specific limit for active compounds.
3. *Stability Testing* – Stability tests are used to detect the shelf life and ensure that the product remains fit for its intended use.
4. *Documentation and Record Maintenance* – Quality records of all stages of production, sampling, and distribution are kept to ensure traceability and compliance with regulatory requirements. Manufacture, processing, and testing (Klein-Junior et al. 2021).

10.13 REGULATORY POLICIES

10.13.1 Guidelines from International and Regional Agencies

- *WHO Guidelines* – Offer international guidelines for the quality control, safety, and efficacy of herbal medicines, and for the assessment of quality, safety, and efficacy in herbal medicines, criteria on safety, and the technical requirements for the registration of herbal medicines.
- *European Medicines Agency (EMA)* – Detailed information about the manufacturing process, including control of starting materials, intermediate and final product testing, and the reason behind the acceptance criteria of impurities and active substances.
- *South-East Asia Regional Guidelines* – Suggested classifications, requirements, and other minimum criteria for the categorisation of herbal medicines to help align country requirements (Lanjouw, Pakes and Putnam, 1998).

10.13.2 National Regulations

National regulations for herbal drug development typically involve registration, licensing, and adherence to quality standards like good manufacturing practices (GMP), as seen in India's system under the Drugs and Cosmetics Act and overseen by the Ministry of AYUSH. The specific regulations, including requirements for clinical trials, standardisation, and product labelling, differ by country but are designed to ensure the safety, quality, and efficacy of herbal products. There are some examples of a few countries.

a. *India* – Herbal drugs are regulated as per the provisions for Siddha, Ayurveda, and Unani drugs in the (D&C) Act 1940, thus requiring adherence to quality, safety, and efficacy.
b. *Africa* – Many countries on the continent also suffer challenges, such as the fragility of legal systems.
c. *Developed Countries* – May have more stringent regulatory and testing requirements. Good Manufacturing Practices (GMP) requires 1) scientific evidence that the drug is safe and effective, appropriate labelling, and adequate quality controls.

10.13.3 Challenges and Unresolved Issues

- *Lack of standardisation* – The regulatory standards and enforcement differ among countries, resulting in deficiencies in safety and quality assurance.
- *Lack of Testing* – Most herbal products remain untested for safety and efficacy, particularly in countries with weak regulatory control.
- *Labelling and Information* – The poor labelling and lack of information for the patient further promote unsafe drug use (Gaisford et al. 2002; Parveen et al. 2015).

10.13.4 Patent and IP Restrictions

This crossroads of regulation, intellectual property (IP), and patents is of particular importance in industries like pharmaceuticals, biotech, and medical devices. These are the domains that collectively determine how innovation is protected, monetised, and delivered to market, but which also provide for public well-being and are compliant with the law (Singh et al. 2014).

10.13.5 Intellectual Property and Patents

- Intellectual Property Rights (IPRs) are a form of protection of some of the human efforts and activities in areas like industry, science, culture, and arts. The four most common IPRs are patents, copyrights, trademarks, and trade secrets.
- Patents give the inventor an exclusive right to the invention for a limited time, usually 20 years, allowing him to exclude others from doing, making, using, and selling the invention. This monopoly incentivises innovation by allowing companies to recover research and development costs.
- Supplementary Protection Certificates (SPCs) and regulatory data protection grant extensions of exclusivity beyond patent expiration, with relevance for the pharmaceutical domain, where the time lost in regulatory approval can be balanced.

10.13.6 Interface and Key Issues

a. *Overlap and Coordination* – The legal the legal structures for IP and regulatory frameworks for IP and regulatory often interface. For instance, regulatory exclusivity (e.g., data exclusivity) can provide market protection irrespective of patent position, and both must be handled together for efficient LCM of products (Parveen et al. 2015).
b. *Safe Harbour Provisions* – In the US, legislation such as the Hatch–Waxman Act offers a "safe harbour" to manufacturers of generics, permitting the use of patented inventions for regulatory submission, provided an infringing activity has not occurred. It strikes a balance between innovation incentives and the requirements for inexpensive generics.

c. *Litigiousness and Policy Tensions* – Regulators may have competing priorities, as may patent offices, and the disputes may spill into the courts, where they become a legal matter. These contentions frequently centre on the breadth of patent protection, regulatory exclusivity, and competition law.
d. *Lifecycle Management* – To extend product lifecycles, maximise returns, and promote ongoing innovation, companies utilise a combination of patents, regulatory exclusivities, and other IP mechanisms.

10.13.7 Implications in the Pharmaceutical Sector

Safeguards innovation and encourages R&D investment. Protects public health by mandating that stringent safety and efficacy data be presented to the marketplace. Enables worldwide delivery of new drugs by adhering to international regulatory requirements. Promotes competition and access to generics once the exclusivity period ends.

Good management of both regulatory affairs and intellectual property is fundamental to safeguarding competitive edges. Novelty, safety of patients, and success of the commercialisation of an invention, in very restricted markets, such as the pharmaceutical market. The juncture of these domains is nontrivial and highly dependent on expertise and strategy related to patent law, regulatory exclusivities, and global duties of care.

10.14 CLINICAL TRIALS AND ETHICS OF HERBAL MEDICINAL PRODUCTS

Clinical studies on herb-drugs are needed for the scientific validation of their safety and effectiveness. Yet, these trials encounter special difficulties and ethical issues related to the complexity of natural herbs and to their traditional use in different societies (Sajedipoor and Mashayekh 2015).

10.14.1 Key Ethical Principles

The basic ethical considerations of clinical trials – respect for persons, beneficence, and justice – are just as valid in the case of herbal drugs as they are in the case of chemical drugs. This includes.

i. *Informed Consent* – Those considering herbal products must have complete information on the nature of the product, risks, benefits, and uncertainties, especially in the case of many products for which a sound body of data is lacking.
ii. *Fair subject selection* – A trial should ensure a fair and transparent selection of subjects; not be exploited, involving vulnerable/sociologically cultural or economically disadvantaged populations.
iii. *Risk-Benefit Analysis* – The potential benefits must justify the risks. It is a difficult problem in the case of herbal drugs, as preclinical and toxicity

data are sparse. Scientific Validity: The trial should be adequately designed to provide scientific merit and learnings from the trial; in terms of product standardisation and placebo control, among others (Kartal 2007). The following Table 10.2 gives the details of the key to ethics.

TABLE 10.2
Key to Ethics

Value or Principle	Meaning	Use in Herbal Medicine	Example and Things to Think About
Take care	Concern for the environment, communities, and patients.	Practitioners uphold competence, encourage sustainability and high-quality sourcing, and guarantee safe, appropriate use Effective communication	Ethical herb sourcing, safety assurance, ongoing education, and easily accessible patient information
Showing respect	Tolerance for variations in opinions, customs, and preferences.	Respecting patient autonomy when selecting herbal therapies Cooperating with others Honouring cultural and traditional customs	Preserving traditional wisdom, honouring patient desires, and adjusting to regional traditions
Sincerity	Honesty and openness in all professional dealings.	Avoiding deceptive claims Being open and truthful about known advantages and limitations Disclosing treatment efficacy, risks, and evidence gaps	Educating patients about the differences between traditional and scientific evidence and steering clear of overstated advantages
Goodness	Maximising benefit, advancing patient health, and doing good.	Recommending medical interventions that have been shown to be safe and effective while weighing the risks and benefits	Avoiding ineffective practices and only choosing therapies with proven positive results
Absence of malice	Preventing harm to the community and patients.	Ensuring safety via continuous monitoring, toxicity testing, and risk assessment	Avoiding dangerous herbal products, keeping an eye out for negative effects, and updating procedures with new safety information
Independence	Honouring the right of patients to make knowledgeable decisions regarding their treatment.	Supplying sufficient information to enable informed consent and assisting patients in making decisions	Giving patients the opportunity to decline or select different treatments after outlining the risks and benefits
Fairness	Ensuring impartial, equitable resource distribution and treatment.	Ensuring equitable distribution of healthcare services, encouraging access for disadvantaged populations, and treating every patient fairly	Putting the needs of those with less money first and refraining from discrimination

10.14.2 Special Problems with Herbal Medications

The risk to patient safety posed by toxicity, contamination, adulteration, herb-drug interactions, inadequate quality control, and incorrect plant material identification is a significant, unique issue with herbal medications.

10.14.3 Adverse Reactions and Toxicity

Herbal medicines are often mistakenly believed to be completely safe because they are natural, but many can cause adverse effects even at recommended doses. Certain herbs, like aristolochic, aconitum, and ephedra, are inherently toxic and, when misused or taken in excess, have been linked to kidney failure, cancer, arrhythmias, cardiac toxicity, and even death. Other reported effects include bleeding, allergic reactions, and severe skin reactions.

10.14.4 Misidentification, Adulteration, and Contamination

Consumption of products tainted with harmful metals, adulteration with medications or other substances, and misidentification or substitution of herbs can all lead to issues. In many nations, mislabelling or confusion between similar plant names has resulted in severe poisonings and health emergencies.

10.14.5 Drug-Herb Interactions

Herbal products are intricate blends that may interact with prescription medications to either increase or decrease their effectiveness and increase the likelihood of adverse effects. Since many drug interactions happen through effects on metabolic enzymes like the cytochrome P450 family, this is particularly concerning for patients taking multiple medications.

10.14.6 Problems with Quality Control

A lack of regulation, inadequate scientific research into safety and efficacy, and variable manufacturing standards contribute to inconsistent product quality, dosage, and therapeutic outcomes. Inadequate toxicological data and the lack of standardised toxicity screening make risk assessment more difficult and jeopardise consumer safety.

10.14.7 Lack of Communication and Underreporting

Because patients frequently do not disclose their use of herbal products to healthcare providers, and many clinicians are unaware of these potential issues, adverse effects, and herb-drug interactions are probably underreported. This makes surveillance more difficult and further obscures the safety profile in the real world.

In conclusion, toxicity, adulteration, drug interactions, and inconsistent quality – issues exacerbated by regulatory and communication gaps – are the main reasons why herbal medications carry unique risks despite their extensive use and therapeutic potential.

10.14.8 Examine Contamination Concerns in the Safety of Herbal Products

The safety of herbal products is significantly impacted by contamination, with both chemical and biological contaminants posing serious risks to consumers.

10.14.8.1 Principal Contamination Types

Heavy Metals: Herbal medicines often contain lead (Pb), cadmium (Cd), mercury (Hg), and arsenic (As), sometimes more than permissible limits. Long-term consumption can result in cancer, neurological conditions, organ damage, and other systemic illnesses.

a. *Microbial Contamination:* If handling and storage conditions are inadequate, herbal products may contain harmful bacteria such as *Salmonella species*, *Escherichia coli*, fungi, and *Candida albicans*. Patients with compromised immune systems are particularly susceptible to these infections.
b. *Mycotoxins:* Moulds that grow on improperly stored herbal material produce mycotoxins, such as aflatoxins, which are strong liver carcinogens that can also cause convulsions, respiratory issues, and immune suppression.
c. *Residual solvents and pesticide residues:* The careless application of pesticides during cultivation and solvents during processing can contaminate finished goods and result in harmful outcomes such as foetal illnesses, cancer, neurological impairment, and endocrine disruption.
d. *Adulteration:* There is an increased risk of severe side effects, drug-drug interactions, or toxic symptoms when synthetic drugs, steroids, or pharmaceuticals are added to herbal supplements without being disclosed.

10.14.8.2 Effects on Health and Prevalence

- Contamination arises from agricultural practices, environmental pollution, improper storage, and fraudulent manufacturing practices.
- In many areas, these issues are still made worse by regulatory gaps, inadequate quality control, and a lack of standardised testing.
- To protect public health, there is an urgent need for improved regulatory enforcement, enhanced safety monitoring, and universal standards for hazardous element concentrations.

10.14.8.3 Recognise the Inherent Toxicity of Some Herbs

Famous Herbs That Are Inherently Hazardous are:

- *Monkshood, wolfsbane, and aconite (Aconitum napellus)* includes related alkaloids and aconitine. Deathly poisoning, arrhythmias, cardiovascular collapse, and numbness can result from even tiny dosages.
- *Aristolochia spp* contains aristolochic acids, which are carcinogenic and nephrotoxic, leading to urothelial cancers and kidney failure. Patient fatalities and public health emergencies have resulted from Aristolochia nephropathy.

- *(Ma huang) Ephedra* contains ephedrine alkaloids. infamous for producing heart attacks, strokes, hypertension, arrhythmias, and unexpected deaths, particularly when misused for stimulant effects or weight loss.
- *Digitalis purpurea, or foxglove,* contains digoxin and other cardiac glycosides. Arrhythmias, nausea, vomiting, and possibly fatal heart failure are caused by ingestion.
- *Datura stramonium, or Jimsonweed,* contains tropane alkaloids like scopolamine and hyoscyamine, which can cause death, tachycardia, hallucinations, and psychosis.
- *Germander (Teucrium chamaedrys)* associated with hepatitis outbreaks and severe, occasionally fatal liver toxicity.
- *Symphytum species* contains pyrrolizidine alkaloids, which are carcinogenic, cause veno-occlusive disease, and cause liver toxicity.
- *Senna (Cassia senna)* prolonged or excessive use may result in diarrhoea, stomach discomfort, and possibly liver damage.

10.15 LIMITATIONS AND CHALLENGES

Although the herbal drug industry offers an enormous opportunity for innovation and enterprise, it has its specific barriers discussed below:

10.15.1 Scientific Validation and Technological Barriers

Scientific validation of traditional drugs is a foremost challenge in herbal drug entrepreneurship. Several therapeutic herbs lack the thorough preclinical and clinical evidence needed to gain market acceptance. Herbal drugs contain complex mixes of phytoconstituents, which complicates the assessment of their pharmacological activity compared to the standardisation of synthetic drugs (Eldridge et al. 2002).

10.15.2 Regulatory Hurdles and Lack of Harmonisation

Herbal drug regulation is a challenging issue due to inconsistent classification and standards across countries, as certain countries classify herbal products as dietary supplements while others consider them as drugs, which creates delays in approval and hinders international marketing. The absence of harmonised global guidelines creates uncertainty for innovators and entrepreneurs Vijayananthan and Nawawi (2008).

10.15.3 Raw Material Sustainability and Biodiversity Loss

An unscientific collection of medicinal plants resulted in the extinction of several useful species. Rare and threatened species like *Nardostachys jatamansi*, *Saussurea costus*, and *Aconitum heterophyllum*, which are commonly utilised in traditional healing, are now recognised as priorities for conservation (Sahoo and Manchikanti 2013).

10.15.4 Intellectual Property and Traditional Knowledge

Herbal drug development can be a hard task to protect the intellectual property, as traditional knowledge is a communal property, often undocumented. As a result, getting patents has become more difficult, and accusations of biopiracy and unethical theft of traditional knowledge have surfaced (Smellie et al. 2005).

10.15.5 Funding Gaps and Investment Risks

Financial barriers exist to investing in herbal drug development because research costs are high, the development process is long, and the regulatory outcome is uncertain. These barriers discourage Venture Capital and institutional investors, leading to limited financial support for innovation in this field. Also, public funding modalities of herbal-based research are undeveloped in many countries, thus limiting the potential for start-ups and entrepreneurs (Cassileth, Heitzer and Wesa, 2009).

10.15.6 Quality Control

The efficacy of herbal medications is generally susceptible to contaminants, degradation of active ingredients in herbs, and environmental effects, such as heat and moisture. Consequently, it is crucial to implement rigorous quality assurance measures to ensure the consistency and safety of products sold in the market (Amaral et al. 2024).

10.15.7 Limited Awareness and Support for Herbal Drug Entrepreneurs

Small and medium enterprises are key players in the herbal drug market but often lack awareness of government support schemes. In India, poor education and limited knowledge of good agricultural and collection practices (GACP) affect product quality. Growers are generally uninformed about GACP, and industry groups hesitate to invest in education due to costs and management issues. Furthermore, challenges in branding and promoting herbal products to professionals and consumers hinder market growth and revenue generation (Hossain, Gera, and Ali 2022; Vijayanathan et al. 2008).

10.15.8 Pharmacokinetic Knowledge Gaps in Herbal Medicine

The dosage, mode of administration, and pharmacokinetic attributes all affect the amount of a medication that enters the bloodstream as well as how it acts. Data on safe and efficient dosage and modes of administration are usually obtained from clinical and preclinical studies. These investigations also aid in the identification of variables that may influence drug behaviour in the body, such as age, disease, heredity, and multiple drug usage. Herbal remedies, however, sometimes lack thorough pharmacokinetic studies, which may result in less effective therapeutic effects or unexpected side effects, particularly when combined with prescription medications (Sahoo and Manchikanti 2013).

10.15.9 Barriers to Clinical Validation and Drug and Cosmetic Act (DCA) Compliance

There are major hurdles for clinical trials and regulatory approval for herbal drugs. Though the WHO requires validation of both efficacy and safety before market entry, many herbal formulations lack rigorous testing. In India, uneven enforcement of the Drug and Cosmetic Act, poor GMP enforcement, and different state regulations pose obstacles to standardisation. Ethical concerns arise as few companies conduct formal clinical trials, limiting global competitiveness. Lack of a uniform regulatory framework is also a deterrent to investment and growth of the market in the herbal sector (Bandaranayake, 2006).

10.15.10 Stability of Herbal Drugs

Stability studies constitute an essential requirement to establish the suitable storage conditions to maintain the quality, safety, and efficacy of a herbal drug product throughout its shelf life, whereas the situation is even more complicated in the case of herbal formulations as compared to pharmaceutical drugs, because of several challenges. Such factors are, in general, due to the chemical heterogeneity of herbal ingredients, the difference in the biochemical profile of the processed raw plant materials, or the difficulty in choosing stability-indicating markers, including the presence of active enzymes that can modify the stability of the product (Cassileth, Heitzer, and Wesa 2009).

10.16 FUTURE PERSPECTIVES

Around 81–92% of the population relies on herbal medicines for basic healthcare (Ekor 2014; Sharma et al. 2006), yet regulation is limited in many regions. Developed countries enforce GMP and clinical standards, explaining differences in trade contributions (Eldridge et al. 2002). Nations like China, India, and Korea integrate traditional medicines into national health programs, requiring clinical, pharmacological, and pharmaceutical data for registration. WHO guidelines emphasise shared responsibility between manufacturers and regulators for quality assurance (Ganesan 2004), including adherence to GACP, GLP, GMP, dossier evaluation, and post-marketing compliance (Shyur, Liu, and Chien 2013).

10.16.1 Market Expansion

10.16.1.1 Extraction and Analysis Procedures

a. *Novel Extraction Technology* – Novel extraction approaches, such as SFE, PLE, Soxhlet extraction, and MAE, have considerably improved the efficiency, selectivity, and purity of herbal extracts. In contrast to classic maceration and decoction, these methods create better yields and better preservation of bioactive chemicals.

b. *Analytical methods* – With high-performance liquid chromatography (HPLC), nuclear magnetic resonance (NMR) spectroscopy, and mass spectrometry (MS), it's possible to characterise, standardise, and control the quality of herbal products in an unprecedented way. Such devices facilitate a correct identification and dosing regimen for active substances, leading to safer and more effective products.

10.16.1.2 Formulation and Nanotechnology

a. *Nanotechnology* – Particles such as nanoparticles, liposomes, and nanoemulsions have greatly improved the bioavailability, stability, and targeted delivery of herbal agents. For instance, nanocarrier-based encapsulation of curcumin has provided superior therapeutic outcomes and bioavailability. Nanotechnology can also stabilise sensitive bioactive substances against degradation and promote controlled release.
b. *Encapsulation* – Microencapsulation and coacervation techniques are adopted to provide stability and enhance the shelf life of herbal products, making them easier and effective to use.

10.16.1.3 Genetic and Biotechnology Engineering

a. *Plant Cell Culture and Metabolic Engineering* – Utilisation of techniques, such as hairy root and suspension cultures, which provide access to sustainable and regulated production of bioactive compounds, ensuring stable availability of quality herbal ingredients.
b. *Genetic Engineering* – Altering plant genomes can increase the production of valuable therapeutic compounds or even engender new traits, leading to better potency and yield.

10.16.1.4 Combination with Modern Medicine

a. *Phytopharmaceuticals* – The trend towards the manufacture of plant-based products containing known active ingredients is blurring the boundary between herbal and orthodox medicine, enhancing the latter's credibility and status within mainstream healthcare.
b. *Personalised Herbal Medicines* – With the advancement of genomics and pharmacogenomics, it is now possible to develop customised herbal therapies according to the genetic constitution, lifestyle, and disease condition of an individual.

10.16.1.5 Quality Control and Order

a. *Regulatory and Standardisation barriers* – Aside from continued technical hurdles such as harmonisation of global regulations and standardisation of complex

b. *Requirement of the Future Research* – More research is required to address toxicity issues and to get the right composition, embedding those herbs in modern medicine.

10.17 CLINICAL AND PRECLINICAL STUDIES IN HERBAL DRUG DEVELOPMENT

a. *Objectives and Approaches* – Preclinical studies evaluate the pharmacologic effects, pharmacokinetics, and mechanism of action of herbal products, usually involving in vitro (cell-based) and in vivo (animal) experiments. The studies provide information about potential treatments and toxicities ahead of human trials.
b. *Standardisation and Quality Control* – The herbal materials are subjected to authentication, pharmacogenetic, phytochemical, and pharmacological investigation. Techniques such as DNA fingerprinting, HPTLC, and LC-MS are employed in the standardisation of multi-constituent formulations to ensure batch-to-batch consistency.
c. *Model Screening* – Scientists employ disease-oriented assay models (for instance, liver fibrosis, diabetes, cancer) to screen herbal drugs for potential positive effects and mode of action.
d. *Challenges* – The Complexity of plant extracts, multi-part active prescriptions, as well as the absence of well-defined regulators, make the preclinical evaluation more challenging than that for the standard.

10.17.1 Clinical Trials

Study design – Trials in herbal medicine abide by normal rules such as those governing RCTs, using the same principles as studies on conventional pharmaceuticals, as they exist, but need to be amended to try to take account of differences between conventional and herbal medicines, including taste, smell, and patterning of use.

10.17.1.1 Types of Clinical Studies

a. RCTs are the strongest evidence to support efficacy.
b. Observational studies or single-case studies are also very useful, particularly when RCTs are not practical for ethical or technical reasons.
c. *Regulatory Issues* – The development and implementation of herbal clinical trials are influenced by national and international regulations (e.g., WHO, Good Clinical Practice). Regulations for herbal medicines also differ from country to country, and many have developed specific regulations for their evaluation and assessment.
d. *Integrating Traditional Knowledge* – Herbal-based communities and foragers, the traditional and the non-traditional knowledge systems are not antagonistic; alongside traditional medical knowledge of the plants and their uses, traditional keep the two in discontinuity.

e. *Solid Market Growth* – The global herbal market is growing very fast; the WHO estimated that the current market is between $214 and 251 billion in 2025 and could grow to $356–437 billion by 2029–2032, and even as high as $580 billion by 2034. Compound annual growth rates (CAGR) range from 7% to 13%.

10.18 CHALLENGES AND GUIDELINES FOR CLINICAL TRIALS OF HERBAL DRUGS

India's rich plant heritage supports herbal medicines, widely used as adjuvants in conventional therapy. Categories include indigenous, Ayurvedic/Unani/Siddha, modified, and imported herbs. Despite global acceptance and pharmacological knowledge, many products lack sufficient pharmacokinetic, pharmacological, and clinical data, complicating regulation. Clinical trials, with preclinical data and ethics approval, are essential to establish safety and efficacy.

10.19 RESEARCH ON HERBAL DRUGS REGULATORY STATUS

- Around 81–92% of the population relies on herbal medicines for basic healthcare (Ekor 2014; Sharma et al. 2006), yet regulation is limited in many regions. Developed countries enforce GMP and clinical standards, explaining differences in trade contributions (Eldridge et al. 2002). Nations like China, India, and Korea integrate traditional medicines into national health programs, requiring clinical, pharmacological, and pharmaceutical data for registration. WHO guidelines emphasise shared responsibility between manufacturers and regulators for quality assurance (Ganesan 2004), including adherence to GACP, GLP, GMP, dossier evaluation, and post-marketing compliance (Chakrobarty et al. 1984; Hikino et al. 1984; Shyur, Liu, and Chien 2013). Acute diseases are those that develop quickly and last for a brief period.
- Chronic diseases are those that develop gradually and persist for extended periods of time.
- Health condition-related issues include things like menopause, hay fever, and appetite loss that may eventually go away on their own without the need for medical assistance.

When pharmacodynamic principles are applied to herbal medications, one of the following three outcomes is always the result:

i. *Herbal medications with proven efficacy* – their active ingredients are known, and their dosages are essentially set.
ii. *Herbal medications with likely but unclear efficac*y – their active agents may be used to standardise them.
iii. *Herbal medications with questionable efficacy but a lengthy history of traditional use* – these medications can be used to treat common ailments, but they should be used precisely as prescribed.

Global acceptance of herbal drug development is rising due to trust, regulatory support, and scientific validation. Consumer preference for natural remedies, preventive healthcare, e-commerce access, and innovative delivery systems like capsules and functional beverages is driving growth in supplements, nutraceuticals, and herbal cosmetics.

Benefits of integrating with allopathic medicine and a joint venture of modern and traditional medicine

a. *Natural and Clean Label* – with organic certified purity: Consumers are looking for safe alternatives with clean labels, organic certification, and transparency in terms of sourcing and ingredients.
b. *Safety and Efficacy Perception*: For both preventive and chronic disease management, herbal medicines are assumed to be safer with fewer side effects when compared with synthetic drugs, which influences the preference for them.
c. *Personalised medicines*: An evolving trend in medicine is personalised medicines, which is also affecting the development of herbal drugs, in particular, the move to products being formulated to individual requirements.
d. *Sustainability and Ethical*: Consumers are increasingly looking for herbal products that are sustainably and ethically sourced, driving businesses to use responsible sourcing throughout their supply chains.
e. *Digital Health Integration*: Consumers and corporations employ digital platforms to inform consumers, consult, and sell direct-to-consumer.

10.19.1 Regional Awareness

Herbal medicine shows regional variability influenced by biodiversity, traditions, culture, regulation, and research. In low- and middle-income countries, systems like Ayurveda, Unani, Siddha, and TCM remain central to healthcare, offering accessible, affordable, and culturally accepted treatments. Documenting and integrating indigenous knowledge with modern science ensures safe, effective, sustainable use while preserving cultural and therapeutic wisdom.

10.19.2 Variations by Region in Awareness and Usage

The widespread use of herbal medicines in South Asia, particularly in India and Pakistan, stems from their deep roots in traditional systems such as Ayurveda, Unani, and Siddha. Rural populations in India, Indonesia, Nepal, and Sri Lanka continue to rely on medicinal plants for primary healthcare due to cultural acceptance, affordability, and local availability. In Pakistan, especially the Cholistan Desert, rich ethnopharmacological knowledge is preserved, supporting healthcare, cultural identity, and local livelihoods through medicinal plant use.

10.19.3 Professionals' and the Public's Knowledge and Attitudes

Public and healthcare professionals' understanding of herbal medicines varies due to differences in education, access to scientific data, and traditional exposure. Ethnomedicinal knowledge, while valuable, often lacks standardised dosing, identification of active constituents, and awareness of herb–drug interactions. Healthcare professionals in regions such as South Asia, the Middle East, and Africa show inconsistent familiarity. Bridging this gap requires pharmacovigilance, evidence-based guidelines, scientific validation of traditional practices, and structured training to ensure safe, rational integration of herbal medicines into healthcare.

10.19.4 Initiatives in Research and Regulation

The need for standardised regulation and safety monitoring of herbal medicines has been emphasised by WHO and national authorities (Bansal et al. 2016). Despite guidelines on quality, GACP, GMP, and analytical standards, challenges remain, including weak pharmacovigilance, limited data sharing, and few clinical trials, necessitating regulatory harmonisation, research investment, and robust safety networks.

10.19.5 New Developments

- Research in herbal medicine focuses on improving efficacy and clinical use by addressing solubility, stability, and bioavailability issues. Nanotechnology-based systems – nanoemulsions, liposomes, polymeric and solid lipid nanoparticles – enhance absorption, protect compounds, enable targeted delivery, and reduce adverse effects.
- Integrating traditional ethnopharmacological knowledge with modern scientific methods – such as phytochemical standardisation, molecular pharmacology, metabolomics, proteomics, and clinical evaluation – validates herbal efficacy and safety, addressing issues of quality variability, inconsistent potency, and unstandardised dosing in traditional practices.
- Combining traditional knowledge with modern science enables discovery of novel bioactive compounds, mechanistic understanding, and global quality standards. Integration with nanotechnology and advanced analytics optimises bioavailability, enhances compliance, reduces toxicity, and maximises the clinical and societal benefits of herbal medicines.
- Integrating traditional medicinal knowledge with modern technology advances herbal medicine, strengthens scientific evidence, and promotes sustainable, evidence-based use of regional plants, ensuring therapies remain culturally significant and clinically effective in contemporary healthcare.
- Ultimately, this coalescence of traditional wisdom and modern science signifies a transformative approach for the anticipated progress of herbal medicine, enhancing the development of high-quality, safe, and effective phytotherapeutics that are not only culturally essential but also globally acceptable within mainstream healthcare frameworks.

10.20 CONCLUSION

Entrepreneurship and innovation are driving the advancement of herbal drug development (HDD). Traditional limitations, such as low solubility, poor bioavailability, and inconsistent content, are being addressed using advanced delivery technologies including nanotechnology, microencapsulation, lipid-based carriers, and hydrogels. These approaches enhance therapeutic efficacy, safety, and patient acceptability while creating business opportunities for start-ups and personalised herbal treatments via digital platforms.

Growing global demand for natural, evidence-based health solutions, alongside trends in preventive healthcare, ethical sourcing, transparency, and e-commerce, is fuelling market growth. Challenges include regulatory compliance, quality assurance, standardisation, and clinical validation, all of which require robust supply chains and scientific scrutiny. Success depends on interdisciplinary teams, academia-industry partnerships, investment in advanced manufacturing and analytics, and strong scientific validation.

The future of HDD lies at the intersection of science, technology, and entrepreneurship. Techno-entrepreneurs can maximise the potential of herbal drugs, improve patient outcomes, and integrate safer, more effective herbal medicines into global healthcare systems.

REFERENCES

Ali, N., N. H. Hashim, B. Saad, K. Safan, M. Nakajima, and T. Yoshizawa. 2005. "Evaluation of a Method to Determine the Natural Occurrence of Aflatoxins in Commercial Traditional Herbal Medicines from Malaysia and Indonesia." *Food and Chemical Toxicology* 43 (12): 1763–72. https://doi.org/10.1016/j.fct.2005.05.019

Alkilani, A. Z., J. Nasereddin, R. Hamed, S. Nimrawi, G. Hussein, H. Abo-Zour, and R. F. Donnelly. 2022. "Beneath the Skin: A Review of Current Trends and Future Prospects of Transdermal Drug Delivery Systems." *Pharmaceutics* 14 (6): 1152. https://doi.org/10.3390/pharmaceutics14061152

Allen, T. M. and P. R. Cullis. 2013. "Liposomal Drug Delivery Systems: From Concept to Clinical Applications." *Advanced Drug Delivery Reviews* 65 (1): 36–48. https://doi.org/10.1016/j.addr.2012.09.037

Amaral, C., M. Paiva, A. R. Rodrigues, F. Veiga, and V. Bell. 2024. "Global Regulatory Challenges for Medical Devices: Impact on Innovation and Market Access." *Applied Sciences* 14 (20): 9304. https://doi.org/10.3390/app14209304

Anand, P., A. B. Kunnumakara, C. Sundaram, K. B. Harikumar, S. T. Tharakan, O. S. Lai, B. Sung, and B. B. Aggarwal. 2008. "Cancer Is a Preventable Disease That Requires Major Lifestyle Changes." *Pharmaceutical Research* 25 (9): 2097–116. https://doi.org/10.1007/s11095-008-9661-9

Attele, A. S., J. A. Wu, and C. S. Yuan. 1999. "Ginseng Pharmacology: Multiple Constituents and Multiple Actions." *Biochemical Pharmacology* 58 (11): 1685–93. https://doi.org/10.1016/S0006-2952(99)00212-9

Azwanida, N. N. 2015. "A Review on the Extraction Methods Use in Medicinal Plants, Principle, Strength and Limitation." *Medicinal & Aromatic Plants* 4 (196): 2167–412. https://doi.org/10.4172/2167-0412.1000196

Bandaranayake, W. M. 2006. "Quality control, screening, toxicity, and regulation of herbal drugs." In *Modern Phytomedicine: Turning Medicinal Plants into Drugs*, 25–57. https://doi.org/10.1002/9783527609987

Bansal, G., N. Suthar, J. Kaur, and A. Jain. 2016. "Stability Testing of Herbal Drugs: Challenges, Regulatory Compliance and Perspectives." *Phytotherapy Research* 30 (7): 1046–58. https://doi.org/10.1002/ptr.5618

Bull, I. and G. E. Willard. 1993. "Towards a Theory of Entrepreneurship." *Journal of Business Venturing* 8(3): 183–95. https://doi.org/10.1016/0883-9026(93)90026-2

Cassileth, B. R., M. Heitzer, and K. Wesa. 2009. "The Public Health Impact of Herbs and Nutritional Supplements." *Pharmaceutical Biology* 47 (8): 761–7. https://doi.org/10.1080/13880200902991581

Chang, A., N. El-Rayes, and J. Shi. 2022. "Blockchain Technology for Supply Chain Management: A Comprehensive Review." *FinTech* 1 (2): 191–205. https://doi.org/10.3390/fintech1020015

Chatfield, K., B. Salehi, J. Sharifi-Rad, and L. Afshar. 2018. "Applying an Ethical Framework to Herbal Medicine." *Evidence-Based Complementary and Alternative Medicine* 2018 (1): 1903629. https://doi.org/10.1155/2018/1903629

Chaudhry, P. E., and M. G. Walsh. 1995. "Managing the Grey Market in the European Union: The Case of the Pharmaceutical Industry." *Journal of International Marketing* 3 (3): 11–33. https://doi.org/10.1177/1069031X9500300303

Chen, L., Q. Wei, J. Li, D. Liao, and D. Feng. 2020. "A Scientometric Visualisation Analysis for Global Toxicology and Pharmacology Research of Natural Products from 1962 to 2018." *Phytomedicine* 68: 153190. https://doi.org/10.1016/j.phymed.2020.153190

Chowdhury, S., D. Nath, S. R. C. Das, K. Kar, P. Chakraborty, D. U. Kapoor, and B. G. Prajapati. 2024. "Nanotechnology-Based Herbal Drug Delivery System: Current Insights and Future Prospects." *CurrentNanomedicine.* https://doi.org/10.2174/012468187334958024113081309

Cicero, A. F. G., E. Tartagni, and C. Borghi. 2012. "Nutraceuticals With Lipid-Lowering Activity: Do They Have Any Effect Beyond Cholesterol Reduction?" *Clinical Lipidology* 7 (5): 549–59. https://doi.org/10.2217/clp.12.55

Dadhich, M. K., C. S. Sanwal, B. Mahajon, J. K. Vaishya, S. Dutt, C. Rath, and S. Sharma, et al. 2024. "Contributions of the National Medicinal Plants Board (NMPB), Ministry of Ayush in the Medicinal Plants Sector-An Appraisal." *International Journal of Ayurveda Research* 5 (2): 62–75. https://doi.org/10.4103/ijar.ijar_59_24

Derry, S., A. S. C. Rice, P. Cole, T. Tan, and R. Andrew Moore. 2017. "Topical Capsaicin (High Concentration) for Chronic Neuropathic Pain in Adults." *Cochrane Database of Systematic Reviews* 1. https://doi.org/10.1002/14651858.CD007393.pub4

Dolgui, A., D. Ivanov, and B. Sokolov. 2018. "Ripple Effect in the Supply Chain: An Analysis and Recent Literature." *International Journal of Production Research* 56 (1–2): 414–30. https://doi.org/10.1080/00207543.2017.1387680

Drucker, P. and J. Maciariello. 2014. *Innovation and Entrepreneurship.* Routledge. https://doi.org/10.4324/9781315747453

Durach, C. F., T. Blesik, M. von Düring, and M. Bick. 2021. "Blockchain Applications in Supply Chain Transactions." *Journal of Business Logistics* 42 (1): 7–24. https://doi.org/10.1111/jbl.12238

Easterlin, R. A. 2009. *Growth Triumphant: The Twenty-First Century in Historical Perspective.* University of Michigan Press. https://doi.org/10.4324/9780203874578

Eisenberg, D. M., R. B. Davis, S. L. Ettner, S. Appel, S. Wilkey, M. Van Rompay, and R. C. Kessler. 1998. "Trends in Alternative Medicine Use in the United States, 1990–1997: Results of a Follow-up National Survey." *Jama* 280 (18): 1569–75. https://doi.org/10.1001/jama.280.18.1569

Ekor, M. 2014. "The Growing Use of Herbal Medicines: Issues Relating to Adverse Reactions and Challenges in Monitoring Safety." *Frontiers in Pharmacology* 4: 177. https://doi.org/10.3389/fphar.2013.00177

El-Atroush, H., M. Magdy, and O. Werner. 2015. "DNA Barcoding of Two Endangered Medicinal Plants from Abou Galoom Protectorate." *Life Science Journal* 12 (9): 101–9.

Eldridge, G. R., H. C. Vervoort, C. M. Lee, P. A. Cremin, C. T. Williams, S. M. Hart, M. G. Goering, M. O'Neil-Johnson, and L. Zeng. 2002. "High-Throughput Method for the Production and Analysis of Large Natural Product Libraries for Drug Discovery." *Analytical Chemistry* 74 (16): 3963–71. https://doi.org/10.1021/ac025534s

Erdem, S. A. and A. Chandra. 2003. "E-Commerce in Healthcare and Pharmaceutical Marketing—Opportunities and Concerns." *Clinical Research and Regulatory Affairs* 20 (4): 399–407. https://doi.org/10.1081/CRP-120026121

Ernst, E. 2002. "The Risk–Benefit Profile of Commonly Used Herbal Therapies." *Annals of Internal Medicine* 137 (12): 1008. https://doi.org/10.7326/0003-4819-137-12-200212170-00024

Fatehi, F. and R. Wootton. 2012. "Telemedicine, Telehealth or e-Health? A Bibliometric Analysis of the Trends in the Use of These Terms." *Journal of Telemedicine and Telecare* 18 (8): 460–4. https://doi.org/10.1258/jtt.2012.gth10

Gaisford, J. D., R. Tarvydas, J. E. Hobbs, and W. A. Kerr. 2002. "Biotechnology Piracy: Rethinking the International Protection of Intellectual Property." *Canadian Journal of Agricultural Economics/*Revue Canadienne d'Agroeconomie 50 (1): 15–34. https://doi.org/10.1111/j.1744-7976.2002.tb00379.x

Ganesan, A. 2004. "Natural Products as a Hunting Ground for Combinatorial Chemistry." *Current Opinion in Biotechnology* 15 (6): 584–90. https://doi.org/10.1016/j.copbio.2004.09.002

Gangwal, A. and A. Lavecchia. 2025. "Artificial Intelligence in Natural Product Drug Discovery: Current Applications and Future Perspectives." *Journal of Medicinal Chemistry* 68(4): 3948–69. https://doi.org/10.1021/acs.jmedchem.4c01257

Gertsch, J. 2009. "How Scientific Is the Science in Ethnopharmacology? Historical Perspectives and Epistemological Problems." *Journal of Ethnopharmacology* 122 (2): 177–83. https://doi.org/10.1016/j.jep.2009.01.010

Gilani, A. H. 2005. "Trends in Ethnopharmacology." *Journal of Ethnopharmacology* 100 (1–2). https://doi.org/10.1016/j.jep.2005.06.001

Goyanes, A., J. Wang, A. Buanz, R. Martínez-Pacheco, R. Telford, S. Gaisford, and A. W. Basit. 2015. "3D Printing of Medicines: Engineering Novel Oral Devices With Unique Design and Drug Release Characteristics." *Molecular Pharmaceutics* 12 (11): 4077–84. https://doi.org/10.1021/acs.molpharmaceut.5b00510

Grunwald, A. 2009. "Vision Assessment Supporting the Governance of Knowledge–The Case of Futuristic Nanotechnology." In *The Social Integration of Science*, 147–70. Nomos Verlagsgesellschaft mbH & Co. KG. https://doi.org/10.1016/j.lfs.2020.118173

Hasenclever, L., J. Paranhos, C. R. Costa, G. Cunha, and D. Vieira. 2017. "The Brazilian Phytotherapics Industry: Challenges and Opportunities." *Ciencia & Saude Coletiva* 22: 2559–69. https://doi.org/10.1590/1413-81232017228.29422016

Hewlings, S. J. and D. S. Kalman. 2017. "Curcumin: A Review of Its Effects on Human Health." *Foods* 6 (10): 92. https://doi.org/10.3390/foods6100092

Hossain, C. M., M. E. E. T. A. Gera, and K. A. Ali. 2022. "Current Status and Challenges of Herbal Drug Development and Regulatory Aspect: A Global Perspective." *Asian Journal of Pharmaceutical and Clinical Research* 15: 31–41. https://doi.org/10.22159

Hosseinzadeh, S., A. Jafarikukhdan, A. Hosseini, and R. Armand. 2015. "The Application of Medicinal Plants in Traditional and Modern Medicine: A Review of Thymus vulgaris." *International Journal of Clinical Medicine* 6 (9): 635–42. https://doi.org/10.4236/ijcm2015.69084

Kamsu-Foguem, B. and C. Foguem. 2014. "Could Telemedicine Enhance Traditional Medicine Practices?" *European Research in Telemedicine/*La Recherche Européenne En Télémédecine 3 (3): 117–23. https://doi.org/10.1016/j.eurtel.2014.08.001

Kaplan, A. M. and M. Haenlein. 2010. "Users of the World, Unite! The Challenges and Opportunities of Social Media." *Business Horizons* 53 (1): 59–68. https://doi.org/10.1016/j.bushor.2009.09.003

Kartal, M. 2007. "Intellectual Property Protection in the Natural Product Drug Discovery, Traditional Herbal Medicine and Herbal Medicinal Products." *Phytotherapy Research: An International Journal Devoted to Pharmacological and Toxicological Evaluation of Natural Product Derivatives* 21 (2): 113–9. https://doi.org/10.1002/ptr.2036

Kasilo, O. M. J. and J. B. Nikiema. 2014. "World Health Organization Perspective for Traditional Medicine." In *Novel Plant Bioresources: Applications in Food, Medicine and Cosmetics*, 23–42. https://doi.org/10.1002/9781118460566.ch2

Kellerman, A. 2023. "Which Country Recorded the Highest Digital Penetration in 2022–2023?" *Netcom. Réseaux, Communication et Territoires* 37 (3–4). https://doi.org/10.4000/netcom.8633

Kessler, R. C., R. B. Davis, D. F. Foster, M. I. Van Rompay, E. E. Walters, S. A. Wilkey, T. J. Kaptchuk, and D. M. Eisenberg. 2001. "Long-Term Trends in the Use of Complementary and Alternative Medical Therapies in the United States." *Annals of Internal Medicine* 135 (4): 262–8. https://doi.org/10.7326/0003-4819-135-4-200108210-00011

Kilbourne, A. M., D. Hynes, T. O'Toole, and D. Atkins. 2018. "A Research Agenda for Care Coordination for Chronic Conditions: Aligning Implementation, Technology, and Policy Strategies." *Translational Behavioral Medicine* 8 (3): 515–21. https://doi.org/10.1093/tbm/ibx084

Klein-Junior, L. C., M. R. de Souza, J. Viaene, T. M. B. Bresolin, A. L. de Gasper, A. T. Henriques, and Y. V. Heyden. 2021. "Quality Control of Herbal Medicines: From Traditional Techniques to State-of-the-Art Approaches." *Planta Medica* 87 (12/13): 964–88. https://doi.org/10.1055/a-1529-8339

Kumar, R. 2023. "Nanotechnology in Herbal Medicine: Challenges and Future Perspectives." In *Nanotechnology in Herbal Medicine*, 515–48. Woodhead Publishing. https://doi.org/10.1016/B978-0-323-99527-6.00008-2

Kumar, S., and R. Ramasamy. 2024. *Textbook on Digital Marketing for PG Students.* Akinik Publication. https://doi.org/10.2139/ssrn.4870982

Kunle, O. F., H. O. Egharevba, and P. O. Ahmadu. 2012. "Standardization of Herbal Medicines-A Review." *International journal of biodiversity and conservation* 4 (3): 101–12. https://doi.org/10.5897/IJBC11.163

Kushwah, S., N. S. Maurya, S. Kushwaha, L. Scotti, A. Chawade, and A. Mani. 2023. "Herbal Therapeutics for Alzheimer's Disease: Ancient Indian Medicine System from the Modern Viewpoint." *Current Neuropharmacology* 21 (4): 764–76. https://doi.org/10.2174/1570159X21666230216094353

Kwahk, K. Y. and X. Ge. 2012. "The effects of social media on e-commerce: A perspective of social impact theory." In *2012 45th Hawaii International Conference on System Sciences*, 1814–23. IEEE. https://doi.org/10.1109/HICSS.2012.564

Lanjouw, J. O., A. Pakes, and J. Putnam. 1998. "How to Count Patents and Value Intellectual Property: The Uses of Patent Renewal and Application Data." *The Journal of Industrial Economics* 46 (4): 405–32. https://doi.org/10.1111/1467-6451.00081

Liang, Y. Z., P. Xie, and K. Chan. 2004. "Quality Control of Herbal Medicines." *Journal of Chromatography B* 812 (1–2): 53–70. https://doi.org/10.1016/j.jchromb.2004.08.041

Liu, Y. F., N. Ai, A. Keys, X. H. Fan, and M. J. Chen. 2015. "Network Pharmacology for Traditional Chinese Medicine Research: Methodologies and Applications." *Chinese Herbal Medicines* 7 (1): 18–26. https://doi.org/10.1016/S1674-6384(15)60015-6

Longhi, C. and S. Rochhia. 2015. "Long Tails in the Tourism Industry: Towards Knowledge Intensive Service Suppliers." *International Journal of Leisure and Tourism Marketing* 4 (3–4): 241–59. https://doi.org/10.1504/IJLTM.2015.072119

Mak, K. K., Y. H. Wong, and M. R. Pichika. 2024. "Artificial intelligence in drug discovery and development." In *Drug Discovery and Evaluation: Safety and Pharmacokinetic Assays*, 1461–98. https://doi.org/10.1016/j.drudis.2020.10.010

Mangold, W. G. and D. J. Faulds. 2009. "Social Media: The New Hybrid Element of the Promotion Mix." *Business Horizons* 52 (4): 357–65. https://doi.org/10.1016/j.bushor.2009.03.002

Memon, M., S. S. Hussain, U. A. Bajwa, and A. Ikhlas. 2018. "Blockchain beyond bitcoin: Blockchain technology challenges and real-world applications." In *2018 International Conference on Computing, Electronics & Communications Engineering (iCCECE)*, 29–34. IEEE. https://doi.org/10.1109/iCCECOME.2018.8658518

Morris, C. A. and J. Avorn. 2003. "Internet Marketing of Herbal Products." *Jama* 290 (11): 1505–9. https://doi.org/10.1001/jama.290.11.1505

Mukherjee, P. K., R. K. Harwansh, S. Bahadur, S. Banerjee, A. Kar, J. Chanda, S. Biswas, S. M. Ahmmed, and C. K. Katiyar. 2017. "Development of Ayurveda–Tradition to Trend." *Journal of Ethnopharmacology* 197: 10–24. https://doi.org/10.1016/j.jep.2016.09.024

Mukherjee, P. K., S. Banerjee, B. Das Gupta, and A. Kar. 2022. "Evidence-based validation of herbal medicine: Translational approach." In *Evidence-Based Validation of Herbal Medicine*, 1–41. Elsevier. https://doi.org/10.1016/B978-0-323-85542-6.00025-1

Narang, J. K., A. Dogra, J. Ali, S. Baboota, and H. Singh. 2022. "Drug regulatory affairs: an introduction." In *Regulatory Affairs in the Pharmaceutical Industry*, 1–30. Academic Press. https://doi.org/10.1016/B978-0-12-822211-9.00001-0

Neekhra, S., J. A. Pandith, N. A. Mir, A. Manzoor, S. Ahmad, R. Ahmad, and R. A. Sheikh. 2022. "Innovative Approaches for Microencapsulating Bioactive Compounds and Probiotics: An Updated Review." *Journal of Food Processing and Preservation* 46 (11): e16935. https://doi.org/10.1111/jfpp.16935

Nurhandayani, A., R. Syarief, and M. Najib. 2019. "The Impact of Social Media Influencer and Brand Images to Purchase Intention." *Jurnal Aplikasi Manajemen* 17 (4): 650–61. https://doi.org/10.21776/ub.jam.2019.017.04.09

Othman, C. N. and M. Farooqui. 2015. "Traditional and Complementary Medicine." *Procedia-Social and Behavioural Sciences* 170: 262–71. https://doi.org/10.1016/j.sbspro.2015.01.036

Palmberg, C. 2004. "The Sources of Innovations–Looking Beyond Technological Opportunities." *Economics of Innovation and New Technology* 13 (2): 183–97. https://doi.org/10.1080/10438590410001628143

Panossian, A., G. Wikman, and J. Sarris. 2010. "Rosenroot (Rhodiola rosea): Traditional Use, Chemical Composition, Pharmacology and Clinical Efficacy." *Phytomedicine* 17 (7): 481–93. https://doi.org/10.1016/j.phymed.2010.02.002

Parveen, A., B. Parveen, R. Parveen, and S. Ahmad. 2015. "Challenges and Guidelines for Clinical Trial of Herbal Drugs." *Journal of Pharmacy and Bioallied Sciences* 7 (4): 329–33. https://doi.org/10.4103/0975-7406.168035

Politzer, R. M., J. Yoon, L. Shi, R. G. Hughes, J. Regan, and M. H. Gaston. 2001. "Inequality in America: The Contribution of Health Centers in Reducing and Eliminating Disparities in Access to Care." *Medical Care Research and Review* 58 (2): 234–48. https://doi.org/10.1177/

Pradhan, D., G. Tripathy, R. K. Pradhan, T. Dasmohapatra, and S. P. Pradhan. 2016. "A Review on Cuminosides Nanomedicine-: Pharmacognostic Approach to Cancer Therapeutics." *Journal of Young Pharmacists* 8 (2): 61. https://doi.org/10.5530/jyp2016.2.2

Qiraouani Boucetta, K., Z. Charrouf, H. Aguenaou, A. Derouiche, and Y. Bensouda. 2015. "The Effect of Dietary and/or Cosmetic Argan Oil on Postmenopausal Skin Elasticity." *Clinical Interventions in Aging* 10: 339–349. https://doi.org/10.2147/CIA.S71684

Rana, T., M. Fatima, A. Q. Khan, Z. Naeem, S. Javaid, N. Sajid, and A. Habib. 2020. "Hydrogels: A Novel Drug Delivery System." *Journal ISSN* 2766: 2276. https://doi.org/10.37871/jbres1176

Rastogi, V. and P. Yadav. 2012. “Transdermal Drug Delivery System: An Overview.” *Asian Journal of Pharmaceutics (AJP)* 6 (3). https://doi.org/10.22377/ajp.v6i3.51

Rehman, F. U., K. U. Shah, S. U. Shah, I. U. Khan, G. M. Khan, and A. Khan. 2017. “From Nanoemulsions to Self-Nanoemulsions, with Recent Advances in Self-Nanoemulsifying Drug Delivery Systems (SNEDDS).” *Expert Opinion on Drug Delivery* 14 (11): 1325–40. https://doi.org/10.1080/17425247.2016.1218462

Ritchie, M. R. 2007. “Use of Herbal Supplements and Nutritional Supplements in the UK: What Do We Know About Their Pattern of Usage?” *Proceedings of the Nutrition Society* 66 (4): 479–82. https://doi.org/10.1017/S0029665107005794

Rowley, J. 2002. “Information Marketing in a Digital World.” *Library Hi Tech* 20 (3): 352–8. https://doi.org/10.1108/07378830210444540

Sahoo, N. and P. Manchikanti. 2013. “Herbal Drug Regulation and Commercialization: An Indian Industry Perspective.” *The Journal of Alternative and Complementary Medicine* 19 (12): 957–63. https://doi.org/10.1089/acm.2012.0275

Sajedipoor, A. A. and S. Mashayekh. 2015. “Investigation on Supply and Demand in Medicinal Plants Market (Case Study: Hamedan, Markazi and Lorestan Provinces).” *Iranian Journal of Medicinal and Aromatic Plants Research* 31 (4): 626–36. https://doi.org/10.22092/ijmapr.2015.102680

Savant, P. B. and M. S. Kareppa. 2022. “A Review: E-pharmacy vs Conventional Pharmacy.” *Asian Journal of Pharmacy and Technology* 12 (1): 84–8. https://doi.org/10.52711/2231-5713.2022.00014

Schork, N. J. 2019. “Artificial Intelligence and Personalized Medicine.” *Precision Medicine in Cancer Therapy*: 265–83. https://doi.org/10.1007/978-3-030-16391-4_11

Senthil Kumar, P. and E. Gunasundari. 2018. “Nanocomposites: Recent Trends and Engineering Applications.” *Nano Hybrids and Composites* 20: 65–80. https://doi.org/10.4028/www.scientific.net/NHC.20.65

Shahriari, M., P. Kesharwani, T. P. Johnston, and A. Sahebkar. 2023. “Anticancer Potential of Curcumin-Cyclodextrin Complexes and Their Pharmacokinetic Properties.” *International Journal of Pharmaceutics* 631: 122474. https://doi.org/10.1016/j.ijpharm.2022.122474

Shankar, V. and S. Balasubramanian. 2009. “Mobile Marketing: A Synthesis and Prognosis.” *Journal of Interactive Marketing* 23 (2): 118–29. https://doi.org/10.1016/j.intmar.2009.02.0

Sharkie, R. 2003. “Knowledge Creation and Its Place in the Development of Sustainable Competitive Advantage.” *Journal of Knowledge Management* 7 (1): 20–31. https://doi.org/10.1108/13673270310463590

Sharma, G., S. Anabousi, C. Ehrhardt, and M. N. V. Ravi Kumar. 2006. “Liposomes as Targeted Drug Delivery Systems in the Treatment of Breast Cancer.” *Journal of Drug Targeting* 14 (5): 301–10. https://doi.org/10.1080/10611860600809112

Shyur, L. F., C. P. Liu, and S. C. Chien. 2013. “Metabolomics in herbal medicine research.” In *The Handbook of Plant Metabolomics*, 155–74. https://doi.org/10.1002/9783527669882

Singh, H. P., S. Sharma, S. B. Chauhan, and I. Kaur. 2014. “Clinical Trials of Traditional Herbal Medicines in India: Current Status and Challenges.” *International Journal of Pharmacognosy* 1 (7): 415–21. https://doi.org/10.13040/IJPSR.0975-8232

Sireesha, B. and S. Haritha. 2024. “A Review on Regulatory Affairs and Regulatory Requirements for New Drug Approval.” *International Journal of Health Care and Biological Sciences*: 22–8. https://doi.org/10.46795/ijhcbs.V5i4.643

Smellie, W. S. A., D. I. Finnigan, D. Wilson, D. Freedman, C. A. M. McNulty, and G. Clark. 2005. “Methodology for Constructing Guidance.” *Journal of Clinical Pathology* 58 (3): 249–53. https://doi.org/10.1136/jcp.2004.018374

Soman, S., R. K. Sharma, V. Sharma, V. Gautam, K. Shrman, R. Kewat, and S. Kour. 2020. “Phytosomes-A Novel Approach for Herbal Drug Delivery.” *Journal of Animal Research* 10 (4): 479–85. https://doi.org/10.30954/2277-940X.04.2020.1

Steinhoff, B. 1999. "Monographs for Herbal Medicinal Products: A Way to Reach Harmonization." *Pharmaceuticals Policy and Law* 2 (1): 37–46. https://doi.org/10.3233/PPL-1999-00014

Straus, S. E. 2002. "Herbal Medicines—What's in the Bottle?" *New England Journal of Medicine* 347 (25): 1997–8. https://doi.org/10.1056/NEJMp020148

Sun, M., Y. Gao, Y. Pei, C. Guo, H. Li, F. Cao, A. Yu, and G. Zhai. 2010. "Development of Nanosuspension Formulation for Oral Delivery of Quercetin." *Journal of biomedical nanotechnology* 6 (4): 325–32. https://doi.org/10.1166/jbn.2010.1133

Swathi, J. and D. N. Venkatesh. 2022. "A Review of Herbal Regulations in India and Worldwide." *Research Journal of Pharmacy and Technology* 15 (3): 1348–52.

Taft, C. A. and S. R. de Lazaro, eds. 2024. *Progress in Hydrogen Energy, Fuel Cells, Nano-Biotechnology and Advanced, Bioactive Compounds.* Springer Nature. https://doi.org/10.1007/978-3-031-75984-0

Thakkar, S., E. Anklam, A. Xu, F. Ulberth, J. Li, B. Li, and M. Hugas, et al. 2020. "Regulatory Landscape of Dietary Supplements and Herbal Medicines from a Global Perspective." *Regulatory Toxicology and Pharmacology* 114: 104647. https://doi.org/10.1016/j.yrtph.2020.104647

Thompson, R. 2018. "Gardening for Health: A Regular Dose of Gardening." *Clinical Medicine* 18 (3): 201–5. https://doi.org/10.7861/clinmedicine.18-3-201

Tian, Q., N. D. Price, and L. Hood. 2012. "Systems Cancer Medicine: Towards Realization of Predictive, Preventive, Personalized and Participatory (P4) Medicine." *Journal of Internal Medicine* 271 (2): 111–21. https://doi.org/10.1111/j.1365-2796.2011.02498.x

Torchilin, V. P. 2014. "Multifunctional, Stimuli-Sensitive Nanoparticulate Systems for Drug Delivery." *Nature Reviews Drug Discovery* **13** (11): 813–827. https://doi.org/10.1038/nrd4333

Uddin, M. A., A. Khatun, and M. A. Mannan. 2023. "Antinociceptive Effect of Methanol Extract of *Diospyros malabarica* (Desr.) Kostel Leaves in Mice." *Pharmacology & Pharmacy* 14 (9): 388–406. https://doi.org/10.4236/pp.2023.149025

Uppoor, V. R. S. 2001. "Regulatory Perspectives on In Vitro (dissolution)/In Vivo (Bioavailability) Correlations." *Journal of Controlled Release* 72 (1–3): 127–32. https://doi.org/10.1016/S0168-3659(01)00268-1

Usman, M., M. Nakagawa, and S. Cheng. 2023. "Emerging Trends in Green Extraction Techniques for Bioactive Natural Products." *Processes* 11 (12): 3444. https://doi.org/10.3390/pr11123444

Van't Riet, A., M. Berg, F. Hiddema, and K. Sol. 2001. "Meeting Patients' Needs with Patient Information Systems: Potential Benefits of Qualitative Research Methods." *International journal of medical informatics* 64 (1): 1–14. https://doi.org/10.1016/S1386-5056(01)00185-X

Vijayananthan, A., and O. Nawawi. (2008). "The Importance of Good Clinical Practice Guidelines and Its Role in Clinical Trials." *Biomedical Imaging and Intervention Journal* 4, no. (1): e5. https://doi.org/DOI: 10.2349/biij.4.1.e5

Yuan, H., Q. Ma, L. Ye, and G. Piao. 2016. "The Traditional Medicine and Modern Medicine from Natural Products." *Molecules* 21 (5): 559. https://doi.org/10.3390/molecules21050559

11 Nano-Interventions for Herbal Drug Development

A Decade Journey

Jhansi Lakshmi Parimi, Mousumi Das, Deepayan Kar, Manonayan Singha, and Santanu Dhara

11.1 INTRODUCTION

Traditional herbal medicines are defined by their long-standing history, with some systems dating back nearly 5000 years. In India, the three primary systems are: Ayurveda, Unani, and Siddha. These systems utilize a holistic approach, focusing on the integration of body, mind, and spirit through nutrition and herbal remedies to prevent and treat illness (Morgan 1994).

Currently, approximately 80% of the world's population uses herbal formulations as their primary form of healthcare. This widespread adoption is driven by increasing concerns over the toxicity and side effects of conventional allopathic drugs. Herbal products are recognized for a vast range of medical applications, treating both acute and chronic conditions such as chronic diseases: diabetes, arthritis, Alzheimer's, and cirrhosis; mental health: depression, anxiety, and cognitive enhancement; and anti-aging, anti-inflammatory, and immune support (e.g., anti-HIV). (Singh and Sharma 2023). Despite their popularity and therapeutic potential, the text notes that traditional herbal medicines still face several significant challenges.

Global regulatory inconsistency is a major hurdle in the herbal industry is the lack of uniform classification. Depending on the country, the same plant may be legally categorized as food or functional food, dietary supplement, and herbal medicine. This fragmentation makes establishing global standards difficult and leads to significant confusion for both patients and consumers (Netzer, Balmith, and Flepisi 2021). In the United States, herbal products are governed by the Dietary Supplement Health and Education Act (DSHEA) of 1994. Key features include; (i) Definition: Herbs and botanicals are classified as "dietary ingredients" intended to supplement the diet. (ii) Safety Testing: Toxicological testing is generally not required for products that were already on the market before 1994. (iii) Burden of Proof: Unlike pharmaceutical drugs, the responsibility lies with the FDA to prove a product is harmful before it can be restricted or removed (Balogun et al. 2021).

DOI: 10.1201/9781003623748-11

Herbal medicines can interact with conventional drugs, particularly those with narrow therapeutic windows. For example, garlic and ginger may increase bleeding risk with warfarin, while combinations involving ephedrine and beta-blockers or MAO inhibitors may lead to hypertensive crises. Furthermore, valerian can enhance CNS depression when mixed with sedatives, and ginseng may affect neurotransmitters alongside MAO inhibitors. Specific combinations, such as acetaminophen and kava, have been linked to hepatotoxicity. The quality of herbal products is influenced by genetic, environmental, and procedural factors, with guidelines like GACP and SOPs under GMP and GLP aiding in raw material handling (Zhang et al. 2012). Quality assurance remains problematic, particularly in polyherbal formulations, as WHO's call for national GMP and labelling systems faces challenges due to poor practices. Safety and efficacy require stringent regulations.

Conducting clinical trials for herbal medicines is particularly challenging because of issues related to study design, standardization, ethical considerations, cost, and regulatory compliance. To address these challenges, the World Health Organization issued specific guidelines for clinical evaluation of herbal medicines in 2005 (Kunle 2012). Although double-blind randomized controlled trials remain the gold standard for establishing efficacy, their application in herbal medicine is limited by the holistic, individualized, and multimodal nature of these therapies. Consequently, single-blind study designs are often considered more feasible (Misra 2012). Additional obstacles include difficulties in developing appropriate placebos and defining standardized inclusion and exclusion criteria due to the distinctive sensory properties and personalized use of herbal formulations.

11.2 NANOTECHNOLOGY-DRIVEN HERBAL MEDICATION DELIVERY SYSTEMS

Herbal nanocarriers are advanced drug delivery systems that encapsulate or incorporate herbal extracts, phytochemicals, or plant-derived compounds into nanoscale carriers, typically ranging from 1 to 100 nanometres in size. These carriers can be made from various materials, including lipids, polymers, or other biocompatible substances, and are designed to overcome the limitations of conventional herbal medicines. Nanocarriers have transformed herbal drug delivery by enhancing stability, bioavailability, and therapeutic efficacy while enabling targeted and controlled release (Jalili et al. 2023).

11.2.1 Incremental Advantages of Herbal Nano Materials

1. **Enhanced Bioavailability**
   a. **Improved Solubility and Stability:** Herbal nanocarriers increase the solubility of poorly water-soluble plant compounds, making them more readily absorbable in the gastrointestinal tract. This is achieved by encapsulating the herbal actives in nanoscale carriers, which can disperse better in biological fluids and protect the payload from degradation.

   b. **Protection from Degradation:** Nanocarriers shield herbal compounds from enzymatic and environmental degradation, improving their stability and allowing more of the active ingredient to reach systemic circulation.
   c. **Increased Absorption:** By reducing particle size to the nanoscale, these carriers enhance the absorption of herbal drugs across biological membranes, leading to higher concentrations in the bloodstream and improved therapeutic efficacy.
   d. **Overcoming First-Pass Metabolism:** Some nanocarrier systems can bypass or reduce first-pass metabolism in the liver, further increasing the amount of active drug available in the body.
2. **Improved Targeting**
   a. **Targeted Delivery:** Nanocarriers can be engineered to deliver herbal drugs directly to specific tissues or cells, such as tumour sites or inflamed areas. This targeted approach enhances the concentration of the drug at the desired site while minimizing exposure to healthy tissues.
   b. **Controlled and Sustained Release:** Many nanocarriers offer controlled or sustained release of herbal actives, maintaining therapeutic levels over extended periods and reducing the frequency of dosing.
   c. **Surface Modification:** The surface of nanocarriers can be modified with ligands or antibodies that recognize and bind to specific cellular markers, further improving site-specific delivery and reducing off-target effects.

11.2.2 Major Classes of Nanotechnology-Driven Herbal Medication Systems

There are four types of classification of nanotechnology-driven herbal medication.

11.2.2.1 Lipid-Based Nanocarriers

11.2.2.1.1 Solid Lipid Nanoparticles (SLNs)

These nanoparticles offer excellent stability and protection for herbal actives. Recent studies show SLNs loaded with curcumin, berberine, and silybin demonstrate enhanced anticancer, antioxidant, and hepatoprotective effects, as well as improved brain delivery and oral absorption.

11.2.2.1.2 Liposomes

Liposomes are versatile nanocarriers that efficiently encapsulate both hydrophilic and lipophilic herbal drugs due to their lipid bilayer structure (Liu et al. 2023), offering high biocompatibility, biodegradability, and low toxicity. Recent formulations of phytochemicals like curcumin and capsaicin have shown enhanced bioavailability and therapeutic effects. Liposomes are used for controlled drug release, stability enhancement, and targeted delivery, evolving significantly since their inception in 1964 and finding applications in pharmaceutical and biomedical fields, including paediatric drug delivery (Kim et al. 2023; Asl et al. 2023; Maurya et al. 2010).

11.2.2.1.3 Phytosomes

These are complexes of phospholipids and plant actives, offering greater stability and absorption than conventional liposomes. Commercialized phytosome products (e.g., for quercetin, silybin, and ginsenosides) have demonstrated higher bioavailability and efficacy in clinical and preclinical studies. PLGA and chitosan nanoparticles are being used to encapsulate herbal actives, providing controlled release and enhanced delivery across biological barriers, such as the blood-brain barrier. Inorganic nanoparticles like gold and silica nanoparticles are being explored for their ability to deliver herbal drugs with improved targeting and imaging capabilities, particularly in cancer therapy and self-targeting and stimuli-responsive nanocarriers.

Recent advances include the design of nanocarriers that respond to specific biological stimuli (e.g., pH, enzymes) or possess inherent targeting capabilities, further increasing the precision of herbal drug delivery.

11.2.3 Synthesis Techniques for Nano Herbals

Nanoformulations are advanced nanoscale systems (1–300 nm) designed for targeted drug delivery and diagnostics due to their unique physicochemical properties (Di et al. 2021; Khulbe et al. 2023). These include liposomes, dendrimers, polymeric nanoparticles, nanoemulsions, and inorganic carriers like gold or quantum dots (Baig et al. 2023; Panja and Patra 2023). They enhance bioavailability, stability, and therapeutic efficacy while enabling controlled release and site-specific delivery (Prasad and Selvaraj 2024; Sanhueza et al. 2023).

11.2.3.1 Nanosuspensions

Nanosuspensions are aqueous colloidal dispersions of solid nanoparticles stabilized by surfactants or polymers, used to enhance the solubility and bioavailability of poorly soluble herbal extracts (Rahman 2023). Techniques like high-pressure homogenization and nanoprecipitation reduce particle size, increasing surface area and improving absorption (Awlqadr et al. 2025).

11.2.3.2 Nanopolymerization

Nanotechnology is used for herbal nanopolymerization technique to increase the bioavailability and bioactivity of the herbal compounds. By encasing, entrapping, or adsorbing plant extracts into nanoscale systems made of biodegradable and biocompatible polymers, their effectiveness can be increased (Idrees et al. 2020). In order to overcome the limitations of herbal medicines, such as their limited solubility, bioavailability, and durability and to increase the bioactivity of phytochemical compounds used in herbal medicines, researchers are actively creating SLNs, polymeric nanoparticles, and other nanosystems. They are also using certain polymers to create nanoherbals. This method increases overall activity, reduces adverse effects, and lowers dosage requirements (Alharbi et al. 2021).

11.2.3.3 Herbal Nanofibers

Nanofibers produced from plant extracts via electrospinning techniques enhance the solubility, bioavailability, and therapeutic performance of herbal compounds. The use of natural polymers such as chitosan and cellulose acetate enables extracellular matrix–mimicking structures, supporting applications in wound healing, drug delivery, and tissue regeneration (Stoyanova, Nachev, and Spasova 2023).

11.2.3.4 Herbal Nanoemulsions

Nanoemulsions are effective delivery systems for herbal extracts, offering enhanced ingredient penetration, increased surface area, reduced droplet size, and improved stability (Gupta et al. 2016). Prepared using low- and high-energy emulsification methods, herbal nanoemulsions have gained wide application in drug delivery, gene therapy, and the food industry. The growing incorporation of medicinal plants, such as Trigonella foenum-graecum (fenugreek), into nanoemulsion formulations highlights the expanding role of plant-based nanotechnology in enhancing therapeutic efficacy (Jampílek et al. 2019).

11.2.3.4.1 Nanoemulsification of Herbal Extracts Through Homogenization

A common method in nano emulsification for the extraction of herbal components is homogenization. High-pressure homogenization is a forceful method that uses extreme homogenization to make droplets in nanoemulsions smaller. This method has produced a large number of nanoemulsions for biotechnological, medicinal, and culinary substances. It enables accurate control over the emulsion's colour, rheology, stability, and particle size (Jo et al. 2025). Small-molecule surfactants are advised for use as emulsifiers in high-pressure homogenizers for optimal results. High-pressure homogenizers are essential for creating minuscule particle sizes and ensuring consistency in nanoemulsions. Furthermore, low-energy methods that create nanoemulsions using internal chemical energy and mild churning include phase inversion and self-nanoemulsification (Sokolov 2014).

11.2.3.4.2 Nanoprecipitation in Herbal Medicine

Nanoprecipitation is a widely used technique for producing nanoparticles from herbal extracts, enhancing their absorption and therapeutic efficacy (Zhou et al. 2023). Bioactive compounds are first dissolved in a solvent and then rapidly mixed with a non-solvent like water, causing particle formation. Stabilizers such as PVA or PEG-PLGA prevent aggregation and ensure uniform particle size (Zhou et al. 2023). The process parameters—solvent volume, mixing rate, and polymer concentration—influence particle characteristics. It accommodates both hydrophilic and hydrophobic compounds and improves pharmacokinetics, as shown with curcumin and β-carboline alkaloids (Ciuca and Racovita 2023).

11.2.3.4.3 SLNs for Herbal Medicine

SLNs are biocompatible and biodegradable carriers, commonly used for herbal extract delivery, with sizes between 50 nm and 1000 nm (Deshmukh 2014). They are created through emulsification, hot melt homogenization, and ultrasonication, resulting in stable drug delivery systems. SLNs can encapsulate both hydrophilic and hydrophobic compounds, facilitating controlled release, enhanced uptake, improved bioavailability, and targeted delivery through surface modification. Their low toxicity and stability enhance the oral absorption of herbal formulations, including those with *Hibiscus rosa-sinensis*.

11.2.3.4.4 Nanostructured Lipid Carriers for Herbal Medicine

For herbal formulations, nanostructured lipid carriers have emerged as viable medication delivery method. There are particular procedures and factors to take into account while adding herbal actives to NLCs. Furthermore, a melt-emulsification method using solid lipids, liquid oils, and emulsifiers is usually used to produce NLCs for herbal formulations. Depending on the exact composition and preparation process, NLCs can range in size from 100 to 1000 nm. By improving the solubility, stability, and bioavailability of herbal actives, the use of NLCs in herbal preparations increases the medicinal efficacy (Kaur, Nautiyal, and Singh 2015) of NLCs synthesis.

11.2.3.4.5 Liposomes

For both natural and herbal medications, liposomes are being studied as versatile nanocarriers to solve problems with stability, limited solubility, and customized distribution. By encapsulating bioactive compounds with different solubilities, lipid-based systems can improve bioavailability and decrease oxidation (Shamsuddin and Zulfakar 2023). Among the many advantages of liposomes are their natural material composition, the presence of aqueous compartments, and their defence against free radical-induced ingredient oxidation. Herbal extracts can increase the pharmacological advantages of liposomes at lower doses while reducing toxicity by improving stability, controlled release, and bioavailability (Sogut, Sezer, and Sezer 2021). Studies have shown that liposomes, microspheres, and phytosomes are used to deliver drugs by trapping or encasing medications.

11.2.4 Biosynthesis of Nanoparticles Loaded with Herbal Extracts

The use of herbal extracts for nanoparticle synthesis is gaining research interest due to its biocompatibility and rapid synthesis capabilities, allowing for controlled particle morphology (Mohamad et al. 2014). This green synthesis approach utilizes plant-derived biomolecules as reducing and stabilizing agents for metal ions, offering advantages over conventional methods. The fabrication process typically involves extraction and reaction with metal salts. Additionally, advanced nanotechnology delivery systems, such as herbal nanogels, enhance the solubility and therapeutic efficacy of herbal compounds (Dewi et al. 2022). These crosslinked polymeric networks enable controlled drug release and various administration routes, showing potential in managing diabetes, skin disorders, and cancer while improving patient safety and compliance.

Advanced nanotechnology-based delivery systems, such as herbal nanogels, further enhance the solubility, dispersion, bioavailability, and therapeutic efficacy of herbal compounds (Dewi et al. 2022). Nanogels are water-swollen, crosslinked polymeric networks capable of controlled drug release and multiple administration routes, including oral, transdermal, parenteral, and pulmonary delivery. Formulated using polymers such as chitin, chitosan, PLGA, and PEG, herbal nanogels show promise in managing diabetes, skin disorders, and cancer while improving patient safety and compliance (Chaudhari and Paithankar 2020).

11.3 CURRENT TRENDS IN NANOTECHNOLOGY INTERVENTIONS FOR HERBAL MATERIALS

Herbal medicine might undergo a transformation thanks to the intriguing and quickly developing science of nanotechnology. Even while traditional herbal medicines have many therapeutic advantages, their efficacy may be limited by issues including poor solubility, low bioavailability, and instability in the body. Herbal compounds can be encapsulated into nanoparticles, nanocarriers, and nanoemulsions through the use of nanotechnology, which improves their stability, absorption, and targeted administration for better therapeutic results. Herbal chemicals may be delivered precisely and with controlled release thanks to nanoparticles which lessen unwanted effects and increase the efficacy of therapies. Additionally, this technology opens the door for novel herbal formulations that make natural medicines more effective and accessible, such as topical applications, oral capsules, and transdermal patches (Chakraborty, Shivakumar, and Ramachandran 2016; Karaś et al. 2017; Fangueiro et al. 2016; Souto et al. 2019) (Figure 11.1).

11.3.1 Biological Activity Augmentation

Phytosomes are lipid-based systems that enhance the solubility and bioavailability of phytochemicals, addressing issues like poor absorption. Polymer-based systems, such as dendrimers, micelles, and polymersomes, also improve phytochemical delivery by allowing controlled drug release (Souto et al. 2019). Common polymers include PLGA, alginate, albumin, chitosan, starch, and polylactic acid, which offer protection to drugs through various methods while ensuring biocompatibility and biodegradability.

11.3.2 Phytosomes Increasing the Activity of Phytochemical Compounds

Phytoconstituents often have limited clinical utility due to issues like poor solubility and low bioavailability. Phytosome-based delivery systems are effective in enhancing lipophilicity and absorption. Sinigrin, a glucosinolate from the Brassicaceae family with wound-healing properties, faces therapeutic limitations due to its solubility. A sinigrin–phytosome complex was developed by Mazumder et al., demonstrating significantly improved wound-healing efficacy compared to pure sinigrin, achieving complete wound closure, unlike the partial healing seen with free sinigrin (Mazumder et al. 2016).

Taxifolin (dihydroquercetin), a flavonoid with antioxidant and anticancer properties, faces limitations due to its hydrophilicity and high molecular weight. To overcome these issues, taxifolin from *Cedrus deodara* bark was integrated into a

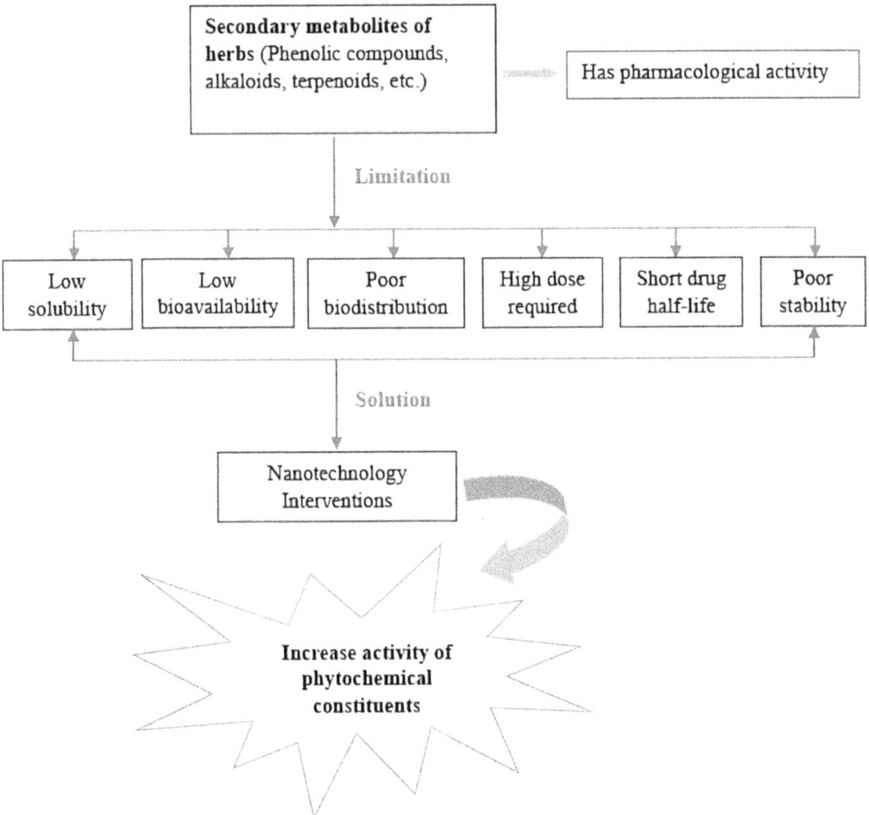

FIGURE 11.1 Insights for nanotechnology advancement benefits to the existing shortcomings.

phytosome system, resulting in improved antioxidant activity and enhanced anticancer efficacy against MCF-7 breast cancer cells, as demonstrated by lower IC_{50} values and greater cytotoxicity compared to the non-encapsulated extract (Kumar, Baldi, and Sharma 2021).

Silybin and apigenin are notable phytochemicals with hepatoprotective properties, but both face limitations in absorption and bioavailability (Chi et al. 2020). Silybin's incorporation into phytosome nanosuspensions enhances its therapeutic effects in liver injury models, while the phytosome delivery system for apigenin significantly improves its solubility and pharmacokinetics, leading to better antioxidant activity in hepatotoxicity models (Telange et al. 2017).

Oral absorption of quercetin, a phytoestrogen with anti-inflammatory and antioxidant properties, is poor. In ovariectomized mice models, phytosome-based administration greatly enhances its oestrogenic, antioxidant, and hypolipidaemic properties, demonstrating more therapeutic efficiency than free quercetin, particularly at higher dosages (Abd El-Fattah et al. 2017). Low solubility and bioavailability also affect other bioactive substances, such as gingerols from Zingiber officinale (Kubra and Jaganmohanrao 2012 ; Singh et al. 2017), wound-healing agents from Moringa oleifera (Bagdas et al. 2015; Chin et al. 2018; Gopalakrishnan et al. 2016; Muhammad et al. 2016), and antidiabetic components from Murraya koenigii (Yankuzo et al. 2011). When compared to traditional extracts, phytosome formulations of these extracts dramatically lower blood glucose levels while also boosting wound healing and antioxidant activity (Rahmani and Aly 2014; Rani, Kumar, and Khar 2022).

11.3.3 Utilization of Polymeric Nanoparticles to Boost Phytochemical Compounds Activity

Nanotechnology-based delivery systems, such as piperine-loaded polymeric nanocapsules (PNCs), have shown potential in enhancing the efficacy of bioactive phytochemicals like piperine, which has limitations due to poor solubility and rapid metabolism. Rani et al. (2020) found that PNCs improved anti-trypanosomal activity against Trypanosoma evansi, significantly outperforming free piperine, evidenced by an IC_{50} value three times lower than its free form. This highlights the effectiveness of nanotechnology in overcoming the challenges of phytochemical delivery.

Jatropha pelargoniifolia (JP) is a medicinal plant noted for its phenolics and flavonoids, contributing to oxidative defence and traditional medicine. However, its therapeutic application is limited due to low solubility, stability, bioavailability, and sensitivity to gastric pH. Research by Gudise, Chowdhury, and Manjappa (2021) indicated that nano-micellar formulations of plant extracts enhance antidiabetic, antihyperlipidaemic, and antioxidant activities compared to non-nano forms, attributed to improved delivery and bioavailability of beneficial phytoconstituents like terpenoids, flavonoids, and phenols.

11.3.4 Solubility Enhancement

Polar or water-soluble chemicals make up most of the physiologically active ingredients found in plants. Because, their large size of molecule, that resist passively diffusion, although poor lipid solubility of these, it mostly prohibits their ability of flowing through lipid-rich biological membranes, water-soluble phytoconstituents such as tannins, terpenoids, and flavonoids are poorly absorbed. As a result, bioavailability is low (Obodozie 2012). These medications fall within the biopharmaceutical classification system's high solubility/low permeability (BCS Class III) classifications. It has been shown that complexation with a few more therapeutically beneficial nutrients significantly increases the bioavailability of these extracts and each of their constituents. Phospholipids are among the nutrients that are highly beneficial for improving absorption (Seema 2014).

Vesicular delivery methods called phytosomes can improve the bioavailability and absorption of active compounds with poor solubility (Bhattacharya 2009). The phospholipids and phytochemical components that make up phytosomes react with that polar groups of phosphatidylcholine and that plant extracts in non-protonated solvents to produce hydrogen bonds (Chang et al. 2025).

11.3.5 Bioavailability and Biodistribution Improvements

Seventy percent of the active compounds derived from plants are hydrophobic, according to the literature (Martins-Gomes, Souto, and Silva 2022). One tactic to improve the bioavailability and bioactivity of phytochemical substances is the introduction of new technologies. The capacity to create appropriate drug delivery formulations is crucial for the development of novel nanotechnology-based therapeutics. Effective disease prevention and therapy depend on the administration of phytochemicals. The bioactivity of phytochemical substances may be enhanced by these delivery methods, which include polymer-based and lipid-based delivery systems (Paroha, Dewangan, and Sahoo 2020; Souto et al. 2019).

Effective herbal medicine delivery has led to the development of new nanotechnology-based drug delivery technologies (Tapadiya 2017). Lipid-based carrier systems, which include lipid particulates (SLN and NLC), vesicular systems (liposomes, phytosomes, transfersomes, ethosomes, and niosomes), and nanoemulsions, have attracted special attention for the delivery of phytochemicals with the goal of enhancing their stability, bioactivity, and bioavailability (Martins-Gomes, Souto, and Silva 2022).

The phytochemicals curcumin and tetrandrine have low absorption due to poor solubility (Song et al. 2022). To improve this, a liposome carrier method was employed, which successfully enhanced their solubility and bioavailability without significant harm to zebrafish in *in-vivo* studies (Blasi et al. 2007). However, liposomes can face self-leakage and stability issues. To address this, the stabilizer DSPE–MPEG 2000 (DP) has been used, significantly increasing the loading dosage and encapsulation effectiveness of curcumin and tetrandrine, as well as improving particle size distribution. Lipids are considered safe and effective for drug delivery.

11.4 INFLUENCE OF NANOENCAPSULATION ON THE RELEASE PROFILE OF FRAGILE HERBAL INGREDIENTS

Nanoencapsulation significantly alters the release profile of sensitive herbal compounds, offering several key advantages for their stability and therapeutic effectiveness. It allows for precise control over how and when fragile herbal ingredients are released, protecting them from degradation and enabling both immediate and sustained therapeutic effects tailored to specific needs.

11.4.1 Controlled and Sustained Release

a. **Regulated Delivery:** Nanoencapsulation enables controlled, often sustained, release of active ingredients, ensuring that fragile herbal compounds are delivered gradually rather than all at once. This controlled release helps maintain therapeutic concentrations over extended periods and reduces the risk of rapid degradation or clearance.
b. **Polymer Properties:** The choice of encapsulating material (such as chitosan, pectin, or proteins) directly influences the release rate. For example, chitosan-based nanocapsules can provide a slow, sustained

release, while pectin-coated nanoliposomes offer both burst and sustained release phases.

11.4.2 Protection from Environmental and Physiological Degradation

a. **Stability Enhancement:** Encapsulation shields herbal actives from harsh environmental conditions (e.g., oxidation, light, temperature) and physiological challenges (e.g., stomach acid, digestive enzymes), allowing for a more predictable and prolonged release profile.
b. **Improved Retention:** Studies have shown that nano encapsulated herbal compounds retain their bioactivity and are released more effectively at physiologically active sites, enhancing their functional performance.

11.4.3 Targeted and Stimuli-Responsive Release

a. **Site-Specific Delivery:** Surface modifications and smart materials enable nanocarriers to release their payload in response to specific triggers (such as pH changes or enzymes), allowing for targeted delivery to desired tissues or organs.
b. **Bio-adhesiveness:** Nanoscale particles can adhere to mucosal surfaces, prolonging retention time and enabling more efficient, localized release.

11.4.4 Burst and Dual-Phase Release

Initial Burst Release: Some nanoencapsulation systems are designed to provide an initial rapid release (burst) followed by a slower, sustained release, which is beneficial for compounds requiring immediate and prolonged action.

11.5 NANOCARRIERS MEDIATED PROTECTION HERBAL DRUGS FROM DEGRADATION IN THE BODY

Herbal drugs are often sensitive to the harsh conditions within the body, such as acidic pH, enzymatic activity, and oxidative environments, which can lead to their degradation and reduced therapeutic efficacy. Nanocarriers offer several mechanisms to protect these sensitive herbal compounds.

Nanocarriers employ multiple strategies to safeguard herbal compounds from degradation, ensuring their stability and efficacy. Below are the key mechanisms, which collectively ensure that sensitive herbal compounds remain intact and bioactive until reaching their target sites, significantly improving therapeutic outcomes

11.5.1 Physical Encapsulation

11.5.1.1 Barrier Formation

Herbal compounds are encapsulated within a nanoscale matrix (e.g., lipids, polymers, or polysaccharides), creating a physical barrier against degrading factors like stomach acid, enzymes, or oxidation. *Naringenin* encapsulated in Eudragit/PVA nanoparticles resisted degradation in the gastrointestinal tract, enhancing hepatoprotective effects. Chitosan nanoparticles protected clove essential oil from oxidation, improving its antimicrobial activity.

11.5.1.2 Protection from Enzymatic and Biological Degradation

Nanocarriers shield herbal compounds from metabolic enzymes in the liver or digestive system.

PLGA-encapsulated *Polygala senega* extract (NEEPS) showed enhanced anticancer activity by avoiding enzymatic breakdown in the bloodstream. Liposomes prevented premature hydrolysis of berberine, maintaining its bioactivity until reaching target cells.

11.5.1.3 Controlled and Sustained Release

Gradual release mechanisms minimize sudden exposure to harsh environments, reducing degradation risks. Chitosan-folic acid-quercetin nanoparticles released their payload over 56 days, ensuring prolonged therapeutic action. Self-assembled cinnamic acid-berberine nanoparticles (CA-BBR NPs) demonstrated sustained release, enhancing antibacterial efficacy.

11.5.1.4 Enhanced Solubility and Stability

Nanosizing hydrophobic herbal compounds improves their dispersibility in biological fluids, reducing aggregation and degradation. Berberine phytantriol liquid crystal nanoparticles (BP-LCNs) increased solubility and bioavailability while reducing toxicity. Yeast cells and pollen shells encapsulated hydrophobic compounds, shielding them from oxidation and temperature fluctuations.

11.5.1.5 Surface Modifications for Immune Evasion

Functionalizing nanocarrier surfaces (e.g., PEGylation) reduces recognition by the immune system, prolonging circulation time. Glycyrrhetinic acid-modified chitosan nanoparticles actively targeted liver cancer cells, minimizing off-target exposure.

11.5.1.6 Material-Specific Protection

a. **Polymeric Carriers:** Polymers like PLGA or chitosan form stable, biodegradable shields.
b. **Lipid-Based Carriers:** Liposomes and SLNs protect thermolabile compounds from heat and pH changes.

11.6 MECHANISMS BY WHICH NANOCARRIERS ENHANCE STABILITY OF HERBAL COMPOUNDS

Nanocarriers enhance stability through physical encapsulation, material-specific protection, and controlled release. During storage, they shield compounds from oxygen, light, and heat. In digestion, they resist acidic pH, enzymes, and rapid clearance, ensuring herbal actives reach systemic circulation intact. Advances in lipid, protein, and polysaccharide-based carriers continue to optimize these protective effects.

11.6.1 During Storage

11.6.1.1 Protection from Environmental Degradation

i. **Oxidation Prevention:** Encapsulation in lipid-based carriers (e.g., liposomes, SLNs) or plant protein matrices (e.g., zein, soy protein) shields herbal compounds from oxygen exposure, reducing oxidative breakdown. For example, chitosan nanoparticles protected clove essential oil from oxidation, preserving its antimicrobial activity.
ii. **Temperature and Light Resistance:** Lipid and polymeric nanocarriers buffer compounds against heat and UV light. Curcumin encapsulated in soy protein isolate (SPI) showed improved thermal stability during storage.

11.6.1.2 Material-Specific Stabilization

a. **Lipid Matrices:** SLNs and nanostructured lipid carriers (NLCs) form rigid matrices that immobilize herbal actives, preventing aggregation or crystallization. SLNs increased the shelf life of puerarin by maintaining its amorphous state.
b. **Polysaccharide Coatings:** Polysaccharide-based nanocarriers (e.g., chitosan, starch) create hydrophilic shells that resist moisture-induced degradation.

11.6.1.3 Surface Modifications

Additives like cholesterol in liposomes stabilize lipid bilayers, reducing leakage and maintaining structural integrity during storage.

11.6.2 During Digestion

11.6.2.1 Resistance to Gastrointestinal Conditions

i. **Acidic pH Protection:** Lipid and protein-based nanocarriers (e.g., zein, gliadin) resist stomach acid, preventing premature release. For instance, glycyrrhetinic acid-modified chitosan nanoparticles protected berberine from gastric degradation.
ii. **Enzymatic Shielding:** Encapsulation blocks digestive enzymes (e.g., proteases, lipases) from degrading herbal compounds. PLGA nanoparticles preserved the anticancer activity of *Polygala senega* extract by avoiding enzymatic hydrolysis in the gut.

11.6.2.2 Controlled Release Mechanisms

Sustained release from nanocarriers (e.g., polymeric nanoparticles, nanoemulsions) ensures gradual exposure to digestive fluids, minimizing degradation. Chitosan-folic acid-quercetin nanoparticles released their payload over 56 days, maintaining stability in the intestines.

11.6.2.3 Enhanced Mucosal Absorption

Nanocarriers improve mucosal permeability, reducing transit time in the harsh gastrointestinal environment. SLN-puerarin demonstrated faster absorption and higher bioavailability compared to unencapsulated forms.

11.7 PARTICLE SIZE INFLUENCE ON THE RELEASE RATE OF SENSITIVE HERBAL INGREDIENTS

Particle size is a critical factor in determining the release rate of sensitive herbal ingredients from nanocarriers and other formulations.

11.7.1 Surface Area Increase

As particle size decreases, the total surface area available for dissolution or release increases. This enhanced surface area allows for more rapid interaction with the surrounding medium, facilitating faster dissolution and release of bioactive compounds.

11.7.1.1 Diffusion Pathway Reduction

Smaller particles have shorter diffusion pathways, which mean that the active ingredients can migrate out of the carrier more easily and quickly.

11.7.1.2 Enhanced Extraction Efficiency

Precise particle size control is crucial in the pharmaceutical value chain of herbal medicines, enhancing the delivery of bioactive compounds. Fine powders lead to over 80% dissolution of key ingredients, compared to less than 25% for coarse materials, improving absorption and physiological effects like hepatoprotection. Consistent small particle sizes are essential for quality control, ensuring accurate dosing in tablets and capsules while reducing variability in production. Small particles improve efficiency but must be managed carefully to maintain therapeutic effects and production quality.

11.8 RECENT INNOVATIONS IN THE FIELD OF NANOTECHNOLOGY BASED HERBAL DRUG DELIVERY SYSTEMS

Key research highlights various nanoformulation types employed to enhance berberine's bioavailability. Notable technologies include nanoemulsions, which showed up to 440% increased bioavailability in rats and improved permeability in Caco2 cells, resulting in better hypoglycaemic effects. Liposomal formulations enhanced metabolic health and glycaemic control compared to free berberine. Berberine-loaded BSA nanoparticles improved antimicrobial activity and biodistribution in wound infections. Additionally, berberine-baicalin micelles achieved targeted drug release and anticancer efficacy in breast cancer models. Silymarin phytoliposomes demonstrated enhanced oral absorption and hepatocyte targeting with improved hepatoprotection. Curcumin SLNs significantly increased antitumor efficacy in lymphoma models. Finally, hybrid systems and muco-inhalable nanosuspensions demonstrated antiviral and anti-inflammatory activity in models of inflammatory lung injury and SARS-CoV-2.

Patents related to nanoformulations for herbal phytochemicals include liposomal curcumin (US 2007/968115 B2), silymarin formulations (US 2011/0201680 A1), and multi-herbal nanoparticle complexes aimed at stress relief and antiviral applications (e.g., Korean patent KR 102118546 B1). A notable patent, CN105030847, focuses on preparing ultrafine Chinese herbal powders to enhance the absorption and activity of compounds like cordyceps. The domain of nanoformulations continues to grow, addressing the solubility and metabolism issues of plant compounds. From a review of phytochemical patents (2001-2021), 139 patents for antidiabetic applications of berberine were identified, although many included combination compositions that aren't exclusive to nanotechnology. Additionally, patents exist for nanotechnology-based formulations of ashwagandha and berberine.

11.8.1 Commercial Herbal Nanomedicine

A comprehensive list for curcumin, silymarin, quercetin, and resveratrol nanoformulations and insights into their **patent portfolios and innovation strategies** has been shown in Table 11.1.

Recent patents from the past year (2024–2025) in the field of **herbal nanomaterials**, focusing on novel formulations, delivery systems, and applications; A **Herbal Colloidal Composition for Healing of Skin Lesions, Publication Number:** 523588,

TABLE 11.1
Market Leaders in Herbal Nanomedicine

Company	Key Products/Tech	Patent Focus	Market Reach
Sabinsa	CurcuWIN®, Siliphos®	SLNs, Phytosomes, manufacturing	Global nutraceuticals
Mucos Pharma	Lipocurc™	Liposome stabilization, loading	Clinical supplements
Theravalues	Theracurmin®	Polymeric nanoparticles	Asia, N. America
Indena	Resveratrol micelles/liposomes	Micellar/liposomal delivery	Pharma & nutraceuticals
Avesthagen (India)	SLNs/polymeric nanoparticles	Nanoformulation processes	India, emerging markets
Vedic Lifesciences	Silymarin, quercetin nanoforms	Nanocarrier compositions	India
Arjuna Natural Extracts	NANOCURC™	Nanoformulations, natural excipients	India
Biotivia	Liposomal resveratrol/quercetin	Liposome stability & composition	USA, International
Herbalife Nutrition	Nanoemulsions/polymeric NPs	Nanoencapsulation of botanicals	Global
PLT Health Solutions	Nano-lipid curcumin	SLNs, surfactant optimization	N. America, Europe
Himalaya Drug Company	Nanoemulsions, liposomes	Nanocarrier formulations for botanicals	India, global

Filed: March 12, 2024, **by** Dr. Sanjay Darvekar, Dr. Atul Kulkarni, Mrs. Sakshi Shah **from** Symbiosis Centre for Nanoscience and Nanotechnology (SCNN), India.

Cascade Cyclone Separator for Separation of Submicron Biological Material with Publication Number: 522761, **Filed on** March 11, 2024, Symbiosis Center for Nanoscience and Nanotechnology (SCNN), India pertains to a device designed for the separation of submicron biological materials, which can be crucial in the preparation of herbal nanomaterials by isolating fine particles.

Herbal Nutraceutical Formulation to Reduce Oxidative Stress, Viral and Microbial Infections, and Inflammation, U.S. Patent No. 12,090,186, Issued: September 17, 2024, discloses an herbal composition aimed at reducing oxidative stress, combating viral and microbial infections, and mitigating inflammation. The formulation includes a blend of natural ingredients known for their therapeutic properties.

11.8.2 Technology Trends in Herbal Nanomaterials (2015–2025)

Growing focus on bioavailability enhancement challenges of many phytochemicals (e.g., curcumin, berberine, quercetin) suffer from poor water solubility and low oral bioavailability. Patents increasingly focus on nanoformulations (nanoemulsions, liposomes, SLNs, polymeric nanoparticles) to improve: (i) Solubility,(ii) Stability against degradation,(iii) Permeability through biological membranes (iv) Controlled/sustained release

Example: Nanoemulsions of curcumin (US 9,894,221) and berberine nanoparticles (CN107315602A)

Notable patents from the last 10 years (2015–2025) in **herbal nanomaterials and nanoformulations** – covering a wide range of botanicals, delivery platforms, and therapeutic applications. Curcumin's bioavailability for anti-inflammatory and anticancer applications is improved by a stable nanoemulsion. For the treatment of diabetes and cancer, polymeric nanoparticles enhance the solubility and regulated release of herbal extracts. For antioxidant and anti-ageing properties, a liposome-based quercetin formulation provides stability and skin penetration. Herbal extracts such as silymarin have improved oral bioavailability thanks to SLNs. Through enhanced absorption, resveratrol nanoemulsion has neuroprotective and cardiovascular effects. Berberine nanoparticles increase solubility for anti-inflammatory and diabetic applications. Formulations of liposomal curcumin have enhanced bioavailability against inflammation and cancer. Herbal nanoemulsions containing extracts of aloe vera and turmeric aim to hydrate and prevent aging of the skin by combining phytochemicals to treat metabolic disorders.

11.9 POLICY AND REGULATORY SUPPORT IN INDIA

India allows product and process patents for nanoformulations and delivery methods (post-2005 TRIPS compliance).Herbal combinations that are non-traditional or synergistic, especially when nano-enabled, are patentable. Indian Patent Office (IPO) accepts "phytopharmaceutical" applications under Section 3(p) for certain well-characterized herbal drugs.

11.9.1 Regulatory Landscape (CDSCO + AYUSH)

CDSCO regulates nanoformulations of herbal drugs as new drugs under Schedule Y. Ministry of AYUSH is working on evolving a regulatory pathway for Ayurvedic nano-formulations, though still early-stage. There is an urgent need for well-documented safety, stability, and pharmacokinetics studies for nanoscale herbal products.

Herbal nanoformulations developed in India can be positioned for global markets, especially, US (via FDA's botanical drug pathway or dietary supplement category), EU (EMA's traditional herbal medicinal products), and Southeast Asia and Africa (via WHO-traditional medicine frameworks)

11.10 NANOTECHNOLOGY SHIFTING CONTRIBUTIONS FOR THE SAFETY CONSIDERATIONS OF HERBAL THERAPIES

Nanocarriers enable site-specific and controlled drug delivery, minimizing systemic exposure and reducing the risk of side effects and toxicity compared to conventional herbal formulations. By protecting herbal actives from degradation and controlling their release, nanotechnology lowers the frequency of dosing and the risk of adverse reactions related to fluctuating drug levels.

Nanoparticles, due to their small size and high surface area, can interact with biological systems in unpredictable ways, potentially causing, (i) pulmonary inflammation and lung diseases if inhaled, due to the easy inhalability of nanoparticles, (ii) mitochondrial damage, oxidative stress, and interference with cellular functions, including inhibition of macrophage activity and platelet aggregation, which can affect immune and cardiovascular systems, and (iii) accumulation in organs and tissues, potentially leading to long-term toxicity or disruption of normal physiological processes.

Nanoparticles can cross biological barriers more easily than larger particles, leading to (i) Unintended distribution to sensitive organs such as the brain or liver, raising concerns about neurotoxicity or hepatotoxicity and (ii) Local irritation or inflammation at the site of administration, especially for topical or inhaled formulations.

Nanocarriers can increase the absorption not only of beneficial herbal actives but also potentially toxic or allergenic components present in the plant extract, which amplifying side effects that might otherwise be negligible in conventional formulations. Many nano-herbal products reach the market with limited long-term safety data and incomplete toxicological profiling, making it difficult to predict rare or delayed adverse effects.

11.10.1 Nano-Delivery Systems Disrupt Natural Plant Compound Interactions and Effects?

Nano-delivery systems, while enhancing the delivery and efficacy of herbal compounds, can indeed alter the natural interactions among plant-derived constituents by, often exerting their therapeutic effects through the combined action of multiple bioactive compounds. Traditional preparations allow these compounds to interact freely, potentially producing synergistic or additive effects. Encapsulation within nanocarriers can change the pharmacokinetics and spatial-temporal release of individual components, possibly disrupting these natural interactions.

By enabling the precise control over the release rate and the ratio of multiple actives. While this can optimize therapeutic outcomes, it may also inadvertently shift the natural balance of compounds, potentially diminishing or altering the intended synergistic effects of whole-plant extracts. In addition to the above, nanocarriers can be engineered to target specific tissues or cells, which may lead to selective delivery of certain compounds over others. This targeted approach, while beneficial for efficacy, might exclude or underrepresent some constituents that contribute to the holistic effect of herbal medicines.

11.10.2 Gaps and Research Needs

There remain significant gaps in understanding how nanoencapsulation affects the complex interplay of plant constituents at the molecular and pharmacokinetic levels. Standardized protocols and further research are needed to evaluate whether nano-delivery systems consistently preserve, enhance, or disrupt the holistic effects characteristic of traditional herbal therapies.

Nano-delivery systems have the potential both to disrupt and to optimize the natural interactions and effects of plant compounds in herbal therapies. The outcome depends on formulation design, the choice of actives, and the delivery strategy. Co-delivery approaches and careful formulation can help preserve synergistic effects, but more research is needed to fully understand and control these complex interactions. The impact of nano-delivery systems is highly variable and depends on the plant species, type of nanoparticle, formulation method, and application dose.

ACKNOWLEDGEMENT

Authors would like to acknowledge School of Pharmacy, Techno India University, West Bengal, for providing facilities required for this work.

REFERENCES

Abd El-Fattah, A. I., M. M. Fathy, Z. Y. Ali, A. E. El-Garawany, and E. K. Mohamed. 2017. "Enhanced Therapeutic Benefit of Quercetin-Loaded Phytosome Nanoparticles in Ovariectomized Rats." *Chemico-Biological Interactions* 271: 30–38. https://doi.org/10.1016/j.cbi.2017.04.020

Alharbi, W. S., F. A. Almughem, A. M. Almehmady, S. J. Jarallah, W. K. Alsharif, N. M. Alzahrani, A. A. Alshehri, and A. A. Alzahrani. 2021. "Phytosomes as an Emerging Nanotechnology Platform for the Topical Delivery of Bioactive Phytochemicals." *Pharmaceutics* 13 (9): 1475. https://doi.org/10.3390/pharmaceutics13091475

Asl, F. D., M. Mousazadeh, S. Taji, A. Bahmani, P. Khashayar, M. Azimzadeh, and E. Mostafavi. 2023. "Nano Drug-Delivery Systems for Management of AIDS: Liposomes, Dendrimers, Gold and Silver Nanoparticles." *Nanomedicine* 18 (3): 279–302. https://doi.org/10.2217/nnm-2022-0274

Awlqadr, F. H., K. R. Majeed, A. B. Altemimi, A. M. Hassan, S. A. Qadir, M. N. Saeed, A. M. Faraj, T. H. Salih, A. J. Abd Al-Manhel, M. A. A Najm, E. Tsakali, J. F. M. Van Impe, A. A. Abd El-Maksoud, and T. G. Abedelmaksoud. 2025. "Nanotechnology-Based Herbal Medicine: Preparation, Synthesis, and Applications in Food and Medicine." *Journal of Agriculture and Food Research*: 101661. https://doi.org/10.1016/j.jafr.2025.101661

Bagdaş, D., B. C. Etoz, Z. Gul, S. Ziyanok, S. Inan, O. Turacozen, N. Y. Gul, A. Topal, N. Cinkilic, S. Tas, M. O. Ozyigit, and M. S. Gurun. 2015. "In Vivo Systemic Chlorogenic Acid Therapy Under Diabetic Conditions: Wound Healing Effects and Cytotoxicity/Genotoxicity Profile." *Food and Chemical Toxicology* 81: 54–61. https://doi.org/10.1016/j.fct.2015.04.001.

Baig, M. S., R. M. Suryawanshi, M. Zehravi, H. S. Mahajan, R. Rana, A. Banu, M. Subramanian, A. K. Kaundal, S. Puri, F. A. Siddiqui, and R. Sharma. 2023. "Surface Decorated Quantum Dots: Synthesis, Properties and Role in Herbal Therapy." *Frontiers in Cell and Developmental Biology* 11: 1139671. https://doi.org/10.3389/fcell.2023.1139671

Balogun, S. T., Okon, K.O., Akanmu, A.O., Paul, L.M., and O. A. Sodipo. 2021. Safety Monitoring of Herbal Medicines in Nigeria: Worrying State of Pharmacovigilance System Based on WHO Core Pharmacovigilance Indicators. *Journal of Herbmed Pharmacology* 10 (2): 202–208. https://doi.org/10.34172/jhp.2021.22

Bhattacharya, S. 2009. "Phytosomes: The New Technology for Enhancement of Bioavailability of Botanicals and Nutraceuticals." *International Journal of Health Research* 2 (3): 225–32. https://doi.org/10.4314/ijhr.v2i3.55421

Blasi, P., S. Giovagnoli, A. Schoubben, M. Ricci, and C. Rossi. 2007. "Solid Lipid Nanoparticles for Targeted Brain Drug Delivery." *Advanced Drug Delivery Reviews* 59 (6): 454–77. https://doi.org/10.1016/j.addr.2007.04.011

Chakraborty, K., A. Shivakumar, and S. Ramachandran. 2016. "Nano-Technology in Herbal Medicines: A Review." *International Journal of Herbal Medicine* 4 (3): 21–7. https://doi.org/10.22271/flora.2016.v4.i3.05

Chang, C., Y. Zhang, Y. C. Kuo, M. Du, K. Roh, R. Gahler, A. Ibi, and J. Solnier. 2025. "Novel Micellar Formulation of Silymarin (Milk Thistle) With Enhanced Bioavailability in a Double-Blind, Randomized, Crossover Human Trial." *Pharmaceutics* 17 (7): 880. https://doi.org/10.3390/pharmaceutics17070880

Chaudhari, P. M., and A. V. Paithankar. 2020. "Herbal Nanogel Formulation: A Novel Approach." *Journal of Science and Technology* 5 (5): 149–53. https://doi.org/10.46243/jst.2020.v5.i5.pp149-153

Chi, C., C. Zhang, Y. Liu, H. Nie, J. Zhou, and Y. Ding. 2020. "Phytosome-Nanosuspensions for Silybin-Phospholipid Complex With Increased Bioavailability and Hepatoprotection Efficacy." *European Journal of Pharmaceutical Sciences* 144: 105212. https://doi.org/10.1016/j.ejps.2020.105212

Chin, C. Y., J. Jalil, P. Y. Ng, and S. F. Ng. 2018. "Development and Formulation of *Moringa oleifera* Standardised Leaf Extract Film Dressing for Wound Healing Application." *Journal of Ethnopharmacology* 212: 188–99. https://doi.org/10.1016/j.jep.2017.10.016.

Ciuca, M. D., and R. C. Racovita. 2023. "Curcumin: Overview of Extraction Methods, Health Benefits, and Encapsulation and Delivery Using Microemulsions and Nanoemulsions." *International Journal of Molecular Sciences* 24 (10): 8874. https://doi.org/10.3390/ijms24108874

Deshmukh, A. S. 2014. "Solid Lipid Nanoparticles." *Research Journal of Pharmaceutical Dosage Forms and Technology* 6 (4): 282. https://doi.org/10.37285/ijpsn.2017.10.3.1

Dewi, M. K., A.Y. Chaerunisaa, M. Muhaimin, and I. M. Joni. 2022. "Improved Activity of Herbal Medicines Through Nanotechnology." *Nanomaterials* 12 (22): 4073. https://doi.org/10.3390/nano12224073

Di, J., X. Gao, Y. Du, H. Zhang, J. Gao, and A. Zheng. 2021. "Size, Shape, Charge and 'stealthy' Surface: Carrier Properties Affect the Drug Circulation Time in Vivo." *Asian Journal of Pharmaceutical Sciences* 16 (4): 444–58. https://doi.org/10.1016/j.ajps.2020.11.006

Fangueiro, J. F., Souto, E. B., and A. M. Silva. 2016. Encapsulation of Nutraceuticals in Novel Delivery Systems. In *Nutraceuticals*, edited by A. M. Grumezescu, 305–342. https://doi.org/10.1016/B978-0-12-804305-9.00009-1.

Gopalakrishnan, A., M. Ram, S. Kumawat, S. K. Tandan, and D. Kumar. 2016. Quercetin Accelerated Cutaneous Wound Healing in Rats by Increasing Levels of VEGF and TGF-β1. *Indian Journal of Experimental Biology* 54 (3): 187–95. https://pubmed.ncbi.nlm.nih.gov/27145632/

Gudise, V., B. Chowdhury, and A. S. Manjappa. 2021. "Antidiabetic and Antihyperlipidemic Effects of *Argyreia pierreana* and *Matelea denticulata*: Higher Activity of the Micellar Nanoformulation Over the Crude Extract." *Journal of Traditional and Complementary Medicine* 11 (3): 259–67. https://doi.org/10.1016/j.jtcme.2021.04.003

Gupta, A., H. B. Eral, T. A. Hatton, and P. S. Doyle. 2016. "Nanoemulsions: Formation, Properties and Applications." *Soft Matter* 12 (11): 2826–41. https://doi.org/10.1039/C5SM02958A

Idrees, H., S. Z. Zaidi, A. Sabir, R. U. Khan, X. Zhang, and S. U. Hassan. 2020. "A Review of Biodegradable Natural Polymer-Based Nanoparticles for Drug Delivery Applications." *Nanomaterials* 10 (10): 1970. https://doi.org/10.3390/nano10101970

Jalili, A., R. Bagherifar, A. Nokhodchi, B. Conway, and Y. Javadzadeh. 2023. "Current Advances in Nanotechnology-Mediated Delivery of Herbal and Plant-Derived Medicines." *Advanced Pharmaceutical Bulletin* 13 (4): 712–22. https://doi.org/10.34172/apb.2023.056

Jampílek, J., K. Kráľová, E. V. R. Campos, and L. F. Fraceto. 2019. "Bio-Based Nanoemulsion Formulations Applicable in Agriculture, Medicine, and Food Industry." In *Nanobiotechnology in Bioformulations*, 33–84. https://doi.org/10.1007/978-3-030-17061-5_2

Jo, Y.-J., S.-B. Lee, J.-K. Lee, and Y.-J. Kwon. 2025. "Preparation of Nanoemulsions Containing Curcumin by High Pressure Homogenization." *Food Engineering Progress* 18 (4). https://doi.org/10.13050/fep.2025.18.4.341

Karaś, M., A. Jakubczyk, U. Szymanowska, U. Złotek, and E. Zielińska. 2017. "Digestion and Bioavailability of Bioactive Phytochemicals." *International Journal of Food Science & Technology* 52 (2): 291–305. https://doi.org/10.1111/ijfs.13323

Kaur, K., U. Nautiyal, and D. Singh. 2015. "Nanostructured Lipid Carrier for Bioavailability Enhancement." *International Journal of Advanced Science and Technology* 2 (1): 1–9. https://doi.org/10.30750/ijast.211

Khulbe, P., D.M. Singh, A. Aman, E. D. Ahire, and R. K. Keservani. 2023. "The Emergence of Nanocarriers in the Management of Diseases and Disorders." *Community Acquired Infection* 10. https://doi.org/10.54844/cai.2022.0139

Kim, S. J., N. Puranik, D. Yadav, J.O. Jin, and P. C. Lee. 2023. "Lipid Nanocarrier-Based Drug Delivery Systems: Therapeutic Advances in the Treatment of Lung Cancer." *International Journal of Nanomedicine* 18: 2659–76. https://doi.org/10.2147/IJN.S414429

Kubra, R., and L. Jaganmohanrao. 2012. "An Overview on Inventions Related to Ginger Processing and Products for Food and Pharmaceutical Applications." *Recent Patents on Food, Nutrition & Agriculture* 4 (1): 31–49. https://doi.org/10.2174/2212798411204010031

Kumar, S., A. Baldi, and D. K. Sharma. 2021. "In Vitro Antioxidant Assay Guided Ex Vivo Investigation of Cytotoxic Effect of Phytosomes Assimilating Taxifolin-Rich Fraction of *Cedrus deodara* Bark Extract on Human Breast Cancer Cell Lines (MCF-7)." *Journal of Drug Delivery Science and Technology* 63: 102486. https://doi.org/10.1016/j.jddst.2021.102486

Kunle, O. F., Egharevba, H.O., and P. O. Ahmadu. 2012. Standardization of herbal medicines: A review. *International Journal of Biodiversity and Conservation* 4 (3): 101–12. https://doi.org/10.5897/IJBC11.163.

Liu, M., Wei, X., Zheng, Z., Li, Y., Li, M., Lin, J., and L. Yang. 2023. Recent advances in nano-drug delivery systems for the treatment of diabetic wound healing. *International Journal of Nanomedicine* 18: 1537–60. https://doi.org/10.2147/IJN.S401540.

Martins-Gomes, C., E. B. Souto, and A. M. Silva. 2022. Nanophytosomes: A Novel Approach for the Delivery of Herbal Drugs. In *Systems of Nanovesicular Drug Delivery*, 239–257. https://doi.org/10.1016/B978-0-323-91100-0.00015-1

Maurya, S. D., S. Prajapati, A. Gupta, G. Saxena, and R. C. Dhakar. 2010. "Formulation Development and Evaluation of Ethosome of Stavudine." *International Journal of Pharmaceutical Education and Research* 13 (1): 16–22. https://dx.doi.org/10.2139/ssrn.3774625

Mazumder, A., Dwivedi, A., Du Preez, J. L., and J. Du Plessis. 2016. In vitro wound healing and cytotoxic effects of sinigrin–phytosome complex. *International Journal of Pharmaceutics* 498 (1–2): 283–93. https://doi.org/10.1016/j.ijpharm.2015.12.019

Misra, S. 2012. "Randomized Double Blind Placebo Control Studies, the 'Gold Standard' in Intervention-Based Studies." *Indian Journal of Sexually Transmitted Diseases and AIDS* 33 (2): 131–34. https://doi.org/10.4103/0253-7184.102108

Mohamad, N. A. N., N. A. Arham, J. Jai, and A. Hadi. 2014. "Plant Extract as Reducing Agent in Synthesis of Metallic Nanoparticles: A Review." *Advanced Materials Research* 832: 350–55. https://doi.org/10.4028/www.scientific.net/AMR.832.350

Morgan, K. 1994. *Medicine of the Gods: Basic Principles of Ayurvedic Medicine*. Oxford: Mandrake.

Muhammad, A. A., P. Arulselvan, P. S. Cheah, F. Abas, and S. Fakurazi. 2016. "Evaluation of Wound Healing Properties of Bioactive Aqueous Fraction from *Moringa oleifera* Lam on Experimentally Induced Diabetic Animal Model." *Drug Design, Development and Therapy* 10: 1715–30. https://doi.org/10.2147/DDDT.S113365

Netzer, K., M. Balmith, and B. T. Flepisi. 2021. "An Appraisal of the Regulatory Policies Governing the Use of Herbal Traditional Medicine." *Traditional Medicine Research* 6 (6): 57. https://doi.org/10.53388/TMR20211013249

Obodozie, O. O. 2012. Pharmacokinetics and Drug Interactions of Herbal Medicines: A Missing Critical Step in the Phytomedicine/Drug Development Process. In *Reading in Advanced Pharmacokinetics—Theory, Methods, and Applications*, 127–56. https://doi.org/10.5772/30690

Panja, A., and P. Patra. 2023. "A Review on Quantum Dots (QDs) and Their Biomedical Applications." *4open* 6: 1. https://doi.org/10.1051/fopen/2022020

Paroha, S., R. P. Dewangan, and P. K. Sahoo. 2020. Pharmaceutical Technology for Improving the Bioavailability of Natural Products. In *Sustainable Agriculture Reviews 43: Pharmaceutical Technology for Natural Products Delivery Vol. 1 Fundamentals and Applications*, 1–32. https://doi.org/10.1007/978-3-030-41384-2_1

Prasad, R., and K. Selvaraj. 2024. "Choice of Nanoparticles for Theranostics Engineering: Surface Coating to Nanovalves Approach." *Nanotheranostics* 8 (1): 12. https://doi.org/10.7150/ntno.10982

Rahman, R. 2023. Plant Nanonutrients for Sustainable Agriculture. In *Nanofertilizers for Sustainable Agroecosystems: Recent Advances and Future Trends*, 151–79. https://doi.org/10.1007/978-3-031-41329-2_6

Rahmani, A. H., and S. M. Aly. 2014. "Active Ingredients of Ginger as Potential Candidates in the Prevention and Treatment of Diseases via Modulation of Biological Activities." *International Journal of Physiology, Pathophysiology and Pharmacology* 6 (2): 125–36. https://doi.org/10.2147/IJPP.S60712

Rani, A., S. Kumar, and R. K. Khar. 2022. "*Murraya koenigii* Extract Loaded Phytosomes Prepared Using Antisolvent Precipitation Technique for Improved Antidiabetic and Hypolipidemic Activity." *Indian Journal of Pharmaceutical Education and Research* 56: s326–38. https://doi.org/10.5530/ijper.56.2s.103

Rani, R., S. Kumar, N. Dilbaghi, and R. Kumar. 2020. "Nanotechnology Enabled the Enhancement of Antitrypanosomal Activity of Piperine Against *Trypanosoma evansi*." *Experimental Parasitology* 219: 108018. https://doi.org/10.1016/j.exppara.2020.108018

Sanhueza, C., D. Vergara, C. Chávez-Aravena, F. Gálvez-Jirón, E. Chávez-Angel, and A. Castro-Álvarez. 2023. "Functionalizing Dendrimers for Targeted Delivery of Bioactive Molecules to Macrophages: A Potential Treatment for Mycobacterium Tuberculosis Infection—A Review." *Pharmaceuticals* 16 (10): 1428. https://doi.org/10.3390/ph16101428

Seema, A. 2014. "Recent Development of Herbal Formulation—A Novel Drug Delivery System." *International Ayurvedic Medical Journal* 2 (6): 952–958. https://doi.org/10.46607/iamj2008122014

Shamsuddin, N. A. M., and M. H. Zulfakar. 2023. "Nanostructured Lipid Carriers for the Delivery of Natural Bioactive Compounds." *Current Drug Delivery* 20 (2): 127–43. https://doi.org/10.2174/1567201819666220324094234
Singh, H., and G. Sharma. 2023. "Recent Development of Novel Drug Delivery of Herbal Drugs." *RPS Pharmacy and Pharmacology Reports* 2 (4): rqad028. https://doi.org/10.1093/rpsppr/rqad028
Singh, R. P., H. V. Gangadharappa, R. M. Narke, and P. V. Jadhav. 2017. "Stability-Indicating RP-HPLC Method Development for the Estimation of Gingerol." *International Research Journal of Pharmacy* 8 (8): 56–61. https://doi.org/10.7897/2230-8407.088145
Sogut, O., U. A. Sezer, and S. Sezer. 2021. "Liposomal Delivery Systems for Herbal Extracts." *Journal of Drug Delivery Science and Technology* 61: 102147. https://doi.org/10.1016/j.jddst.2020.102147
Sokolov, Yu. V. 2014. "Nanoemulsion Formation by Low-Energy Methods: A Review." *News of Pharmacy* 3 (79): 16–9. https://doi.org/10.24959/nphj.14.1981
Song, J. W., Y. S. Liu, Y. R. Guo, W. X. Zhong, Y. P. Guo, and L. Guo. 2022. "Nano–Liposomes Double Loaded with Curcumin and Tetrandrine: Preparation, Characterization, Hepatotoxicity and Anti–Tumor Effects." *International Journal of Molecular Sciences* 23 (12): 6858. https://doi.org/10.3390/ijms23126858
Souto, E. B., J. Dias-Ferreira, J. Oliveira, E. Sánchez-Lopez, A. Lopez-Machado, M. Espina, M. L. Garcia, S. B. Souto, C. Martins-Gomes, and A. M. Silva. 2019. "Trends in Atopic Dermatitis—From Standard Pharmacotherapy to Novel Drug Delivery Systems." *International Journal of Molecular Sciences* 20 (22): 5659. https://doi.org/10.3390/ijms20225659
Souto, E. B., A. R. Fernandes, C. Martins-Gomes, T. E. Coutinho, A. Durazzo, M. Lucarini, S. B. Souto, A. M. Silva, and A. Santini. 2019. "Nanomaterials for Skin Delivery of Cosmeceuticals and Pharmaceuticals." *Applied Sciences* 10 (5): 1594. https://doi.org/10.3390/app10051594
Souto, E. B., A. C. Sampaio, J. R. Campos, C. Martins-Gomes, A. Aires, and A. M. Silva. 2019. Polyphenols for Skin Cancer: Chemical Properties, Structure-Related Mechanisms of Action and New Delivery Systems. In *Bioactive Natural Products, Studies in Natural Products Chemistry* Vol. 63, 21–42. https://doi.org/10.1016/b978-0-12-817901-7.00002-2
Stoyanova, N., N. Nachev, and M. Spasova. 2023. "Innovative Bioactive Nanofibrous Materials Combining Medicinal and Aromatic Plant Extracts and Electrospinning Method." *Membranes* 13 (10): 840. https://doi.org/10.3390/membranes13100840
Tapadiya, G. G. 2017. "Impact of Nanotechnology on Global Trade of Herbal Drugs: An Overview." *International Journal of Green Pharmacy (IJGP)* 11 (3): 161–5. https://doi.org/10.22377/ijgp.v11i03.1490
Telange, D. R., A. T. Patil, A. M. Pethe, H. Fegade, S. Anand, and V. S. Dave. 2017. "Formulation and Characterization of an Apigenin-Phospholipid Phytosome (APLC) for Improved Solubility, In Vivo Bioavailability, and Antioxidant Potential." *European Journal of Pharmaceutical Sciences* 108: 36–49. https://doi.org/10.1016/j.ejps.2016.10.030
Yankuzo, H., Q. U. Ahmed, R. I. Santosa, S. F. Akter, and N. A. Talib. 2011. "Beneficial Effect of the Leaves of *Murraya koenigii* (Linn.) Spreng (Rutaceae) on Diabetes-Induced Renal Damage In Vivo." *Journal of Ethnopharmacology* 135 (1): 88–94. https://doi.org/10.1016/j.jep.2011.03.073
Zhang, L., J. Yan, X. Liu, Z. Ye, X. Yang, R. Meyboom, K. Chan, D. Shaw, and P. Duez. 2012. "Pharmacovigilance Practice and Risk Control of Traditional Chinese Medicine Drugs in China: Current Status and Future Perspective." *Journal of Ethnopharmacology* 140(3): 519–25. https://doi.org/10.1016/j.jep.2012.01.056
Zhou, F., T. Peterson, Z. Fan, and S. Wang. 2023. "The Commonly Used Stabilizers for Phytochemical-Based Nanoparticles: Stabilization Effects, Mechanisms, and Applications." *Nutrients* 15 (18): 3881. https://doi.org/10.3390/nu15183881

12 Way Forward
Strategies and Prospects in Herbal Drug Development

Sonal Dubey, J. Josephine Leno Jenita, Shanaz Banu, and Seema S. Rathore

12.1 INTRODUCTION

Herbal medicine/phytotherapy refers to the use of plants or plant extracts for therapeutic purposes. It encompasses a wide range of practices that utilize various parts of plants, including leaves, roots, flowers, and seeds, to prevent, treat, or cure health conditions. Herbal drug development has gained significant momentum due to its therapeutic potential, technological advancements, and expanding market opportunities. Herbal drug development stands at the intersection of ancient wisdom and modern innovation, offering immense potential for healthcare, industry, and entrepreneurship. Throughout this book, *Herbal Drug Development: Insight into Techno-Entrepreneurial Opportunities*, the authors have explored the multifaceted journey of herbal medicines, from the rich legacy of traditional knowledge systems to the cutting-edge advancements in extraction technologies, standardization, regulatory frameworks, and commercialization strategies.

The central themes woven through the book highlight the importance of integrating time-tested ethnobotanical practices with rigorous scientific validation and contemporary business models. The book has examined the challenges of ensuring quality, efficacy, and safety in herbal products, the complexities of navigating regulatory landscapes, and the need for sustainable sourcing and conservation of medicinal plant resources. Furthermore, the book has illuminated the expanding global market for herbal products and the rising opportunities for innovation-driven entrepreneurship in this sector.

As the herbal drug industry evolves, a forward-looking perspective becomes essential. Rapid technological advancements, shifting consumer preferences, and increasing demand for evidence-based natural therapies are reshaping the landscape. Stakeholders-including researchers, entrepreneurs, policymakers, and traditional practitioners-must anticipate emerging trends, adapt to new challenges, and seize opportunities for responsible growth.

The aim of this concluding chapter is to synthesize the key insights presented in the preceding sections and to chart actionable directions for all stakeholders involved in herbal drug development. By reflecting on lessons learned and envisioning future possibilities, this chapter seeks to provide a strategic roadmap that fosters innovation, sustainability, and inclusive growth in the herbal sector. Ultimately, the goal is

DOI: 10.1201/9781003623748-12

to inspire continued collaboration and proactive engagement, ensuring that herbal drug development remains a dynamic and impactful field for years to come.

This chapter explores the integration of modern scientific techniques such as nanotechnology, biotechnology, and AI-driven drug discovery to enhance herbal formulations. It also examines the importance of standardization, quality control, and regulatory frameworks in ensuring efficacy and safety. Furthermore, entrepreneurial opportunities in herbal pharmaceuticals are analyzed, highlighting innovation, sustainable practices, and successful business models in the industry. With increasing global demand for natural remedies, techno-entrepreneurial initiatives in herbal drug development present promising avenues for both research and commercialization.

This chapter also presents the critical insights throughout the book and outlines a strategic vision for the future of herbal drug development. It underscores integrating traditional medicinal knowledge with modern scientific approaches, emphasizing the need for robust phytochemical standardization, clinical validation, and regulatory harmonization. Technological advancements—particularly the emergence of nano-formulations, targeted delivery systems, and green extraction techniques—are highlighted as transformative tools for enhancing herbal drug bioavailability, efficacy, and safety. The chapter also explores the role of digital technologies, such as artificial intelligence (AI), blockchain for supply chain traceability, and e-commerce platforms, in reshaping product development and market dynamics. Persistent challenges, including regulatory inconsistencies, gaps in clinical evidence, and sustainability concerns, are addressed with actionable recommendations, such as fostering interdisciplinary R &D, promoting public-private partnerships, and implementing supportive policy reforms. Looking ahead, it identifies emerging trends and future opportunities shaped by technological innovation, globalization, and evolving consumer preferences (Figure 12.1). Furthermore, it identifies emerging opportunities in nutraceuticals, functional foods, and personalized herbal therapeutics. By offering a techno-entrepreneurial roadmap, this

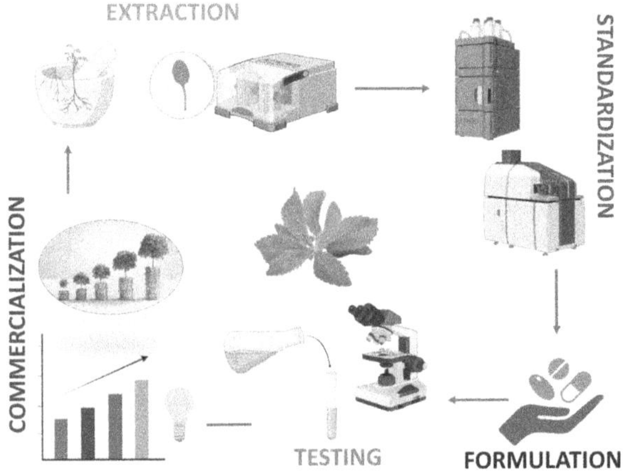

FIGURE 12.1 Overview of herbal journey from extraction to market.

chapter envisions a resilient, innovation-driven, and globally competitive herbal sector that bridges tradition with technology and science with commerce.

12.2 HERBAL DRUG DISCOVERY AND DEVELOPMENT

Discovery of herbal drugs and development focuses on identification, isolation, and testing the compounds from herbs for their therapeutic potential. This chapter merges traditional herbal knowledge with modern scientific approaches for the development of novel medications (Figure 12.2).

12.2.1 Sources of Herbal Drugs: Traditional vs. modern approaches

Traditional approaches: Ethnobotany draws the traditional knowledge from indigenous communities about medicinal plants and their applications (Xiao et al. 2021; NCBI 2025). They rely on ancient texts and practices that detail herbal remedies for different health conditions. Empirical knowledge relies on centuries of hands-on experience and anecdotal observations that have guided herbal usage over the ages. Ethnobotanical Knowledge is traditional healing systems like Ayurveda, Siddha, Unani, homeopathy, and Chinese medicine have historically depended upon plant knowledge, acquired through practical

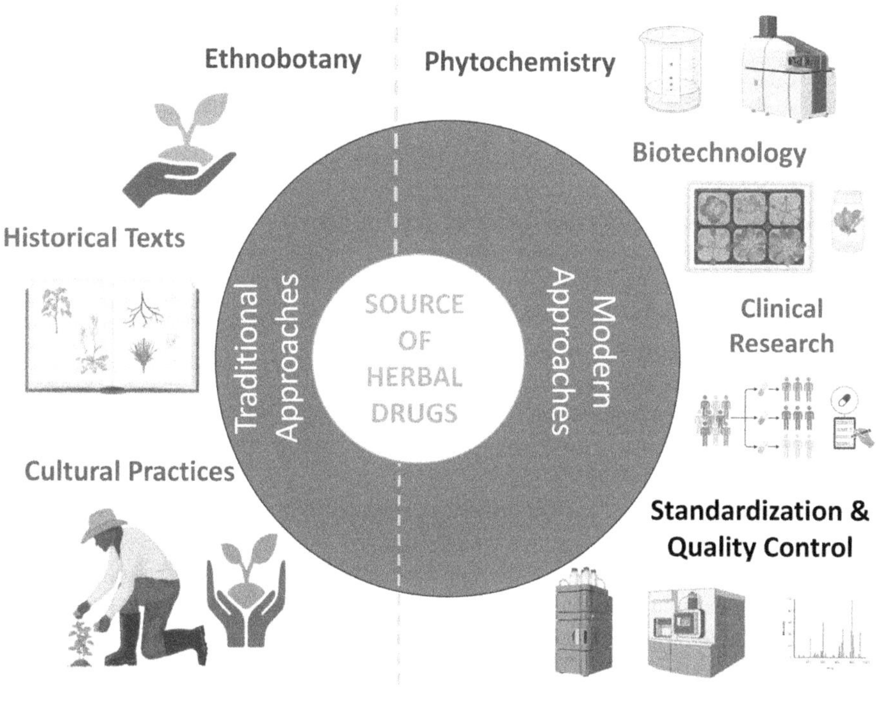

FIGURE 12.2 Comparison of ethnobotanical and phytochemical approaches.

use and empirical observations. Historical Texts, such as the Huangdi Neijing in traditional Chinese medicine and the Charaka Samhita and Sushruta Samhita in Ayurveda, offer us valuable insights into the medicinal properties of plants and plant-based remedies in ancient healing practices. Cultural rituals involving herbal practices, reflecting both spiritual beliefs and ancestral customs, are often handed down through storytelling and lived experience across generations (Ouoba, Fofana, and Semdé 2022).

Modern Approaches: Phytochemistry focuses on identifying and analyzing the bioactive phytocompounds with potential therapeutic applications. Advanced analytical techniques like chromatography and mass spectrometry enable the isolation and identification of active compounds in medicinal plants. This offers insights into their chemical composition and pharmacological properties. Biotechnology utilizes advanced techniques like gene editing and micropropagation to improve the production and potency of herbal medicines. Plant tissue culture and genetic engineering tools can increase the yield of specific therapeutic compounds in plants, enabling the cultivation of varieties with higher concentrations of beneficial compounds. Through rigorous systematic research assessing herbal products for safety and efficacy by clinical trials, it contributes to creating standardized and effective formulations. This approach merges traditional healing with modern medicine to meet the current healthcare standards. By the fusion of traditional insights and contemporary scientific validation, a holistic and effective approach is achieved in herbal drug development. Thereby bridging the gap between modern medicine and ancient healing (Li et al. 2023a).

12.2.2 Bioactive Compounds and Their Therapeutic Potential

Naturally occurring chemical substances in plants are the bioactive compounds that affect the biological systems in a meaningful way. Their therapeutic effects make them a major focus in drug discovery and development (Hao et al, 2024).

Alkaloids are basic nitrogenous compounds that have pronounced effects on the nervous system (Krishna, Ulhas, and Malaviya 2024). Examples involve morphine from Opium poppy capsules, quinine from Cinchona bark, potentially used for analgesic, antimalarial activity, respectively.

Flavonoids are polyphenolic compounds abundantly found in fruits, vegetables, plant-based foods, and herbs and are valued for their antioxidant activity. They exhibit anti-inflammatory, antiviral effects, lower cancer risk, and improve heart function (Wenbo et al. 2024).

Terpenoids are a diverse class of organic compounds that are produced through plant metabolic pathways. These are isoprenoids derived from isoprene units and widely found in plants, in essential oils, resins, and pigments. Examples include saponins from ginseng and menthol from mint, among others. These compounds are known for their anti-inflammatory, antimicrobial, and anticancer effects (Li et al. 2023b).

Glycosides are secondary metabolites with a sugar moiety and a non-sugar moiety linked to each other with a glycosidic linkage, enhancing the pharmacological properties of plant-derived drugs. Therapeutic potential effects of it on heart disease

TABLE 12.1
Examples of Synergistic Effects in the Therapeutic Potential of Bioactive Compounds in Herbal Drug Discovery

Synergy Type	Herbal Components/ Examples	Observed Effect
Synergistic antimicrobial activity	*Vitellaria paradoxa* (Shea butter tree) leaf extract + Ampicillin	Enhanced activity against methicillin-resistant *Staphylococcus aureus* by targeting beta-lactamase enzymes (Thomas, Deming, and Sarkar 2022)
Synergistic antioxidant and antimicrobial	*Triticum aestivum* (Wheatgrass) + *Ocimum sanctum* (Tulsi) + *Tinospora cordifolia* (Giloy)	Greater antioxidant and antimicrobial effects due to combined phenolic and bioactive compounds (Laureanti et al. 2023)
Synergistic anticancer effects	*Wedelia chinensis* compounds: indole-3-carboxyaldehyde, wedelolactone, luteolin, apigenin	Stronger anticancer activity at natural ratios; removal of any one compound reduces overall efficacy (Shan et al. 2021)
Synergistic herb-pair combinations	*Salvia miltiorrhiza* (Danshen) + *Pueraria lobata* (Gegen)	Enhanced anti-inflammatory and anti-atherogenic effects for coronary heart disease treatment (Wang et al. 2007)
Enhancement of drug efficacy	Polyphyllin I, Evodiamine + chemotherapy agents (platinum or 5-Fluorouracil)	Stronger inhibition of gastric cancer cells than chemotherapy alone (Yang et al. 2025)
Improved bioavailability and multi-targeting	Piperine (black pepper) + Curcumin (turmeric) Quercetin and Kaempferol (*Ginkgo biloba* extract)	Increased bioavailability and combined anti-inflammatory/neuroprotective action via complementary mechanisms (Ehambarampillai and Wan 2025)

management and as a diuretic. Phenolic Compounds are a diverse group of compounds, like tannins and phenolic acids, with strong antioxidant activity (Wang et al. 2024).

Enlisted below in Table 12.1 are some of the examples of synergistic effects in the therapeutic potential of bioactive compounds in herbal drug discovery.

In Personalized Medicine, detailed knowledge of bioactive compounds and an individual's biological characteristics opens new avenues for the development of personalized therapies aligning with patients' unique metabolic and genetic profiles.

12.2.3 Preclinical Research and Efficacy Studies

Preclinical research and efficacy studies of the plant drugs are crucial phases in drug development. Such studies are carried out in labs through *in vitro* and *in vivo* studies, in order to evaluate the safety, efficacy, and mechanism of action of drug formulations (Figure 12.3) (Zhang et al. 2024).

This foundational research is vital in determining the entry of drugs into clinical trial phases. Given below is a detailed overview of these critical steps:

Phytochemical analysis involves isolation, identification, and profiling of bioactives within a herbal formulation, utilizing advanced analytical tools such as

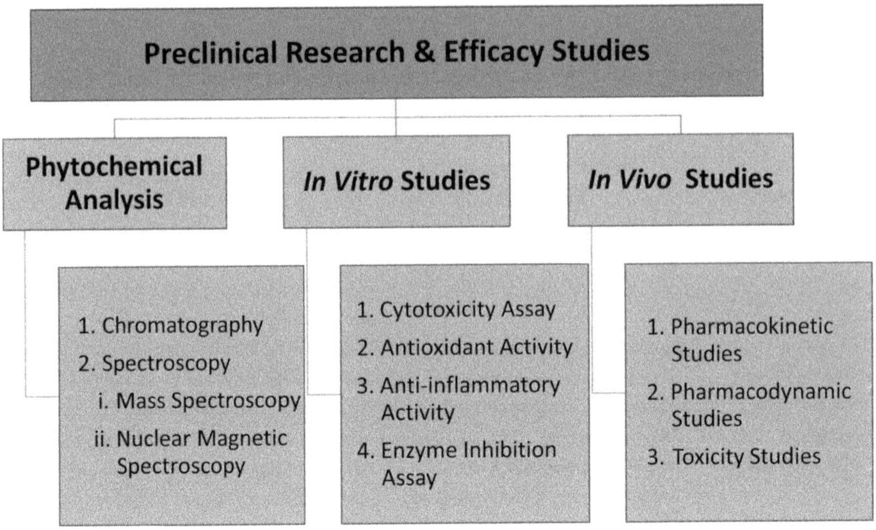

FIGURE 12.3 Requisites for preclinical research and to establish the efficacy of phytopharmaceuticals.

chromatography, mass spectrometry, and nuclear magnetic resonance spectroscopy. Phytochemical profiling is important for linking the bioactives to their medicinal activity (Badoni Semwal et al. 2014). Lab experiments using cell cultures determine the biological effects of herbal extracts (*In Vitro* Studies). High-throughput screening rapidly analyzes the therapeutic potential of multiple herbal components.

Cell culture assays test the herbal drugs on isolated cells to evaluate the cytotoxicity and biological activity. Some typical examples include:

Enzyme Activity Assays: Evaluating the effects of bioactives on cytochrome P450 enzymes (CYPs) is crucial for understanding drug metabolism. Researchers also explored the influence of herbal compounds on transport proteins such as organic anion transporting polypeptides (OATPs) and P-glycoprotein (P-gp) (Wang et al. 2014).

Anti-Inflammatory Assays: Protein denaturation assay assesses the potential of herbal extracts to inhibit protein denaturation, which is a key factor in inflammatory responses. The membrane stabilization assay measures the ability of plant compounds to protect the red blood cells against stress-induced lysis (Joshi et al. 2025). Enzyme inhibition studies analyze the influence of herbal compounds on key enzymes that play vital roles in inflammation such as cyclooxygenase (COX) and lipoxygenase (LOX) (Dubey and Shukla 2023).

In Vivo Studies: The pharmacological effects of the herbal drugs are assessed by using animal models. The parameters evaluated by researchers include optimal dosage, administration routes, pharmacodynamic and pharmacokinetic behavior, i.e., absorption, distribution, metabolism, and excretion. Pharmacodynamic parameters help us to evaluate the biological effects of herbal drugs, including their onset, duration, and mechanism of action, along with the therapeutic potential (Dotsenko and Mosiĭchuk 2012). Such evaluations are essential for understanding the safety and efficacy of herbal drugs in the living system.

Toxicological Assessment: Toxicity studies are evaluated by performing acute, sub-chronic, and chronic toxicity studies to establish the safety of herbal compounds. Once safety is validated through toxicity studies, the herbal formulation is evaluated for its consistent and measurable clinical effects. Mechanism of Action Studies are examined to explore the influence of active constituents on molecular targets, such as interactions with receptors, impact on cellular processes, and modulation of signaling pathways. In Comparative Studies, researchers compare herbal formulations with standard drugs or placebos to evaluate their relative therapeutic value and clinical significance. Such comparisons are very important for clinical validation and to obtain regulatory approval. Clinical validation advances into human trials to confirm the safety and efficacy of herbal formulations. However, the multifaceted nature of herbal formulations presents unique challenges in drug development (Zeng, Song, and Shen 2012). For successful integration of herbal medicines in mainstream healthcare, regulatory compliance, standardization, and reproducibility must be assured.

Long-term efficacy studies of herbal drugs for managing chronic conditions must be monitored to identify the risk of side effects and complications. Systematic assessment of these parameters enables the researchers to validate the herbal therapies for safe human use and creates a bridge between traditional remedies and contemporary clinical practice.

12.3 FORMULATION AND STANDARDIZATION OF HERBAL DRUGS

12.3.1 Techniques for Herbal Extraction and Purification

The fundamental process in herbal drug development is the extraction and purification of bioactive compounds. The techniques used to isolate the phytoconstituents should be such that they maintain their medicinal properties and product quality and consistency are ensured. Efficient extraction, isolation and purification methods are required for preserving the therapeutic effects of plant material (Figure 12.4). The most widely used methods of extraction and purification are:

Extraction Techniques: *Maceration* is the basic and conventional technique that involves the plant material being soaked in a suitable solvent at room temperature for a specific duration, which allows the soluble bioactives to slowly dissolve in the given solvent. *Soxhlet extraction* is a continuous extraction method utilizing the Soxhlet apparatus, in which the plant material is loaded into a thimble and continuously extracted with repeatedly condensed solvent vapors. This continuous process assures effective and

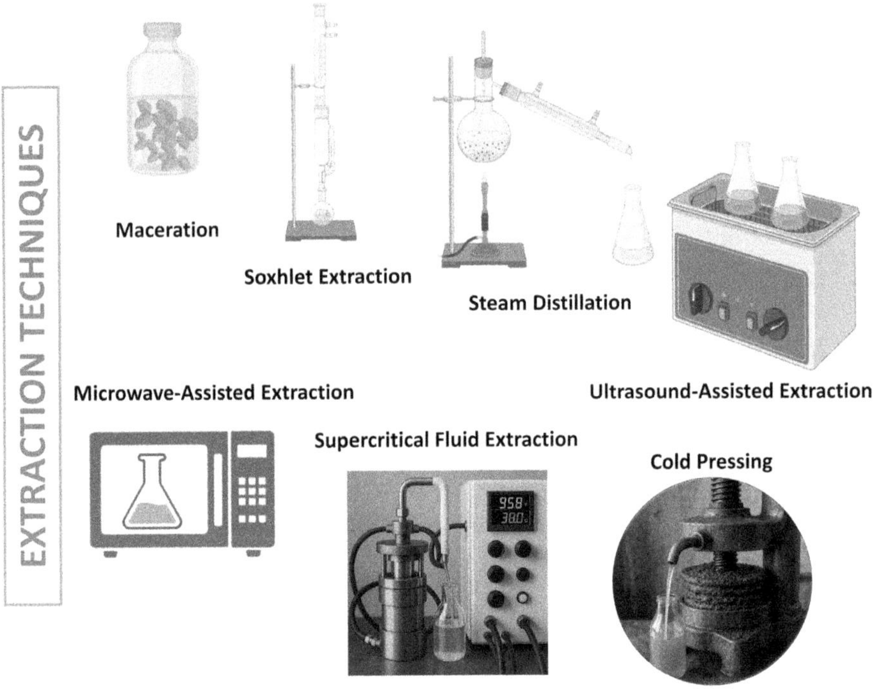

FIGURE 12.4 Various herbal extraction techniques.

exhaustive extraction of herbal phytocompounds (Panontin et al. 2022). *Steam distillation* is a widely used method of isolating essential oils from the aromatic plant material by directing steam through the plant material, where the volatile components will vaporize. The combination of essential oil and steam is then condensed and separated accordingly to obtain the essential oil (Ibrahim et al. 2022). Application of *ultrasound-assisted extraction (UAE)* with high-frequency sound waves to enhance the cell permeability and disrupt the cell walls of plant cells and release the bioactives in the solvent of extraction. This process yields high yields within a shorter time as compared to the conventional methods (Sanou et al. 2023). *Cold Pressing* is a mechanical method mainly used for the extraction of fixed oils from seeds and fruits without the aid of heat, and retains the nutritional value and the natural quality of oils (Rehman, Choe, and Yoo 2016).

Supercritical fluid extraction (SFE) incorporates supercritical CO_2 as a solvent of extraction, which enables efficient extraction of bioactives with minimal thermal degradation (Pastare et al. 2023). This green technology ensures superior purity of extracts and avoids the use of toxic organic solvents. *Microwave-assisted extraction (MAE)* utilizes microwave energy to heat the plant material and the solvent instantly, in order to enhance the extraction of bioactives in minimal time (Tsai et al. 2024).

Purification Techniques: In the standardization of herbal drugs, purification techniques play a crucial role in ensuring the quality and therapeutic efficacy of formulations. Chromatographic methods such as TLC, HPLC, and GC-MS are commonly employed to separate and identify bioactive constituents based on their differential affinities toward stationary and mobile phases. Molecular imprinting polymers (MIPs) offer a sophisticated approach to purification by employing synthetic polymers engineered with selective binding sites for target molecules. Together, these advanced techniques significantly enhance the purity, consistency, and pharmacological reliability of herbal products, making them more suitable for pharmaceutical applications. Electrophoresis is a method that utilizes an electric field to drive the charged molecules through a matrix gel that enables the separation of mixtures based on charge and molecular weight. It's especially used for analyzing complex mixtures like proteins and genetic material. The selection of the extraction technique and purification method depends on the type of plant material, properties of desired components, desired therapeutic purpose. Utilizing multiple extraction and purification techniques often enhances efficiency and results in maximum recovery of components while preserving their therapeutic properties. Table 12.2 summarizes these techniques and some of the references where these techniques are used.

12.3.2 Quality Control and Standardization Parameters

Quality control in herbal drug development is a critical process aimed at ensuring the safety, efficacy, and consistency of herbal formulations, which are inherently variable due to differences in plant sources, cultivation practices, and environmental

TABLE 12.2
Purification Techniques for Ensuring Quality in Herbal Drug Standardization

S.No	Technique	Principle	Application
1.	Chromatography	Separation based on differential affinities toward mobile and stationary phases	TLC, HPLC, and GC-MS help isolate and identify bioactive constituents (Zhao et al. 2025)
2.	Liquid-Liquid Extraction	Partitioning of compounds between two immiscible solvents	Isolates target phytochemicals from complex herbal mixtures
3.	Crystallization	Selective formation of crystals from a supersaturated solution	Purifies bioactive molecules by removing soluble impurities (Takeno et al. 2018)
4.	Membrane Filtration	Size- or pressure-based separation using semi-permeable membranes	Eliminate particulates, microbes, or small impurities from herbal extracts
5.	Molecular Imprinting Polymers (MIPs)	Synthetic polymers are engineered with specific binding sites for target molecules	Offers highly selective purification of specific bioactives (Sahu et al. 2025)

conditions (Yau et al 2015). It begins with proper botanical identification to confirm the correct species and plant parts, followed by phytochemical profiling to verify the presence and concentration of active constituents using techniques like HPTLC and HPLC. Standardization ensures batch-to-batch uniformity, while contaminant testing screens for harmful residues such as heavy metals, pesticides, and microbial contaminants (Wang et al. 2023). Stability studies are conducted to determine shelf life, and appropriate packaging and labeling help preserve product integrity and enhance traceability. This holistic approach integrates pharmacognosy, analytical chemistry, and regulatory frameworks, but challenges remain, including the need for global standardization, risk of adulteration, and sustainable sourcing. Addressing these factors is essential for developing high-quality, scientifically validated herbal products. These quality control parameters and evaluation test for the herbal drug are given in Table 12.3 and 12.4, respectively. The herbal drugs need to be standardized as per specific parameters as stated in Table 12.5.

12.3.3 Innovative Drug Delivery Systems for Herbal Formulations

The rising interest in herbal medicines has spurred innovation in advanced delivery methods to maximize their clinical potential. As the holistic wellness sector gains momentum, advanced technologies open new techno-entrepreneurial avenues in herbal medicine. This section examines innovative drug delivery platforms for herbal products and the business opportunities they present. Nano formulations have significantly transformed herbal drug development by enhancing bioavailability, stability and efficacy. The major techniques include **nanotechnology-based delivery systems** is a powerful tool for transforming herbal drug delivery, by encapsulating

TABLE 12.3
Key Parameters of Quality Control in Herbal Drug Development

Parameter	Purpose/Details
Botanical Identification	Authenticates the right species of plant using morphological, microscopical, and DNA tools (Chen et al. 2014)
Foreign Matter	Ensures that the plant material is free from impurities and adulterants (Kim et al. 2021)
Moisture Content	Ensures no microbial growth, degradation, and spoilage (Chen et al. 2010)
Ash Values	Measures the amount of carbonates, phosphates, silicates, and silica. It detects the impurities present (Chen et al. 2010)
Extractive Values	Defines the amount of phytoconstituents soluble in the respective solvents (Wang et al. 2019)
Phytochemical Screening	Identifies the presence of various phytoconstituents (e.g., glycosides, alkaloids) (Wang et al. 2019)
Microbial Limit Tests	Tests the presence of microorganisms
Heavy Metal Testing	Ensures the extent of heavy metals like lead, arsenic levels are within safe limits
Pesticide Residue Analysis	Detects fertilizer and harmful pesticide residues from cultivation

TABLE 12.4
A Series of Tests and Evaluations to Ensure the Integrity and Safety of Herbal Products

Quality Control Parameter	Description	Techniques Used
Identification of Plant Material	Confirms the right plant species for formulation	Morphological, anatomical, and molecular methods (DNA barcoding)
Purity Testing	Detects for presence of pesticides, microbes and heavy metals.	Chromatographic and Spectroscopic methods are used (HPLC, GC), spectroscopic methods (UV-Vis, FTIR)
Active Ingredient Content	Estimates the bioactive components concentration, which determines the efficacy	Quantitative Determination using HPLC, Spectrophotometry
Physical and Chemical Properties	Identifies the drug for its color, odor, taste, solubility, etc.	Sensory evaluation, testing of physico-chemical parameters
Stability Testing	Assesses the stability and quality changes over time at different temperature conditions	Evaluates the shelf life at varied temperatures, light, and humid conditions

TABLE 12.5
Standardization Parameters in Herbal Drug Formulation

Standardization Parameter	Description	Purpose	Additional Details	Refinements & Insights
Phytochemical Profiling	Defines reference for active constituents	Ensures stability and therapeutic efficacy	Utilizes chromatographic and spectroscopic methods *Example*: Standardization of turmeric for curcumin content	Bioactive components are identified by Advanced metabolomics
Reference Standards	Compare with known concentration of standard bioactive compounds	Maintains uniformity in bioactive ingredient levels	Standard reference compounds are validated through analytical methods. *Example*: Compare the alkaloid content in Rauwolfia serpentina against pharmacopeial standards	Certified Reference Materials (CRMs) enable reproducibility and global standardization
Extraction Methods	Optimizes solvent, temperature and extraction	Ensured uniform yield.	Optimization prevents bioavailability and degradation of active components	Green methods such as ultrasound-assisted extraction (UAE) and Supercritical fluid extraction (SFE) lower environmental impact and boost efficiency

(Continued)

TABLE 12.5 *(Continued)*
Standardization Parameters in Herbal Drug Formulation

Standardization Parameter	Description	Purpose	Additional Details	Refinements & Insights
Formulation Consistency	Ensure uniform granulation, mixing and tablet compression methods	Ensures consistency of product and reproducibility	Involves evaluation of excipients, dissolution, and distribution of particle size uniformly	Nanoformulations enhance the absorption and efficacy of herbal therapies
Labeling and Documentation	Provides an accurate list of ingredients, dosage and safety details	Complies with regulatory compliance	Label claim must include all details about biological source, active constituents, expiry dates and storage conditions	Blockchain ensures product authenticity and secures traceability
Regulatory Compliance	Complies with WHO, EMA, GMP and FDA guidelines	Ensures quality, safety and efficacy of herbal drugs	Should adhere to pharmacopeial standards, including stability and microbial contamination	Herbal trade and acceptance is supported by Global regulatory harmonization

the extracts in nanoformulations to enhance stability, solubility and bioavailability. Through nanocarriers such as solid lipid nanoparticles, liposomes, and niosomes, targeted delivery is achieved, which localizes herbal components to the specific desired sites. By adopting nanotechnology-based strategies, entrepreneurs have the opportunity to develop proprietary herbal products that offer superior therapeutic effects. Encapsulation of herbal compounds in polymeric or metallic nanoparticles protects them from degradation and enhances their bioavailability. The *microencapsulation technique* encases herbal ingredients in a protective layer, enhancing their stability, shelf life and enabling sustained and controlled release for longer therapeutic action. This approach attracts startups desiring to deliver reliable, consistent, and long-lasting herbal solutions. *Transdermal herbal drug delivery system* provides a non-invasive method of delivering the phytoconstituents to penetrate the skin and enter the circulation more efficiently. This approach bypasses the gastrointestinal metabolism and offers the users a convenient and promising route for the treatment of various conditions like hormonal disorders and chronic pain (Raghav and Verma 2024). *Smart drug delivery systems* incorporate sensors and microchips, integrating herbal compounds for dynamic, trigger-based dosing, ensuring timely and efficient physiological activities. This is a personalized approach that offers a transformative path for herbal drugs, facilitating tailored treatments to enhance the outcomes and adherence (Kaspute, Ramanavicius, and Prentice 2025).

Advanced *polymer-based technologies* are used to develop sustained-release oral formulations that allow gradual release of active components, which minimizes the need for frequent dosing. Such innovations enhance patient convenience, offering an opportunity for startups in the herbal formulations sector. Delivery of herbal formulations *through inhalation* by using systems like inhalers or nebulizers enables direct absorption of active phytoconstituents through the respiratory tract, thus improving the bioavailability and providing rapid relief from symptoms.

12.4 TECHNOLOGICAL ADVANCEMENTS IN HERBAL DRUG PRODUCTION

12.4.1 Role of Biotechnology in Herbal Drug Synthesis

Biotechnology plays a vital role in the synthesis of herbal drugs by employing advanced scientific tools to enhance their consistency, quality, and sustainability. One of the key methods employed is Plant tissue culture, in particular micropropagation, which allows rapid multiplication of sterile plants. Tissue culture techniques can also be engineered to increase the biosynthesis of secondary metabolites like alkaloids, glycosides, and flavonoids, which possess core medicinal benefits of medicinal plants. Genetic engineering is another significant advancement in the field of herbal drug development. It uses techniques like CRISPR and recombinant DNA technology to modify the genetic composition of plants. These genetic manipulations could modify the key metabolic pathways and boost the biosynthesis of targeted bioactive compounds. Furthermore, genetically engineered plants exhibit enhanced resistance to diseases and pests, minimizing the dependence on chemical pesticides for higher crop yields, which is essential for sustainability and scaling up of herbal drug manufacturing (Shi et al. 2022).

Fermentation technology has enhanced the efficiency and sustainability of herbal drug development. Microbial fermentation is an eco-friendly method of synthesizing plant-derived components more efficiently and sustainably by decreasing the environmental impact as compared to that of conventional extraction methods. Additionally, biotransformation by using microbes can transform an existing plant component into a more effective and easily absorbed compound with enhanced medicinal value. The field of metabolomics plays a crucial role in herbal medicine research by enabling detailed profiling and analysis of plant-derived metabolites. This advanced approach allows researchers to precisely identify and quantify active constituents, which supports rigorous quality assurance and standardization. By generating chemical fingerprints, metabolomic analysis ensures consistent composition and effectiveness of herbal formulations (Xia et al. 2023).

Bioinformatics plays a crucial role in herbal drug development by enabling the analysis of phytochemical datasets and the vast genomic composition of herbal drugs. These studies help uncover the promising therapeutic compounds, predict the biosynthetic pathways, and provide refinement of both agricultural practices and extraction methods. Furthermore, they support the design of more targeted and efficient production strategies (Yu et al. 2024). Lastly, sustainable practices in herbal drug development are promoted by biotechnology. It has supported conservation of biodiversity and protection of endangered plant species by reducing the dependence on wild harvesting. Further, the combination of hydroponics and aeroponics allows cultivation of medicinal plants utilizing minimal land resources and water for their growth, contributing towards production process for eco-friendly. Biotechnology bridges a vital link between modern scientific innovation and traditional herbal knowledge, offering scalable, sustainable methods for high-quality production of herbal drugs. This integration enhances the safety, effectiveness and availability to meet the global healthcare demands.

12.4.2 Nanotechnology Applications in Herbal Medicine

Nanoparticles are drug carriers of nanoscale operating at 1–100 nanometers. Herbal drugs encapsulated with nanoparticles improve their solubility, permeability and stability. The various herbal nanoformulations include phytosomes, liposomes, nanoemulsions, and other polymeric carrier nanoparticles that are tailored to enhance bioavailability, site-specific and controlled release of herbal drugs, thereby boosting the therapeutic performance of herbal medications (Ibrahim et al. 2018).

Nano-herbal drug delivery systems improve the stability issues and targeted delivery, thereby elevating the herbal therapies to a higher level of reliability and clinical effectiveness. Table 12.6 gives few examples of nano-herbal formulations-types, their merits, applications, and challenges.

TABLE 12.6
Nano-formulations in Herbal Drug Development

Category	Description	Examples	Additional Insights
Types of Nanotechnology-based herbal formulations	Nanotechnology-based approaches enhance herbal drug delivery	Curcumin loaded nanoparticles – anti-inflammatory and anticancer activities, Quercetin liposomes – antioxidant activity (Qi et al. 2022) Silymarin phytosomes are hepatoprotective. Essential oils nanoemulsions – antimicrobial. Polymeric carriers-chitosan-based nanoparticles – antioxidant (de Matos et al. 2019)	Surface modification of nanoparticles offers more precise targeting and minimizes the toxicity risk (Soleymani, Velashjerdi, and Asgari 2021)
Merits of nanotechnology-based herbal formulations	Nanotechnology enhances the stability and efficacy of targeted herbal formulations	*Improved Bioavailability*: Enhances the solubility of poorly soluble herbal drugs *Targeted Delivery*: Herbal drugs are transported precisely to the tissues, minimizing side effects *Controlled Release*: Sustained release herbal formulations exhibit a prolonged therapeutic effect *Improved Stability*: Enhances the shelf life of herbal drugs and protects them from environmental degradation. *Reduced Toxicity*: Due to targeted delivery, adverse effects are minimized	Smart nanocarriers respond to variations in the environment, like pH, temperature, to deliver the herbal drugs on demand for therapy. They are also eco-friendly, as made with biodegradable materials and enhance biocompatibility for safer treatments (Osanloo et al. 2022)

(Continued)

TABLE 12.6 *(Continued)*
Nano-formulations in Herbal Drug Development

Category	Description	Examples	Additional Insights
Applications	Application of nanotechnology in a wide range of therapeutic fields	*Treatment of cancer*: Nanoformulations of polyphenols from green tea, resveratrol and curcumin enhance anticancer activity *Neuroprotection*: Nanoformulations of Ginseng and Bacopa improve cognitive performance and protect against neurodegenerative disorders (Murugesan et al. 2022) *Anti-inflammatory Therapy*: Liposomal formulation of boswellic acid reduces inflammation (Mardiana et al. 2025) *Antimicrobial Agents*: Nanoemulsions with neem extract and tea tree oil improve antimicrobial & antifungal activities (Osanloo et al. 2022). *Cardiovascular Health*: Phytosome-based delivery of flavonoid formulations improves cardiovascular health by increasing their antioxidant activity (Mardiana et al. 2025)	Combination therapies of conventional drugs with nano-herbal formulations enhance therapeutic effectiveness. Personalized nano-herbal medicines are gaining traction to tailor therapies based on patients unique requirements
Challenges and Future Prospects	Various challenges and opportunities faced by herbal nanoformulations	Robust regulatory approvals are essential to ensure standardization and safety evaluation of non-herbal products *Manufacturing Complexity*: Scaling up the manufacturing process without compromising consistency is a big challenge *Cost Considerations*: Production cost may be increased with the manufacturing of advanced nano-formulations. *Long-Term Safety*: Long-term research is essential to evaluate the effects and long-term safety of nano-herbal formulations *Future Research*: Utilizing AI and machine learning facilitates precise optimization of nano-formulations. Green nanotechnology emphasizes sustainable approaches to synthesizing nanoparticles	AI-driven formulation designs improve the precision and efficiency of nano-herbal drug development. Green technology leads to eco-friendly and sustainable production methods

12.4.3 AI and Machine Learning (ML) in Herbal Drug Research

The powerful tools in herbal drug research are artificial intelligence (AI) and ML, which are transforming herbal research for evaluating predictive modeling, data analysis and optimization of the process. One of its major applications is data mining, in which AI algorithms sift through large volumes of scientific literature and the databases related to phytochemicals to unveil the potential of herbal compounds, their pharmacological and therapeutic activities. This can enhance the process of discovery by presenting their hidden patterns and unnoticed relationships. Moreover, ML models can predict the therapeutic activity of herbal drugs on the basis of their molecular structure, supporting the researchers in the identification of promising drug candidates in herbal drug development. The models also help for safety assessments of herbal drugs at a preliminary stage by evaluating ADMET (Absorption, Distribution, Metabolism, Excretion, and Toxicity) profiles and reduce the risk of product failure at a later stage. AI and ML play a vital role in phytochemical analysis and metabolomics by incorporating complex datasets to generate an elaborate chemical fingerprint of herbal extracts. ML can identify the biomarker compounds by correlating them with specific disease conditions and therapeutic actions. This can support standardization and also enhance the accuracy of herbal formulations. In addition, AI-driven optimization of various parameters to be considered in extraction, like the solvent type to be used and its composition, the temperature and the duration of extraction, significantly enhances the yield and quality of bioactive components, resulting in more efficient and sustainable production of herbal formulations. Furthermore, the consistency across the batches can be monitored and parameters could be adjusted in real time by ML models (Niu, Alvarez-Alvarez, and Chen 2025).

In the development of herbal formulations, AI predicts interactions between herbal bioactive components and supervises the design of stable and effective herbal products. The emerging field of personalized medicine using herbals, AI supports and analyzes patient-specific data to tailor the treatments for enhanced efficiency. It also allows the analysis of real-world evidence, providing insights into the long-term safety and effectiveness of herbal medicine across diverse populations (Babu et al. 2025).

AI also excels in the field of safety and toxicity assessment of herbal drugs. It analyzes the structural data and predicts potential toxic effects of these drugs and enhances patient safety. AI can also detect the adverse effects from large datasets, which can improve post-market surveillance and ensure continuous monitoring and regulatory compliance. Through AI and ML, herbal drug research is revolutionized by streamlining its discovery, improving safety and advancing personalized treatment strategies. By integration of AI and ML into every stage of the drug development pipeline, that is, from compound identification to clinical validation. It marks a significant advancement in modernizing herbal medicine. This synergism between traditional knowledge and cutting-edge technology holds immense promise for the global healthcare future (Spanakis et al. 2025).

12.5 REGULATORY FRAMEWORK AND MARKET TRENDS

The regulatory outlook and evolving market trends in herbal drug development present a vibrant environment filled with opportunities for techno-entrepreneurs. The regulatory bodies like the USFDA, the European Medicines Agency and India's Ministry of Ayush, working globally, oversee the safety, efficacy, and quality of herbal medicines. These agencies categorize herbal products into dietary supplements, traditional remedies, or pharmaceuticals – each with distinct regulatory pathways. For example, dietary supplements face less stringent requirements as those of pharmaceutical grade herbal drugs. Adherence to GMP (good manufacturing practices) is vital, assuring consistent production quality and safety. Entrepreneurs should also be ready to generate clinical evidence to support the health claims, which often require collaboration with research institutions. In addition to this, accurate labeling and ethical marketing are vital, as regulatory bodies demand that all claims related to health to be backed by scientific data (Sharma and Pundarikakshudu 2019).

On the market front, demand for natural and organic products by consumers continues to surge, due to growing health awareness and cynicism toward synthetic drugs. This shift opens avenues for entrepreneurs to innovate in novel herbal formulations. The incorporation of advanced technologies like AI, biotechnology, and nanotechnology has extended the scope for innovation in extraction methods, formulation processes, and delivery systems. Personalized herbal medicine is becoming widespread, offering tailored treatments based on individual genotypes and lifestyle profiles. Entrepreneurs mainly capitalize on this aspect by developing digital platforms in order to provide customized herbal solutions. The rise in e-commerce has reformed the broader market reach by transforming distribution models and enabling direct-to-consumer sales. Additionally, sustainability and ethical sourcing have become a key consumer concern, motivating businesses to adopt eco-friendly and transparent supply chains (Rojas et al. 2022).

The fusion of regulatory clarity, market demand technological innovations create a reliable ecosystem for visionary entrepreneurs in the herbal drug sector (Liang et al. 2021).

12.5.1 Global Regulatory Requirements for Herbal Drugs

International regulatory requirements for herbal medicines vary significantly across the regions, but share a common goal by ensuring the safety, quality and efficacy of herbal medicinal products. Each of the Regulatory bodies across the world, like the US FDA (Food and Drug Administration), EMA (European Medicines Agency), and India's Ministry of AYUSH, has frameworks tailored to their national healthcare systems. In the United States, herbal products are regulated as dietary supplements under the Dietary Supplement Health and Education Act (DSHEA), which requires formulators to ensure the safety of the product and proper labeling, but does not require pre-market approval. On the other hand, EMA categorizes herbal medicines as either traditional herbal medicines or well-established used products, each with well-defined documentation and clinical evidence (Shi et al. 2023).

Regulatory bodies in India are governed by the "Drugs and Cosmetics Act of 1940". They classify Ayurveda, Siddha and Unani (ASU) drugs in a separate class, governed by the Central Drugs Standard Control Organization (CDSCO) and AYUSH. These bodies regulate the license approval, manufacturing and quality control, mainly focusing on scientific validation and GMP (CDSCO 2020). To synchronize these diverse approaches, the World Health Organization (WHO) established *the International Regulatory Cooperation for Herbal Medicines (IRCH)*, a global network of 49 member countries and regional bodies as of 2025. This regulatory agency promotes collaboration, best practices, and regulatory alignment to enhance the global management of herbal medicines (WHO 2021). For technology entrepreneurs, understanding these regulatory hues is imperative. Products classified as traditional medicines may face fewer problems but offer limited claims, while those positioned as pharmaceuticals require thorough clinical validation. Steering through these frameworks strategically can enable access to international markets and also build consumer trust (Hossain et al, 2022).

12.5.2 Challenges in Herbal Drug Approval and Commercialization

The roadmap to approval and commercialization of herbal medicine is difficult with multifaceted challenges that can delay or derail their market entry significantly. One of the foremost problems lies in managing the complex and often inconsistent **regulatory guidelines** across different countries. Each region has its own set of rules, documentation requirements, and classification systems for herbal products, which range from dietary supplements to pharmaceuticals, that can be both time-consuming and financially burdensome, specifically for small and medium-sized entrepreneurs. The lack of Universal standardization in assessing the quality, safety, and efficacy of herbal drugs further complicates regulatory compliance, leaving manufacturers unsure about how to meet expectations.

Another major problem is with **limited availability of scientific evidence that** justifies the therapeutic claims of many herbal drugs. Compared to synthetic drugs, herbal products often depend on traditional knowledge passed down through generations, which may not meet the stringent standards of modern clinical validation. Conducting clinical trials to establish the safety and efficacy is not only costly but is complicated due to the presence of multiple-component nature of herbal drugs, making it challenging to isolate and study individual active components. This complexity presents significant challenges in designing studies that meet regulatory requirements (Wu et al. 2025).

Quality control and consistency present additional concerns. Depending on the factors like plant species, harvesting season and geographic origin and processing techniques, the therapeutic potency of herbal drugs varies widely. Ensuring uniformity in each batch is vital but difficult to achieve without strict adherence to GMP. Unfortunately, many smaller manufacturers may fail to have the infrastructure or proficiency to implement these standards effectively (Govindaraghavan and Sucher 2015).

From a **market perspective**, herbal medicines face tough competition from widely recognized synthetic pharmaceuticals that result from many years of research, marketing, and consumer trust. Public suspicion about the efficacy and safety of herbal products often exaggerated by misinformation or lack of awareness that can hinder their acceptability. Enlightening consumers and healthcare professionals about the advantages and demerits of herbal medicine is essential for building recognition and demand. **Intellectual property rights** also present a great challenge. Obtaining patents for herbal formulations is often difficult due to the existence of prior art and the traditional use of many herbal remedies. This can demotivate investment in research and innovation of herbal drugs. Additionally, challenges about **biopiracy**, the exploitation of traditional knowledge and resources without legal permission, raise ethical and legal problems that must be addressed through transparent and mutually beneficial sharing mechanisms. Finally, the economic impact of developing and commercializing herbal medicine cannot be unseen. High costs related to research, quality assurance, regulatory compliance, and marketing can be costly, particularly for startups and smaller entrepreneurs. These economic hurdles, together with scientific and regulatory complexities, emphasize the need for promotive policies, public-private partnerships, and global collaboration to release the full potential of herbal medicines in modern healthcare (Imran et al. 2021).

12.5.3 Market Demand and Emerging Trends in Herbal Therapeutics

The worldwide demand for herbal medicines is experiencing consistent and significant growth, powered by shifting consumer preferences, heightened awareness of wellness, and a strong preference toward holistic and natural healthcare solutions. A major force behind this trend is the growing concern over the adverse effects and drawbacks of conventional pharmaceuticals, prompting many individuals to explore alternative and complementary therapies. As health awareness rises, so does the appeal of herbal medicines and supplements, particularly among those striving for preventive care and sustainable wellness approaches. This demand is further multiplied by the aging global community, which is turning to herbal products to manage chronic conditions and support overall well-being. Particularly, older adults are attracted to herbal treatments for their perceived safety and mild side effect profiles. Simultaneously, herbal medicine is gaining wide acceptance within mainstream healthcare networks. More healthcare professionals are integrating herbal medicines into patient care, and in certain regions, insurance companies are beginning to cover such treatments, boosting their accessibility and legitimacy (Costa, Magalhães, and Di Stasi 2022).

The growth of e-commerce has also played a key role in making herbal products more widely available. Table 12.7 is depicting Global Herbal Medicine Market Forecast till 2030. Furthermore, multidisciplinary approaches that combine herbal and conventional therapies are gaining more popularity, as consumers seek comprehensive solutions that draw from both traditional and modern medicine. There is a growing interest in evidence-based herbal medicine. Consumers are continuously demanding scientific validation of safety and efficacy, motivating companies to invest more in clinical research and robust validation. Sustainability and ethical sourcing have become more important, with eco-conscious consumers favoring brands that focus on responsible harvesting and open supply chains. Technological developments are further shaping the herbal medicine landscape. Innovations such as AI-driven formulation, ML, and digital health services offering personalized

TABLE 12.7
Global Herbal Medicine Market Forecast (2024–2030)

Metric	Details
Market Size (2023)	USD 70.57 billion
Projected Market Size (2030)	USD 279.8 billion
Compound Annual Growth Rate (CAGR)	10.6% to 20.9% depending on region and source
Key Growth Drivers	Health awareness, inclination towards natural remedies, aging population, integration of AI and biotechnology
Top Regions	Asia-Pacific, North America, Europe
Popular Segments	Herbal dietary supplements, functional foods, herbal cosmetics, and traditional medicine systems (Ayurveda, TCM)

medicine, and advanced extraction methods are improving the quality of the product and consumer engagement. The upsurge of nutraceuticals, the functional foods and food products that merge nutrition with therapeutic benefits, is another notable trend, as customers look for convenient ways to support their health through regular selections. Finally, the global integration of herbal medicine is boosting the market potential with diverse traditions and practices. Natural remedies from Ayurveda (Sen and Chakraborty 2015), traditional Chinese medicine (TCM), and Western herbalism are increasingly crossing boundaries, creating a vibrant and global consumer market (Cunningham et al. 2019).

12.6 ENTREPRENEURIAL OPPORTUNITIES IN HERBAL DRUG DEVELOPMENT

12.6.1 Start-ups and Innovations in Herbal Pharmaceuticals

Entrepreneurial opportunities in herbal drug development are expanding rapidly due to increasing global demand for natural and holistic healthcare solutions, technological advancements, and supportive innovation ecosystems. Here is a detailed overview addressing startups and innovations, funding and investment strategies, and case studies of successful herbal drug enterprises:

The herbal pharmaceutical sector is witnessing vibrant start-up activity, particularly in India, where institutions like CSIR-IIIM Jammu foster biotech and herbal health innovations through incubation and strategic partnerships. At the National Startup Festival 2025, over 45 startups were empowered, including those focusing on herbal extracts, topical formulations, nutraceuticals, and personalized herbal therapies (iiim.res.in/2025). Examples of innovative startups include: Vipragen Biosciences Pvt. Ltd. specializing in drug discovery and preclinical testing, JMD Herbal Pharmaceuticals focusing on manufacturing herbal extracts and suppositories, Avtar Health Care developing sustainable, pocket-friendly topical skin formulations, Velanutrition Nutraceuticals, working on heart health supplements combining natural ingredients.

Biotechnological advancements such as tissue culture, genetic engineering, and molecular breeding are enhancing the yield and potency of medicinal plants like Ashwagandha, enabling mass propagation and high-potency extracts (Govindaraghavan and Sucher 2015). Nano-phytopharmaceuticals are also emerging as innovative delivery systems for herbal drugs. The integration of traditional knowledge with modern scientific research is accelerating the development of standardized, effective herbal formulations, exemplified by products like standardized curcumin extracts and high-potency Ashwagandha.

12.6.2 Funding and Investment Strategies for Phytopharmaceutical Research

Effective funding strategies for herbal pharmaceutical research combine institutional grants, public-private partnerships, and innovative business models to address high development costs and regulatory demands. Below are the most impactful approaches:

- *Government and Institutional Grants:* Funding for herbal drug startups often comes from government-supported incubators (e.g., ICMR, CSIR-IIIM's Industrial Biotechnology Park), grants (such as BIRAC), and partnerships with academic and research institutions (iiim.res.in/2025) (Tandon and Yadav 2017).
- *NIH/NCCIH Programs:* The National Center for Complementary and Integrative Health (NCCIH) offer phased funding for natural product trials, (www.nccih.nih.gov) including:
  - *R61/R33 Exploratory Grants:* Supports early-stage clinical development (e.g., dose-finding studies) with milestones for transitioning to larger Phase II trials.
  - *R01 Grants:* Funds mid-phase trials for refining herbal interventions, with awards up to $3 million over five years.
  - *Advance Biotech Grant Program:* Provides biotech startups with products, services, and consultation to accelerate commercialization, prioritizing therapeutic innovations in chronic diseases (www.sigmaaldrich.com).
- *Public-Private Partnerships:* Collaborations between academic institutions, research organizations (e.g., CSIR-IIIM Jammu), and private firms enable access to shared infrastructure, expertise, and co-funding. For example, Cuban herbal medicine development leveraged European university partnerships for technology transfer (CSIR website, June 2025). *Low-Cost Franchise Models* require minimal upfront investment (no manufacturing setup) and grant regional monopoly rights, allowing entrepreneurs to market diverse products (e.g., Ayurvedic supplements, personal care items) with promotional support from suppliers (www.hanisan.com)

Targeting *niche markets* like women's wellness, chronic disease management, and sustainable agriculture to attract focused investment (Thakor et al. 2011). Investment strategies include leveraging *public-private partnerships* to access infrastructure and expertise Collaborating internationally for technology transfer and capacity building- the Cuban herbal medicine development model supported by European universities is a good example of this case.

Companies in the herbal medicine market invest heavily in R&D, quality control, and regulatory compliance to ensure safety and efficacy, which is crucial for attracting investment and consumer trust.

12.6.3 Case Studies of Successful Herbal Drug Enterprises (iiim.res.in)

- *Nirog Street*: Originating as a small start-up incubated at CSIR-IIIM, it developed Diabe 250, herbal diabetes drug that regulates blood sugar and improves sleep and digestion. It has grown to employ over 400 people and gained national recognition, illustrating successful scale-up from research to market.
- *JK Aroma*: Leveraging the 'Purple Revolution' of lavender farming in Jammu & Kashmir, JK Aroma scaled up production of essential oils, soaps, and aromatic products with technical and financial support from CSIR-IIIM, now competing internationally.
- *GAURICO*: A herbal cosmetics brand focused on women's wellness that developed innovative Ayurveda-inspired products with technical support from CSIR-IIIM, preparing for market launch with a strong emphasis on purity and sustainability.
- *Hemp Street Pvt. Ltd*: Developed patented technology for bacterial cellulose membranes used in transdermal patches for pain management, with technology transfer valued at 1.5 crore INR, currently undergoing commercialization.
- Internationally, the Cuban model supported by European academic partnerships demonstrates how capacity building, research, and technology transfer can develop herbal medicine industries in resource-constrained settings, offering a scalable model for other regions.

12.7 SUSTAINABLE PRACTICES AND ETHICAL CONSIDERATIONS

Sustainable and ethical approaches are basic to the responsible development of herbal products, assuring that their production confirms environmental health, social equity, and long-term financial viability. Sustainable and ethical sourcing begins with responsible wild collection that adheres to conservation principles, such as those outlined by CITES (Convention on International Trade in Endangered Species), to protect endangered species and prevent over-exploitation. To further this, the cultivation of medicinal plants using organic farming methods aids in preserving soil fertility, reducing chemical inputs, and ensuring a consistent supply of superior quality raw materials. Agroforestry systems that integrate medicinal plants with food crops not only promote biodiversity but also offer economic stability to agricultural communities. Protecting genetic diversity through the conservation of traditional plant varieties further enhances ecological adaptability and the therapeutic potential of herbal drugs. Efficient resource use is another cornerstone; methods such as drip irrigation conserve water, while recycling plant waste into value-added products reduces environmental impact. Ecologically friendly processing techniques, such as supercritical CO_2 and ultrasound-assisted extraction, minimize solvent usage and energy consumption, and sustainable packaging using biodegradable materials aligning with expanding market demand for green products.

Ethical considerations are likewise crucial. Fair trade practices ensure that local communities and farmers are fairly rewarded for their contributions, while community engagement in resource management encourages shared ownership and benefits. Altogether, these sustainable and ethical practices form the backbone of a responsible herbal drug company that protects natural resources, empowers communities, and meets the growing global demand for safe, effective, and environmentally mindful healthcare solutions (Koonrungsesomboon et al. 2021).

12.7.1 Eco-friendly Extraction and Production Methods Like

Supercritical Fluid Extraction, Ultrasound-Assisted Extraction, Microwave-Assisted Extraction to minimizes processing time and also enhances extraction efficiency and reduces the volume of solvents, making it a more ecofriendly option or using water as solvent for extraction (Giera et al. 2022). For example, the use of biodegradable solvents, closed-loop systems to recycle extraction media, and energy-efficient equipment all facilitate to cleaner production cycles (Zhang et al. 2015). In addition to extraction, ecofriendly approaches expand to post-processing and packaging. Techniques like freeze-drying and vacuum drying are preferred to safeguard the integrity of herbal compounds without much energy usage. Furthermore, adoption of **recyclable or biodegradable packaging materials** aligns with growing consumer needs and demand for environmentally friendly products and helps in reducing plastic waste in association with pharmaceutical packaging. These practices not only foster environmental goals but also enhance the market appeal of herbal medicinal products among eco-conscious consumers.

12.7.2 Ethical Sourcing of Medicinal Plants

Ethical sourcing of medicinal plants is a cornerstone of the responsible herbal drug development process, assuring that botanical sources are sustainably used while respecting the rights, knowledge, and livelihoods of indigenous communities. At its core, ethical sourcing promotes **fair trade practices**, where the contributions of local farmers and harvesters are fairly compensated. Establishing long-term partnerships with these communities enhances mutual trust and economic stability, moving away from overexploitation and short-term procurement models. Similarly important is the commitment to **sustainable harvesting**, which follows scientifically informed protocols that protect biodiversity and allow regeneration of plant populations. This includes adhering to guidelines of International Conservation, like those set by the Convention on Biological Diversity (CBD) and the IUCN, to inhibit overharvesting and habitat degradation (Chen et al. 2016).

Drawing respectful partnerships with indigenous communities is another vital pillar. Ethical sourcing involves **respect for traditional knowledge**, recognizing the cultural and traditional healing wisdom passed down through generations. **Certification and Traceability systems are adopted by the companies to** reinforce clarity and accountability. Certifications from recognized bodies support consumers in identifying ethically sourced products, while traceability mechanisms allow for tracing the source of plant materials throughout the supply chain (Abubakar et al. 2022).

Environmental stewardship is a fundamental aspect of ethical sourcing. This includes backing conservation initiatives to protect ecosystems, where medicinal plants grow and promote agroecological cultivation methods that foster biodiversity, enhance soil fertility and minimize synthetic chemical usage. These practices not only protect the environment but also contribute to the long-term viability of herbal medicine. Finally, **education and training** play an essential role in empowering both producers and consumers. Capacity-building initiatives support local communities with sustainable harvesting skills and ecological knowledge, while consumer education fosters knowledge-based purchases and encourages support for ethically produced herbal medicinal products (Leonti and Casu 2013).

12.8 CONSUMER AWARENESS AND EDUCATION

As interest in herbal medicine increases, it becomes important for consumers to understand these products, what they are, how they work, and how to use them safely. Consumers should be informed about the active ingredients of these herbal medicinal products and the scientific evidence supporting their safety and efficacy for specific health conditions. At the same time, it is important to be guided on proper dosage and administration, as well as potential adverse effects and interactions with other medications, especially for individuals managing chronic illnesses or taking multiple prescriptions (Obahiagbon and Ogwu 2023).

Another critical area of focus is quality assurance. Consumers benefit from understanding the GMP, third-party certifications, and ethical sourcing in ensuring the safety and integrity of herbal medicinal products. Moreover, awareness of consumer rights, like access to clear labeling and the ability to report side effects, strengthens public trust and accountability in the herbal drug industry. Collaboration with healthcare professionals is also important, as they can guide patients in incorporating herbal options into their treatment plans safely and effectively (Kennedy, Hart, and Seely 2009).

Finally, technology plays a groundbreaking role in consumer education. Mobile apps and digital platforms can offer accessible, current information on herbal medicinal products, including safety profiles, guidelines of usage, and quality ratings. By empowering consumers with the knowledge and resources to make informed decisions, the herbal medicine sector can foster a more empowered, mindful, health-oriented, and socially responsible public.

12.9 INTEGRATION OF HERBAL MEDICINE WITH CONVENTIONAL TREATMENT

There is a growing trend toward scientifically integrating traditional herbal medicine, such as Ayurveda and TCM, into evidence-based clinical practice. Herbal medicines are increasingly recognized as complementary to conventional drugs, often used in combination to enhance therapeutic efficacy or reduce side effects like hypertension, arthritis and diabetes (GBD 2019 Acute and Chronic Care Collaborators, 2025). Some Chinese herbal components have demonstrated synergistic anti-cancer effects when combined with chemotherapy agents, indicating potential for integrative oncology. Countries like India have incorporated Ayurveda into their national healthcare systems, and several Western countries (UK, USA, UAE, Sweden) acknowledge and regulate Ayurvedic treatments, reflecting broader acceptance and integration globally (Niemeyer et al, 2013). Table 12.8 is highlighting the Integration Model of Herbal and Conventional Medicine, the experts involved and the outcome.

Multidisciplinary teams consisting of physicians, herbalists, naturopathists working together to formulate comprehensive treatment plans, while where referral pathways between conventional and herbal practitioners are contributing to build a better integrated healthcare system.

TABLE 12.8
Integration Model: Herbal and Conventional Medicine

Component	Herbal Medicine	Conventional Medicine	Integration Outcome
Therapeutic Role	Traditional remedies for chronic illness, symptom relief, immune support	Scientific evidence-based treatments for both acute and chronic diseases	Complementary therapy that boosts efficacy and reduces side effects (Tsang et al. 2015)
Personalization	Customized to individual needs, lifestyle, and cultural beliefs	Based on diagnosis, genetics, and clinical protocols	Tailored medicine combining tradition and precision (Coulter and Oldham 2016)
Practitioner Collaboration	Herbalists, naturopathic practitioners, traditional healers	Physicians, pharmacists, specialists	Interdisciplinary groups, offering holistic care (Steel et al. 2022)
Research & Evidence	Traditional herbal knowledge, ethnobotanical information	Clinical trials, pharmacological evidence	Evidence-based collaboration through integrative research (Ryan 2016)
Patient Engagement	Focusing on self-care, wellness, and prevention	Systematic treatment protocols and monitoring	Inspired patients with more choices and shared decision-making (Iovino, Uchmanowicz, and Vellone 2024)
Regulatory Oversight	Region specific variations; often less standardized	Stringent protocols and quality control	Establishment of harmonized standards and safety validation (Govindaraghavan and Sucher 2015)

12.10 RESEARCH DIRECTIONS AND TECHNOLOGICAL INNOVATIONS

Research is intensifying on understanding the synergistic interactions within multi-herb formulations and between herbal and pharmaceutical drugs. Despite challenges due to the complex multi-component nature of herbal medicines, advances in systems biology, mathematical modeling, and computer-aided analysis are improving the scientific evidence base (Ramawat and Merillon 2019). Technological advancements revolutionizing herbal drug development include:

- *Advanced extraction techniques*: Supercritical fluid extraction, pressurized liquid extraction, and microwave-assisted extraction improve yield, purity, and selectivity of herbal extracts.
- *Analytical tools*: High-performance liquid chromatography (HPLC), nuclear magnetic resonance (NMR) spectroscopy, and other sophisticated methods enable better characterization and standardization of herbal products.
- *Formulation technologies*: Nanotechnology and encapsulation enhance bioavailability, stability, and targeted delivery of herbal compounds. Advancements in nano-phytomedicine have significantly enhanced the therapeutic potential of herbal compounds. Nano-formulations—such as liposomes, phytosomes, and various nanocarriers—have been shown to improve the solubility, chemical stability, bioavailability, and targeted delivery of phytochemicals. Customization to meet individual patient needs, thereby paving the way for more personalized approaches to herbal therapy (Khan and Islam, 2019).
- **AI** and **ML** are also playing a transformative role in herbal drug discovery and development. These technologies facilitate the prediction of biological activities, optimization of extraction techniques, and identification of novel phytochemical synergies. Additionally, AI-driven modeling of multi-target interactions inherent in complex herbal formulations supports a shift from reductionist, single-compound paradigms to more integrative systems biology frameworks. Network pharmacology, in particular, has emerged as a valuable tool for elucidating the synergistic mechanisms underlying traditional polyherbal therapies (Bachunaga et al. 2025).

12.11 CONCLUSION

The global herbal medicine market is expanding at an unprecedented pace. This growth is driven by a confluence of factors: rising consumer demand for natural health solutions, increasing acceptance of traditional therapies, and a wave of innovation that is redefining what herbal products can be.

Emerging markets in Latin America, the Middle East, and Africa are particularly promising. In Africa alone, an estimated 70–80% of the population relies on herbal medicine for primary healthcare, with over 6,000 plant species used in traditional practices. As healthcare infrastructure improves and regulatory frameworks evolve, these regions are poised to become major players in the global herbal economy.

12.11.1 Innovation and Diversification

Product innovation is another key driver. The development of herbal nutraceuticals, functional foods, and hybrid formulations that combine botanical and synthetic ingredients is gaining traction. Personalized herbal healthcare—tailored to individual genetic, metabolic, and lifestyle profiles—is emerging as a frontier in both clinical and consumer markets. Wellness tourism, too, is on the rise, offering immersive experiences centered around traditional healing practices.

In North America and Europe, the market continues to grow steadily. Aging populations, increasing health awareness, and a preference for organic and sustainable products are fueling demand. Countries like Germany, France, and the United Kingdom have well-established herbal markets, supported by stringent quality standards and integration into conventional healthcare systems.

Asia-Pacific remains the fastest-growing region, led by China, India, and Japan. China is expected to become the world's largest producer and consumer of herbal medicines, driven by significant R&D investment, particularly in the wake of the COVID-19 pandemic. India's growth is supported by initiatives like the WHO Global Centre for Traditional Medicine and increasing demand for organic and Ayurvedic products.

12.11.2 Integration and the Path Forward

One of the most promising trends is the growing integration of herbal and conventional medicine. This convergence offers the potential for more comprehensive, personalized, and patient-centered care. Herbal medicines are increasingly being used as complementary therapies in the management of chronic diseases, enhancing both efficacy and safety.

The rise of e-commerce has further democratized access to herbal products, allowing companies—both established and emerging—to reach global audiences with unprecedented efficiency. Governments around the world are also recognizing the economic and cultural value of traditional medicine, promoting exports and strengthening regulatory oversight to ensure quality and safety.

According to Grand View Research, the global herbal medicine market was valued at USD 70.57 billion in 2023 and is projected to grow at a compound annual growth rate (CAGR) of 20.91%, reaching over USD 248 billion by 2030. These figures underscore not only the commercial potential of herbal medicine but also its growing legitimacy within the broader healthcare landscape (www.grandviewresearch.com).

FUTURE IMPACT

Integrating traditional and modern medical systems offers a multifaceted approach to healthcare that can significantly improve treatment outcomes. One of the key advantages lies in the ability to deliver holistic and personalized care. Traditional practices such as Ayurveda and TCM when combined with the diagnostic precision and therapeutic rigor of modern medicine, enables more comprehensive and tailored interventions that address both the underlying causes and the symptoms of disease—ultimately improving efficacy and patient satisfaction.

Therapeutic synergy is another major benefit. The integration of herbal remedies and complementary practices—such as acupuncture, yoga, and meditation—with conventional pharmacological or surgical treatments has shown promise in managing chronic conditions, minimizing side effects, and supporting mental health. For instance, botanicals like Ashwagandha and turmeric, known for their anti-inflammatory and immunomodulatory properties, can enhance the effectiveness of standard therapies (Rajeev Gupta 2024).

In the context of chronic and life-limiting illnesses, integrative approaches offer valuable tools for symptom relief and emotional support. Complementary therapies such as aromatherapy and mindfulness practices can help manage pain, anxiety, and treatment-related side effects, contributing to a more compassionate and holistic model of care.

On the regulatory front, regulatory bodies are expected to enforce more rigorous guidelines to ensure product safety, efficacy, and quality, including standardized testing and transparent labeling. Efforts toward global harmonization of herbal drug regulations may also gain momentum, facilitating smoother global trade and compliance.

Sustainability and ethical sourcing will remain fundamental to Industry's evolution. Today's consumers becoming more conscious about environmental and social issues are choosing brands that prioritize responsible harvesting, biodiversity conservation, and fair trade practices.

Finally, patient engagement and satisfaction tend to increase when cultural values and preferences for natural therapies are acknowledged and respected. This alignment fosters trust, improves adherence to treatment plans, and ultimately leads to better health outcomes.

REFERENCES

Ayush Research Portal – Ayush Research Portal. Accessed June 13, 2025 https://ayushportal.nic.in/ Accessed June 13, 2025 https://hanisan.com/is-it-worth-investing-in-herbal-pcd-franchise/ Accessed June 13, 2025 https://iiim.res.in/the-national-startup-festival-2025-organised-by-csir-iiim-jammu-empowered-45-startups-promoting-biotech-agritech-health-innovation/

Council of Scientific and Industrial Research (CSIR) – CSIR Official Website. https://www.csir.res.in/en/rd-projects-csir

Herbal Medicine Market Size, Share & Trends Analysis Report By Intervention (Ayurveda, Traditional Chinese Medicine), By Product Form (Tablet/Capsules, Powder), By Source, By Distribution Channel, By Region, And Segment Forecasts, 2024 – 2030 Accessed on June 13, 2025 https://www.grandviewresearch.com/industry-analysis/herbal-medicine-market-report Accessed June 13, 2025 https://www.nccih.nih.gov/grants/nccih-clinical-trials-funding-opportunities/natural-products-clinical-trials-funding-opportunities Accessed June 13, 2025 https://www.sigmaaldrich.com/IN/en/applications/pharmaceutical-and-biopharmaceutical-manufacturing/biotech-resource-hub/startup-programs

Abubakar, I. B., S. S. Kankara, I. Malami, J. B. Danjuma, Y. Z. Muhammad, H. Yahaya, D. Singh, U. J. Usman, A. N. Ukwuani-Kwaja, A. Muhammad, S. J. Ahmed, S. O. Folami, M. B. Falana, and Q. O. Nurudeen. 2022. "Traditional Medicinal Plants Used for Treating Emerging and Re-Emerging Viral Diseases in Northern Nigeria." *European Journal of Integrative Medicine* 49: 102094. https://doi.org/10.1016/j.eujim.2021.102094

Babu, B. K. 2025. "Revolutionizing Herbal Medicine: The Role of E-Health Informatics and Network Pharmacology in Personalized Herbal Therapies." *International Journal of Pharmaceutical Investigation* 15(1): 219–27. https://doi.org/10.5530/ijpi.20251754

Badoni Semwal, R., D. K. Semwal, S. Combrinck, C. Cartwright-Jones, and A. Viljoen. 2014. "Lawsonia inermis L. (Henna): Ethnobotanical, Phytochemical and Pharmacological Aspects." *Journal of Ethnopharmacology* 155 (1): 80–103. https://doi.org/10.1016/j.jep.2014.05.042

CDSCO, Traditional drugs [Internet]. India: CDSCO; 2020 [cited 2023 Oct. 2]. https://cdsco.gov.in/opencms/opencms/en/Approval_new/Approved-New-Drugs/

Chen, H., S. Shi, A. Liu, and X. Tang. 2010. "Combined Application of Extrusion-Spheronization and Hot-Melt Coating Technologies for Improving Moisture-Proofing of Herbal Extracts." *Journal of Pharmaceutical Sciences* 99 (5): 2444–54. https://doi.org/10.1002/jps.21990

Chen, S., X. Pang, J. Song, L. Shi, H. Yao, J. Han, and C. Leon. 2014. "A Renaissance in Herbal Medicine Identification: from Morphology to DNA." *Biotechnology Advances* 32 (7): 1237–44. https://doi.org/10.1016/j.biotechadv.2014.07.004

Chen, S. L., H. Yu, H. M. Luo, Q. Wu, C. F. Li, and A. Steinmetz. 2016. "Conservation and Sustainable Use of Medicinal Plants: Problems, Progress, and Prospects." *Chinese Medicine* 11: 37. https://doi.org/10.1186/s13020-016-0108-7

Costa, E. F., W. V. Magalhães, and L. C. Di Stasi. 2022. "Recent Advances in Herbal-Derived Products With Skin Anti-Aging Properties and Cosmetic Applications." *Molecules (Basel, Switzerland)* 27 (21): 7518. https://doi.org/10.3390/molecules27217518

Coulter, A., and J. Oldham. 2016. "Person-Centred Care: What is It and How Do We Get There?" *Future hospital journal* 3 (2): 114–6. https://doi.org/10.7861/futurehosp.3-2-114

Cunningham, A. B., J. A. Brinckmann, X. Yang, and J. He. 2019. "Introduction to the Special Issue: Saving Plants, Saving Lives: Trade, Sustainable Harvest and Conservation of Traditional Medicinals in Asia." *Journal of ethnopharmacology* 229: 288–92. https://doi.org/10.1016/j.jep.2018.10.006

de Matos, S. P., H. F. Teixeira, Á. A. N. de Lima, V. F. Veiga-Junior, and L. S. Koester. 2019. "Essential Oils and Isolated Terpenes in Nanosystems Designed for Topical Administration: A Review." *Biomolecules* 9 (4): 138. https://doi.org/10.3390/biom9040138

Dotsenko, V. A., and L. V. Mosiĭchuk. 2012. Effectiveness of dietary supplements–biologically active substances in patients with obesity. Voprosy pitaniia, *81*(5): 33–6.

Dubey, R. K., and S. Shukla. 2023. "Exploring Novel Herbal Compounds and Formulations for Inflammatory Bowel Disease (IBD) Management." *Zhongguo ying yong sheng li xue za zhi = Zhongguo yingyong shenglixue zazhi = Chinese Journal of Applied Physiology* 39: e20230003. https://doi.org/10.62958/j.cjap.2023.003

Ehambarampillai, D., and M. L. Y. Wan. 2025. "A Comprehensive Review of Schisandra chinensis Lignans: Pharmacokinetics, Pharmacological Mechanisms, and Future Prospects in Disease Prevention and Treatment." *Chinese medicine* 20 (1): 47. https://doi.org/10.1186/s13020-025-01096-z

GBD 2019 Acute and Chronic Care Collaborators. 2025. "Characterising Acute and Chronic Care Needs: Insights from the Global Burden of Disease Study 2019." *Nature Communications 16*(1): 4235. https://doi.org/10.1038/s41467-025-56910-x

Giera, D. S., M. Preisitsch, H. Brevard, and J. Nemetz. 2022. "Quantitative Removal of Pyrrolizidine Alkaloids from Essential Oils by the Hydrodistillation Step in Their Manufacturing Process." *Planta Medica 88*(7): 538–47. https://doi.org/10.1055/a-1534-6928

Govindaraghavan, S., and N. J. Sucher. 2015. "Quality Assessment of Medicinal Herbs and Their Extracts: Criteria and Prerequisites for Consistent Safety and Efficacy of Herbal Medicines." *Epilepsy & Behavior: E&B* 52 (Pt B): 363–71. https://doi.org/10.1016/j.yebeh.2015.03.004

Gupta, R. 2024. "A Prospective Clinical Trial to Find the Efficiency of the Therapeutic Use of a Combination of Transdermal gels Producing CO2 in Treating Varicose Ulcers That Do Not Respond to Primary Treatment." *International Journal of Clinical Case Reports and Reviews* 20 (2). https://doi.org/10.31579/2690-4861/582

Hao, S., H. Yang, J. Hu, L. Luo, Y. Yuan, and L. Liu. 2024. "Bioactive Compounds and Biological Functions of Medicinal Plant-Derived Extracellular Vesicles." *Pharmacological Research* 200: 107062. https://doi.org/10.1016/j.phrs.2024.107062

Hossain, C. M., M. E. E. T. A. Gera, and K. A. Ali. 2022. "Current Status and Challenges of Herbal Drug Development and Regulatory Aspect: A Global Perspective." *Asian Journal of Pharmaceutical and Clinical Research* 15(2): 31–41. https://doi.org/10.22159/ajpcr.2022.v15i12.46134

Ibrahim, M. F., F. S. Robustelli Della Cuna, C. Villa, M. Corti, H. I. M. Amin, P. Faris, P. Grisoli, and G. Brusotti. 2022. "A Chemometric Assessment and Profiling of the Essential Oils from *Hibiscus sabdariffa* L. from Kurdistan, Iraq." *Natural Product Research* 36 (9): 2409–12. https://doi.org/10.1080/14786419.2020.1833198

Ibrahim, S., T. Tagami, T. Kishi, and T. Ozeki. 2018. "Curcumin Marinosomes as Promising Nano-Drug Delivery System for Lung Cancer." *International Journal of Pharmaceutics* 540 (1–2): 40–9. https://doi.org/10.1016/j.ijpharm.2018.01.051

Imran, Y., N. Wijekoon, L. Gonawala, Y. C. Chiang, and K. R. D. De Silva. 2021. "Biopiracy: Abolish Corporate Hijacking of Indigenous Medicinal Entities." *The Scientific World Journal* 2021: 8898842. https://doi.org/10.1155/2021/8898842

Iovino, P., I. Uchmanowicz, and E. Vellone. 2024. "Self-Care: An Effective Strategy to Manage Chronic Diseases." *Advances in Clinical and Experimental Medicine: Official Organ Wroclaw Medical University* 33 (8): 767–71. https://doi.org/10.17219/acem/191102

Joshi, M., M. Pandey, R. Shankar, and A. Ved. 2025. "Emerging Natural Therapies for the Treatment of Inflammatory Bowel Disease." *Current Pharmaceutical Biotechnology* 26 (8): 1175–88. https://doi.org/10.2174/0113892010293150240415143650

Kaspute, G., A. Ramanavicius, and U. Prentice. 2025. "Natural Drug Delivery Systems for the Treatment of Neurodegenerative Diseases." *Molecular Biology Reports* 52 (1): 217. https://doi.org/10.1007/s11033-025-10286-9

Kennedy, D. A., J. Hart, and D. Seely. 2009. "Cost Effectiveness of Natural Health Products: A Systematic Review of Randomized Clinical Trials." *Evidence-Based Complementary and Alternative Medicine: eCAM* 6 (3): 297–304. https://doi.org/10.1093/ecam/nem167

Khan, M. S. S., and S. Islam. 2019. "Technological Advancements in Herbal Medicine: Enhancing Efficacy, Safety, and Accessibility." *Australian Herbal Insight* 1(1): 1–12. https://doi.org/10.25163/ahi.1120053

Kim, S. B., J. Bisson, J. B. Friesen, L. Bucchini, S. Gafner, D. C. Lankin, S. N. Chen, G. F. Pauli, and J. B. McAlpine. 2021. "The Untargeted Capability of NMR Helps Recognizing Nefarious Adulteration in Natural Products." *Journal of Natural Products* 84 (3): 846–56. https://doi.org/10.1021/acs.jnatprod.0c01196

Koonrungsesomboon, N., N. Morakote, and J. Karbwang 2021. "Ethical Considerations and Challenges in Herbal Drug Trials with the Focus on Scientific Validity and Risk Assessment." *Phytotherapy Research, PTR* 35(5): 2396–402. https://doi.org/10.1002/ptr.6962

Krishna, K. V., R. S. Ulhas, and A. Malaviya. 2024. "Bioactive Compounds from *Cordyceps* and Their Therapeutic Potential." *Critical Reviews in Biotechnology* 44 (5): 753–73. https://doi.org/10.1080/07388551.2023.2231139

Laureanti, E. J. G., T. S. Paiva, L. M. de Matos Jorge, and R. M. M. Jorge. 2023. "Microencapsulation of Bioactive Compound Extracts Using Maltodextrin and Gum Arabic by Spray and Freeze-Drying Techniques." *International Journal of Biological Macromolecules* 253 (Pt 4): 126969. https://doi.org/10.1016/j.ijbiomac.2023.126969

Leonti, M., and L. Casu. 2013. "Traditional Medicines and Globalization: Current and Future Perspectives in Ethnopharmacology." *Frontiers in Pharmacology* 4: 92. https://doi.org/10.3389/fphar.2013.00092

Li, J., J. X. Li, H. Jiang, M. Li, L. Chen, Y. Y. Wang, L. Wang, N. Zhang, H. Z. Guo, and K. L. Ma. 2023a. "Phytochemistry and Biological Activities of Corynanthe Alkaloids." *Phytochemistry* 213: 113786. https://doi.org/10.1016/j.phytochem.2023.113786

Li, Y., J. Wang, L. Li, W. Song, M. Li, X. Hua, Y. Wang, J. Yuan, and Z. Xue 2023b. "Natural Products of Pentacyclic Triterpenoids: From Discovery to Heterologous Biosynthesis." *Natural Product Reports* 40(8): 1303–53. https://doi.org/10.1039/d2np00063f

Liang, Z., H. Hu, J. Li, D. Yao, Y. Wang, and C. O. L. Ung. 2021. "Advancing the Regulation of Traditional and Complementary Medicine Products: A Comparison of Five Regulatory Systems on Traditional Medicines with a Long History of Use." *Evidence-Based Complementary and Alternative Medicine: eCAM* 2021: 5833945. https://doi.org/10.1155/2021/5833945

Mardiana, L., T. Milanda, Y. E. Hadisaputri, and A. Y. Chaerunisaa. 2025. "Phytosome-Enhanced Secondary Metabolites for Improved Anticancer Efficacy: Mechanisms and Bioavailability Review." *Drug Design, Development and Therapy* 19: 201–18. https://doi.org/10.2147/DDDT.S483404

Murugesan, M., R. Mathiyalagan, V. Boopathi, B. M. Kong, S. K. Choi, C. S. Lee, D. C. Yang, S. C. Kang, and T. Thambi. 2022. "Production of Minor Ginsenoside CK from Major Ginsenosides by Biotransformation and Its Advances in Targeted Delivery to Tumor Tissues Using Nanoformulations." *Nanomaterials (Basel, Switzerland)* 12 (19): 3427. https://doi.org/10.3390/nano12193427

National Center for Biotechnology Information (NCBI) – Phytochemical Database, Available at https://www.phytochemdb.com/ Accessed on November 27, 2025.

Niemeyer, K., I. R. Bell, and M. Koithan. 2013. Traditional Knowledge of Western Herbal Medicine and Complex Systems Science. *Journal of Herbal Medicine* 3(3): 112–9. https://doi.org/10.1016/j.hermed.2013.03.001

Niu, H., I. Alvarez-Alvarez, and M. Chen. 2025. "Artificial Intelligence: An Emerging Tool for Studying Drug-Induced Liver Injury." *Liver International: Official Journal of the International Association for the Study of the Liver* 45 (3): e70038. https://doi.org/10.1111/liv.70038

Obahiagbon, E. G., and M. C. Ogwu. (2023). Consumer Perception and Demand for Sustainable Herbal Medicine Products and Market. In: Izah, S.C., Ogwu, M.C., and Akram, M. (eds.), *Herbal Medicine Phytochemistry. Reference Series in Phytochemistry.* Cham: Springer. https://doi.org/10.1007/978-3-031-21973-3_65-1

Osanloo, M., S. Firooziyan, A. Abdollahi, S. Hatami, A. Nematollahi, N. Elahi, and E. Zarenezhad. 2022. "Nanoemulsion and Nanogel Containing Artemisia Dracunculus Essential Oil; Larvicidal Effect and Antibacterial Activity." *BMC Research Notes* 15 (1): 276. https://doi.org/10.1186/s13104-022-06135-8

Ouoba, K., S. Fofana, and R. Semdé. 2022. "Factors Limiting Traditional Medicine Integration into Modern Health Systems in Africa: Protocol for a Systematic Review of Qualitative Studies." *Public Health* 211: 62–65. https://doi.org/10.1016/j.puhe.2022.07.011

Panontin, J. F., M. K. D. Rambo, V. Isaac, C. S. Seibert, and E. Scapin. 2022. "New Antioxidant Lauryl-Free Herbal Shampoo Formulation with a Brazilian Plant Extract." *Brazilian Journal of Biology* = Revista brasleira de biologia 82: e264677. https://doi.org/10.1590/1519-6984.264677

Pastare, L., M. Berga, L. Kienkas, M. Boroduskis, A. Ramata-Stunda, D. Reihmane, M. Senkovs, G. Skudrins, and I. Nakurte. 2023. "Exploring the Potential of Supercritical Fluid Extraction of Matricaria chamomilla White Ray Florets as a Source of Bioactive (Cosmetic) Ingredients." *Antioxidants (Basel, Switzerland)* 12 (5): 1092. https://doi.org/10.3390/antiox12051092

Qi, W., W. Qi, D. Xiong, and M. Long. 2022. "Quercetin: Its Antioxidant Mechanism, Antibacterial Properties and Potential Application in Prevention and Control of Toxipathy." *Molecules (Basel, Switzerland)* 27 (19): 6545. https://doi.org/10.3390/molecules27196545

Raghav, R. S., and S. Verma. 2024. "A Comprehensive Review on Potential Chemical and Herbal Permeation Enhancers Used in Transdermal Drug Delivery Systems." *Recent Advances in Drug Delivery and Formulation* 18 (1): 21–34. https://doi.org/10.2174/0126673878272043240114123908

Ramawat, K. G., and J. M. Merillon. 2019. *Phytopharmaceutical Technology: Emerging Opportunities. Mukherjee PK. Quality Control and Evaluation of Herbal Drugs: Evaluating Natural Products and Traditional Medicine.* Elsevier.

Rehman, S. U., K. Choe, and H. H. Yoo. 2016. "Review on a Traditional Herbal Medicine, Eurycoma longifolia Jack (Tongkat Ali): Its Traditional Uses, Chemistry, Evidence-Based Pharmacology and Toxicology." *Molecules (Basel, Switzerland)* 21 (3): 331. https://doi.org/10.3390/molecules21030331

Rojas, P., H. Jung-Cook, E. Ruiz-Sánchez, I. S. Rojas-Tomé, C. Rojas, A. M. López-Ramírez, and A. A. Reséndiz-Albor. 2022. "Historical Aspects of Herbal Use and Comparison of Current Regulations of Herbal Products Between Mexico, Canada and the United States of America." *International Journal of Environmental Research and Public Health* 19 (23): 15690. https://doi.org/10.3390/ijerph192315690

Ryan, E. J. 2016. "Undergraduate Nursing Students' Attitudes and Use of Research and Evidence-Based Practice - An Integrative Literature Review." *Journal of Clinical Nursing* 25 (11–12): 1548–56. https://doi.org/10.1111/jocn.13229

Sahu, D. R., B. M. Sahoo, P. Nandi, C. K. Brahma, and A. Nanda. 2025. "Microwave-Assisted Green Extraction of Flavonoids: An Approach for the Development of Antiepileptic Agents." *Current Drug Research Reviews*, 10.2174/0125899775347899250303041223. Advance online publication. https://doi.org/10.2174/0125899775347899250303041223

Sanou, A., K. Konaté, K. Kabakdé, R. Dakuyo, D. Bazié, S. Hemayoro, and M. H. Dicko. 2023. "Modelling and Optimisation of Ultrasound-Assisted Extraction of Roselle Phenolic Compounds Using the Surface Response Method." *Scientific Reports* 13 (1): 358. https://doi.org/10.1038/s41598-023-27434-5

Sen, S, and R. Chakraborty 2015. "Toward the Integration and Advancement of Herbal Medicine: A Focus on Traditional Indian Medicine." *Botanics: Targets and Therapy* 5: 33–44. https://doi.org/10.2147/BTAT.S66308

Shan, Y., F. Wang, Z. Wei, and Y. Lu. 2021. "Synthetic Lethality Theory Approaches to Effective Substance Discovery and Functional Mechanisms Elucidation of Anti-Cancer Phytomedicine." *Phytomedicine: International Journal of Phytotherapy and Phytopharmacology* 91: 153718. https://doi.org/10.1016/j.phymed.2021.153718

Sharma, A. K., and K. Pundarikakshudu. 2019. "Regulatory Aspects of Traditional Indian Medicines (TIM) in India and in International Purview." *Journal of AOAC International* 102 (4): 993–1002. https://doi.org/10.5740/jaoacint.18-0379

Shi, J., X. Chen, H. Hu, and C. O. L. Ung. 2023. "Application of Implementation Science Framework to Develop and Adopt Regulatory Science in Different National Regulatory Authorities." *Frontiers in Public Health* 11: 1172557. https://doi.org/10.3389/fpubh.2023.1172557

Shi, J., X. Ni, J. Huang, Y. Fu, T. Wang, H. Yu, and Y. Zhang. 2022. "CRISPR/Cas9-Mediated Gene Editing of BnFAD2 and BnFAE1 Modifies Fatty Acid Profiles in Brassica napus." *Genes* 13 (10): 1681. https://doi.org/10.3390/genes13101681

Soleymani, M., M. Velashjerdi, and M. Asgari. 2021. "Preparation of Hyaluronic Acid-Decorated Mixed Nanomicelles for Targeted Delivery of Hydrophobic Drugs to CD44-Overexpressing Cancer Cells." *International Journal of Pharmaceutics* 592: 120052. https://doi.org/10.1016/j.ijpharm.2020.120052

Spanakis, M., E. Tzamali, G. Tzedakis, C. Koumpouzi, M. Pediaditis, A. Tsatsakis, and V. Sakkalis. 2025. "Artificial Intelligence Models and Tools for the Assessment of Drug-Herb Interactions." *Pharmaceuticals (Basel, Switzerland)* 18 (3): 282. https://doi.org/10.3390/ph18030282

Steel, A., R. Redmond, J. Schloss, H. Cramer, J. Goldenberg, M. J. Leach, J. E. Harnett, C. Van de Venter, A. McLintock, R. Bradley, J. Hawrelak, K. Cooley, B. Leung, J. Adams, and J. Wardle. 2022. "International Prevalence of Consultation with a Naturopathic Practitioner: A Systematic Review and Meta-Analysis." *BMJ Open* 12 (7): e056075. https://doi.org/10.1136/bmjopen-2021-056075

Takeno, M., S. Kitagawa, J. Yamanaka, M. Teramoto, H. Tomita, N. Shirai, S. Itoh, S. Hida, K. Hayakawa, K. Onozaki, and T. Takii. 2018. "5-Hydroxy-2-methylpyridine Isolated from Cigarette Smoke Condensate Aggravates Collagen-Induced Arthritis in Mice." *Biological & Pharmaceutical Bulletin* 41 (6): 877–84. https://doi.org/10.1248/bpb.b17-00982P

Tandon, N., and S. S. Yadav. 2017. "Contributions of Indian Council of Medical Research (ICMR) in the Area of Medicinal Plants/Traditional Medicine." *Journal of Ethnopharmacology* 197: 39–45. https://doi.org/10.1016/j.jep.2016.07.064

Thakor, V., M. J. Leach, D. Gillham, and A. Esterman. 2011. "The Quality of Information on Websites Selling St. John's Wort." *Complementary Therapies in Medicine* 19(3): 155–60. https://doi.org/10.1016/j.ctim.2011.05.005

Thomas, P. M., M. A. Deming, and A. Sarkar. 2022. "β-Lactamase Suppression as a Strategy to Target Methicillin-Resistant Staphylococcus aureus: Proof of Concept." *ACS Omega* 7 (50): 46213–21. https://doi.org/10.1021/acsomega.2c04381

Tsai, H. W., N. F. Tampubolon, T. Wu, M. Y. Wu, and Y. W. Lin. 2024. "Characterization of Carbon Dots from Fructus gardeniae (Zhi-zi) and Gardenia charcoal (black Zhi-zi) via Microwave-Assisted Extraction." *Journal of Food and Drug Analysis* 32 (3): 371–81. https://doi.org/10.38212/2224-6614.3513

Tsang, S. W., S. P. Ip, J. C. Wu, S. C. Ng, K. K. Yung, and Z. X. Bian. 2015. "A Chinese Medicinal Formulation Ameliorates Dextran Sulfate Sodium-Induced Experimental Colitis by Suppressing the Activity of Nuclear Factor-kappaB Signaling." *Journal of Ethnopharmacology* 162: 20–30. https://doi.org/10.1016/j.jep.2014.12.035

Wang, H., Y. Chen, L. Wang, Q. Liu, S. Yang, and C. Wang. 2023. "Advancing Herbal Medicine: Enhancing Product Quality and Safety Through Robust Quality Control Practices." *Frontiers in Pharmacology* 14: 1265178. https://doi.org/10.3389/fphar.2023.1265178

Wang J., J. Zou, Y. Shi, N. Zeng, D. Guo, H. Wang, C. Zhao, F. Luan, X. Zhang, J. Sun. 2024. "Traditional Chinese Medicine and Mitophagy: A Novel Approach for Cardiovascular Disease Management." *Phytomedicine* 128:155472. https://doi.org/10.1016/j.phymed.2024.155472.

Wang, Q., P. Han, M. Zhang, M. Xia, H. Zhu, J. Ma, M. Hou, Z. Tang, and W. Ling. 2007. "Supplementation of Black Rice Pigment Fraction Improves Antioxidant and Anti-Inflammatory Status in Patients with Coronary Heart Disease." *Asia Pacific Journal of Clinical Nutrition* 16 (1): 295–301.

Wang, X. J., J. L. Ren, A. H. Zhang, H. Sun, G. L. Yan, Y. Han, and L. Liu. 2019. "Novel Applications of Mass Spectrometry-Based Metabolomics in Herbal Medicines and Its Active Ingredients: Current Evidence." *Mass Spectrometry Reviews* 38 (4–5): 380–402. https://doi.org/10.1002/mas.21589

Wang, Y., Y. M. Jiang, Y. T. Wang, J. W. Kang, T. Yu, H. Y. Zhao, B. L. Bian, M. Huang, and H. C. Bi. 2014. "Inhibiton of Cytochrome P450 Isoenzymes and P-gp Activity by Multiple Extracts of Huang-Lian-Jie-Du Decoction." *Journal of Ethnopharmacology* 156: 175–81. https://doi.org/10.1016/j.jep.2014.08.044

Wenbo, Z., H. Jianwei, L. Hua, T. Lei, C. Guijuan, and T. Mengfei. 2024. "The Potential of Flavonoids in Hepatic Fibrosis: A Comprehensive Review." *Phytomedicine:*

International Journal of Phytotherapy and Phytopharmacology 133: 155932. https://doi.org/10.1016/j.phymed.2024.155932

World Health Organization. 1996. *WHO Expert Committee on Specifications for Pharmaceutical Preparations. World Health Organization Technical Report Series, 863*, 1–194. Geneva: World Health Organization.

World Health Organization. 2021. *Index of World Pharmacopoeias and Pharmacopoeial Authorities [Internet].* WHO [cited 2023 Sep. 11]. https://www.who.int/publications/m/item/QAS-11.453-Rev.12

Wu, Q., K. Chen, L. Ma, and S. Man. 2025. "Pueraria: A Comprehensive Review of Chemical Composition, Health Benefits, and Current Applications." *Comprehensive Reviews in Food Science and Food Safety* 24 (3): e70177. https://doi.org/10.1111/1541-4337.70177

Xia Z. D., B. Sun, J. F. Wen, R.X. Ma, F. Y. Wang, Y. Q. Wang, Z. H. Li, P. Jia, and X. H. Zheng. Research Progress on Metabolomics in The Quality Evaluation and Clinical Study of Panax Ginseng. *Biomedical Chromatography.* 37(7):e5546. https://doi.org/10.1002/bmc.5546.

Xiao, M., M. Yang, X. Ji, D. Li, Y. Xie, Y. Lyu, and Z. Zuo. 2021. "Protective Effect of Glechoma hederacea Extract Against Gallstone Formation in Rodent Models." *BMC Complementary Medicine and Therapies* 21(1): 199. https://doi.org/10.1186/s12906-021-03368-1

Yang, W., J. Liu, T. Hou, X. Lu, and L. Yang. 2025. "Inhibition of Human Gastric Cancer Growth by Cytokine-Induced Killer Cells Plus Chemotherapy With or Without Cadonilimab in a Mouse Xenograft Tumor Model." *Frontiers in Immunology* 16: 1609320. https://doi.org/10.3389/fimmu.2025.1609320

Yau, W. P., C. H. Goh, and H. L. Koh. 2015. Quality Control and Quality Assurance of Phytomedicines: Key Considerations, Methods, and Analytical Challenges. *Phytotherapies: Efficacy, Safety, and Regulation*, 18–48. DOI:10.1002/9781119006039.ch2

Yu, X., W. Qin, H. Cai, C. Ren, S. Huang, X. Lin, L. Tang, Z. Shan, W. H. A. Al-Ameer, L. Wang, H. Yan, and M. Chen. 2024. "Analyzing the Molecular Mechanism of Xuefuzhuyu Decoction in the Treatment of Pulmonary Hypertension with Network Pharmacology and Bioinformatics and Verifying Molecular Docking." *Computers in Bbiology and Medicine* 169: 107863. https://doi.org/10.1016/j.compbiomed.2023.107863

Zeng, Y., J. X. Song, and X. C. Shen. 2012. "Herbal Remedies Supply a Novel Prospect for the Treatment of Atherosclerosis: A Review of Current Mechanism Studies." *Phytotherapy Research: PTR* 26 (2): 159–67. https://doi.org/10.1002/ptr.3587

Zhang, J., Y. Wu, Y. Tian, H. Xu, Z. X. Lin, and Y. F. Xian. 2024. "Chinese Herbal Medicine for the Treatment of Intestinal Cancer: Preclinical Studies and Potential Clinical Applications." *Molecular Cancer* 23 (1): 217. https://doi.org/10.1186/s12943-024-02135-3

Zhang, Q., S. H. Zhao, J. Chen, and L. W. Zhang. 2015. "Application of Ionic Liquid-Based Microwave-Assisted Extraction of Flavonoids from Scutellaria baicalensis Georgi." *Journal of Chromatography. B, Analytical Technologies in the Biomedical and Life Sciences* 1002: 411–7. https://doi.org/10.1016/j.jchromb.2015.08.021

Zhao L. Z., Cao H. L., He Z. Q., Sun Y., Fang L. L., Li W. L. 2025. Recent Advances in Green Solvents-Based Liquid-Phase Microextraction Techniques for Chromatographic Analysis of Active Components in Traditional Chinese Medicine. *J Chromatogr A.* 1741:465604. doi: 10.1016/j.chroma.2024.465604.

Index

M

N

O

P

Q

R

S

T

U

V

W

Y

For Product Safety Concerns and Information please contact our EU
representative GPSR@taylorandfrancis.com
Taylor & Francis Verlag GmbH, Kaufingerstraße 24, 80331 München, Germany

www.ingramcontent.com/pod-product-compliance
Ingram Content Group UK Ltd.
Pitfield, Milton Keynes, MK11 3LW, UK
UKHW021504240926
475702UK00035B/365
* 9 7 8 1 0 4 1 0 3 4 1 9 3 *